Intelligent Systems Reference Library

Volume 273

The aim of this series is to publish a Reference Library, including novel advances and developments in all aspects of Intelligent Systems in an easily accessible and well structured form. The series includes reference works, handbooks, compendia, textbooks, well-structured monographs, dictionaries, and encyclopedias. It contains well integrated knowledge and current information in the field of Intelligent Systems. The series covers the theory, applications, and design methods of Intelligent Systems. Virtually all disciplines such as engineering, computer science, avionics, business, e-commerce, environment, healthcare, physics and life science are included. The list of topics spans all the areas of modern intelligent systems such as: Ambient intelligence, Computational intelligence, Social intelligence, Computational neuroscience, Artificial life, Virtual society, Cognitive systems, DNA and immunity-based systems, e-Learning and teaching, Human-centred computing and Machine ethics, Intelligent control, Intelligent data analysis, Knowledge-based paradigms, Knowledge management, Intelligent agents, Intelligent decision making, Intelligent network security, Interactive entertainment, Learning paradigms, Recommender systems, Robotics and Mechatronics including human-machine teaming, Self-organizing and adaptive systems, Soft computing including Neural systems, Fuzzy systems, Evolutionary computing and the Fusion of these paradigms, Perception and Vision, Web intelligence and Multimedia.

Indexed by SCOPUS, DBLP, zbMATH, SCImago.

All books published in the series are submitted for consideration in Web of Science.

Jozo Dujmović

Graded Logic

Modeling Human Commonsense Reasoning in Decision-Making and AI

Springer

Jozo Dujmović
Department of Computer Science
San Francisco State University
San Francisco, CA, USA

ISSN 1868-4394 ISSN 1868-4408 (electronic)
Intelligent Systems Reference Library
ISBN 978-3-031-88557-0 ISBN 978-3-031-88558-7 (eBook)
https://doi.org/10.1007/978-3-031-88558-7

This Springer imprint is published by the registered company Springer Nature Switzerland AG
The registered company address is: Gewerbestrasse 11, 6330 Cham, Switzerland

Graded Logic is fifty years old. This book is a birthday present for the fiftieth anniversary of Graded Logic. To celebrate, we begin with two poems: a clever piece by a celebrated haiku author and a playful poem by a gifted two-year-old.

A Caterpillar
A caterpillar,
this deep in fall –
still not a butterfly.

—Matsuo Bashō (1644–1694)

The Journey of Graded Logic
Fifty years ago, a caterpillar small,
Graded Logic inched and crawled,
In tender growth, it dreamed its flight,
Through branches dense and skies of night.

Today it spreads wings, bold and bright,
A butterfly of graceful might,
In this book, we celebrate the way,
That Graded Logic flies today.

—ChatGPT (2022–forever)

Preface

Graded Logic (GL) is a continuum-valued propositional logic of human commonsense reasoning and decision making. GL is *fully continuum-valued*; i.e., everything is a matter of degree. It is based on continuum-valued logic variables (graded truth), continuum-valued simultaneity (graded conjunction), continuum-valued substitutability (graded disjunction), and continuum-valued importance of logic variables. The graded conjunction and the graded disjunction are dualized, complementary, and unified in a single continuum-valued, andness-directed, importance-weighted, idempotence-selectable, and annihilator-selectable logic function called Graded Conjunction/Disjunction (GCD). The strict use of continuum-valued concepts and variables is a unique distinctive property introduced in Graded Logic. Based on that property, GL is a seamless generalization of the classical bivalent Boolean logic, fuzzy logic propositional calculus, and non-classical continuum-valued logics.

All properties of GL are derived from the observation, measurement, modeling, and explanation of natural human commonsense logical reasoning with graded (continuum-valued) percepts. Furthermore, GL serves as a fundamental component in quantitative models of natural human decision-making. The primary feature of GCD is the parameterized continuous transition from drastic conjunction (the model of ultimate simultaneity) to drastic disjunction (the model of ultimate substitutability). To adjust the properties of graded conjunction and graded disjunction, GL introduces an adjustable conjunction degree ("andness") and an adjustable disjunction degree ("orness"). The basic GL concepts were introduced in 1973–1974, fifty years ago. This book celebrates GL's 50th anniversary, summarizes its key concepts, and presents its current state.

Our main goal is to use GL as a mathematical infrastructure for professional decision-support systems. Based on a human-centric approach, we assume that decision making is a mental process of a specific stakeholder/decision-maker. This process consists of identifying a set of alternatives (or competitive objects) and selecting the most suitable alternative. The perception of suitability is continuum-valued (graded), and the number of alternatives can even be as low as one. To compute the overall suitability of a single alternative, the decision-making process must include a systematic GL-based evaluation method. Consequently, GL can be regarded as the propositional logic of commonsense evaluation, observable in natural human commonsense reasoning and applicable to decision engineering and AI systems. These constraints are essential for defining the scope and role of GL within a broad spectrum of non-classical logics.

In this book we do not present the history of GL development and use, but its distinctive ideas and its current status in the context of soft computing and computational intelligence. We present both the descriptive and the normative aspects of GL and GL-based decision making. The intention is to offer a condensed and applicable presentation which strongly emphasizes the following three fundamental concepts that were the core of GL both in 1973–1974 and today, fifty years later:

- *A human-centric approach*: logic modeling based on experimentally measurable, and explainable properties of the natural commonsense logical reasoning used in the observable mental process of human decision-making. We use natural human commonsense reasoning as the starting point for developing all GL models. In other words, we always give preference to observable natural human reasoning with graded percepts above formalized axiomatic theoretical approaches and considerations. This approach sometimes reduces the theoretical purity but offers a bigger benefit of broad applicability in the decision-making area.
- *Staying strictly within the framework of logic*: our graded propositional calculus preserves the classical ideas of George Boole and Augustus De Morgan, specified by logic values in the vertices of the unit hypercube $\{0,1\}^n$, but extends them, seamlessly and naturally, within the whole unit hypercube $[0,1]^n$.
- *Applicability in decision engineering*: The development of GL is motivated by the needs of the professional area of *computational intelligence* and specific needs of *decision engineering* to use a mathematical logic infrastructure that is derived from and consistent with the commonsense reasoning and natural human decision making. The current interest in developing machines that behave as humans and can make confident and explainable decisions offers a good opportunity for using GL in AI.

This book stands on the shoulders of the author's earlier book *Soft Computing Evaluation Logic* (Wiley and IEEE Press, 2018). That book emphasized industrial applications of decision models in the area of evaluation, and the use of software tools in that process. The primary goal of the present book is to focus on the human-centric GL-based mathematical infrastructure necessary for developing a spectrum of explainable decision models and their use in professional applications.

This book contains a linear sequence of twenty essays addressing important GL topics. The essays can be studied independently, and to facilitate direct access, they sometimes contain slightly redundant components. Readers who choose to access only selected chapters should have a basic understanding of the topics and the willingness to jump from topic to topic when needed. Readers who decide to read all the chapters sequentially can expect to do so with minimal effort and without prerequisites.

Readers with questions, comments, or proposals for joint work are encouraged to contact the author via email at jozo@sfsu.edu.

San Francisco, CA, USA Jozo Dujmović

Contents

Chapter 1
Graded Propositional Logic in the Context of Decision Making

In a sense, we make a decision every time we do one thing when we could have done something else.
—Jonathan St. B. T. Evans [EVA17]
A system (and this includes an individual) which possesses no purpose has no decision problems.
—Derek W. Bunn [BUN84]

Decision making is one of the most frequent and the most observable human logic-based mental activities [EVA17, KAH11, BUN84]. Yet, the commonsense decision-making logic often stays hidden in plain sight. Logical decision-making activities are visible in many simple everyday household tasks. So, let us start by showing a decision-making process that is well known to everybody who buys food or other products.

Commonsense Reasoning and Decision Making in a Food Store

When we enter a food store to buy apricots, our first action is to evaluate the quality of apricots, Q. We investigate the color, touch the apricots to see whether they are ripe enough, listen to what other shoppers are saying, and eventually we form our percept[1] of the apricot quality Q. Obviously, Q is a percept of a specific buyer, i.e., an intuitive feeling/impression of quality, in the range from the lowest (worthless) to the highest (perfect). If mathematicians buy apricots, then for worthless apricots they use the value $Q = 0$ and for perfect apricots the value $Q = 1$. If you ask them to explain the meaning of Q, the expected answer is "Q is the degree of truth of the statement 'these apricots completely satisfy all my requirements'." So, the apricots' quality is expressed using an appropriate value statement, and its continuously adjustable degree of truth $Q \in [0, 1]$. That degree of truth does not exist in a vacuum – it is created by a specific buyer, and it reflects the buyer's individual needs and interests.

Apricots are not free. Therefore, the next step is to investigate the affordability of apricots, A. We look at the price of apricots, count the money we have in our

[1] In this context, a *percept* is defined as a quantifiable mental result (an impression or sensation) of perceiving and/or reasoning. Generally, we assume that all percepts have a continuously adjustable intensity in the range from zero (or another minimum value) to some specific maximum value.

J. Dujmović, *Graded Logic*, Intelligent Systems Reference Library 273,
https://doi.org/10.1007/978-3-031-88558-7_1

pocket, recall what else we must buy, and eventually form the percept of affordability $A \in [0, 1]$. Again, for mathematicians who like apricots, A is the degree of truth of the statement "these apricots are perfectly affordable for me." Obviously, all buyers of all products form an intuitive impression of affordability and use it in the process of deciding whether to buy or not to buy a product.

Now comes the question that exposes the need for graded propositional logic: "are these apricots suitable for purchasing today?" Understandably, everybody wants *simultaneously* a good quality *and* a good price, not only for apricots, but for all possible products. Thus, everybody has in mind a model of *simultaneity*, which is a form of conjunction: the overall suitability of apricots S is high only if their quality Q *and* their affordability A are *simultaneously high.* So, while standing in front of apricots, our mental process is to intuitively create a percept of the *overall suitability* that our friend mathematician denotes $S = Q\&A$. The symbol "&" denotes some kind of *AND* operation (i.e., a mental conjunction operator that has continuously adjustable strength). Obviously, now come inevitable questions: (1) what is the overall suitability S of apricots, and (2) which among numerous existing logic theories explicitly offers mathematical infrastructure for justifiable solutions of this ubiquitous trivial problem which must be intuitively solved in each human mind in each grocery store? Very surprisingly, there is no right logic theory that is developed with explicit goal to precisely model this type of observable, measurable, and explainable human natural logical reasoning. This is why we need a graded logic.

Variables Q, A, and S are called *perceptual variables* because they quantify the percepts of quality, affordability, and the overall suitability. Perceptual variables are graded, i.e., their magnitude varies in a range from a minimum value to a maximum value. We assume that "grading" means using degrees of truth of *value statements* that assert "the complete satisfaction," i.e., the maximum quality (Q), the fully acceptable affordability (A), and the maximum overall suitability (S). Degrees of truth of value statements vary in the range from the completely false to the completely true. Therefore, all perceptual logic variables are graded: they belong to the interval $I = [0, 1]$. Humans use more graded percepts: both the *importance of a value statement*, and the *strength of simultaneity (and substitutability)* are also graded. That justifies the attribute "graded" to logic that develops mathematical models with graded percepts. That is where the name Graded Logic (GL) comes from.

It is easy to see that there is a spectrum of mental *AND* operations because some people want to penalize low values of Q and/or A stronger than other people. That is the reason why the *AND* operation used in the human mind, even for everyday decisions, must be a matter of degree: sometimes we use a higher penalty and sometimes we use a lower penalty. Thus, the *strength of mental conjunction is adjustable.* Our friend mathematician, who believes that the only right models of conjunction are the t-norms, without hesitation uses the product t-norm $Q\&A = QA$. His sister (still a math student) uses a much lower penalty, $Q\&A = \sqrt{QA}$, and they both justifiably believe that $Q\&0 = 0\&A = 0$, i.e., it is necessary to support the *annihilator* 0 and never buy apricots that have unacceptable quality or unacceptable price. However, both siblings think that low price and good quality are *equally important* (Q and A are *commutative*, a favorite property in the majority of mathematical models). Is that acceptable? Each real decision situation shows that in a general case, commutativity is not acceptable because some real buyers are more sensitive to price, while others are more sensitive to the quality of products. Of course, commutativity can be a special case.

Ignoring the concept of different degrees of importance and using the commutativity of arguments (which are degrees of truth!) is generally unacceptable in all logic problems that include human beings. It is equally unacceptable in the simple intuitive grocery store reasoning as in sophisticated mathematical models for professional decision making (e.g., buying a nuclear submarine, or determining the best time to start a risky therapy for a patient). Indeed, for humans (and their groups, institutions, and organizations) all truths are *not* equally important. Both Q and A have a specific *meaning* (different interpretation, different role, and different importance) for each buyer. In other words, both Q and A have a *semantic identity*,[2] and semantic components must also affect logic models. Unsurprisingly, wealthy buyers emphasize high quality, and poorer buyers emphasize low cost. For example, the mathematician's mother wants to buy 12 pounds of apricots to make preserves. More sugar can usually compensate for a lower quality of apricots if they are not completely ripe, but nothing can compensate for an unreasonably high price. So, in this example, an acceptable intuitive criterion might be that the affordability of apricots for preserves may be assessed by a particular buyer as two times more important than the quality of apricots. In other words, an increment/decrement of affordability should produce two times stronger impact on S than the impact of an increment/decrement in quality. This is expressed (both intuitively and formally) using weights 2/3 and 1/3. So, the preserves maker is expected to evaluate the apricot suitability using logical reasoning like the weighted harmonic mean formula $S = Q\&A = 1/\left(\frac{0.33}{Q} + \frac{0.67}{A}\right)$. In addition, this buyer would not buy apricots if S is less than an appropriate threshold value S_{min} (the mathematician's mother typically uses the threshold $S_{min} = \frac{3}{4}$, or less demanding $S_{min} = \frac{2}{3}$). In this example, Q and A are not commutative: if $0 < Q \neq A > 0$, and we swap Q and A, we get two different values of S. Thus, in natural human logic, commutativity is just an infrequent special case.

What happens if apricots are too expensive? Obviously, the buyer will not buy them. In other words, if $A = 0$ then $S = 0$; so, this form of simultaneity supports the annihilator 0. However, in natural human reasoning, and in GL, the support for annihilators must be selectable because humans often use simultaneity that does not support annihilators. For example, the location of an apartment for a senior citizen is suitable if the apartment is close to a nice park, *and* a good restaurant, *and* a public library. This is again a simultaneity requirement. However, the park, restaurant, and library are desirable but not necessary: the senior citizen will *not* reject an attractive

[2] Generally, semantics denotes a study of meaning, defined as a branch of linguistics and logic. In this book we use semantics as a concept in the context of graded propositional calculus. We define semantics (and "semantic identity") as the interpretation, meaning, and importance of a statement and its degree of truth for a specific stakeholder/decision-maker who has specific goals and requirements. If degrees of truth are defined as anonymous real numbers, then we assume that such a logic operates in a "semantic vacuum".

and affordable apartment only because the public library is too far. The logical reasoning of the senior citizen can be based on simultaneity that does *not* support the annihilator 0. But when the senior citizen buys apricots s/he uses simultaneity that supports annihilators. Thus, in natural human reasoning the *annihilators are selectable* (either indispensable or unacceptable).

The presented examples show the basic logic properties of natural human decision making. Decision making needs justifiable evaluation of objects or alternatives, and evaluation is based on logic. So, it is useful and natural to consider logic in the context of commonsense decision making and evaluation processes.

The ubiquitous trivial examples of buying food or selecting a location of residence show that the simplest forms of natural human (commonsense) reasoning[3] use logic based on the following elementary properties:

- Natural propositional logic is an indispensable part of the *decision-making* process of a specific *stakeholder/decision-maker*. Logical reasoning is a human mental activity. Thus, no justifiable natural logic properties without a known stakeholder.
- The degree of truth of all value statements is *continuously adjustable*. No bivalent truth, except as a special case.
- All degrees of truth must have semantic identity, i.e., stakeholder's interpretation of meaning, explanation of role, and specification of importance, in the context of specific decision-making activity. The degrees of truth are not anonymous real numbers.
- Depending on the specific stakeholder's goals and interests, the degrees of importance of all arguments of value statements are *continuously adjustable*. No commutativity, except as a special case.
- The strength of both simultaneity (conjunction) and substitutability (disjunction) is *continuously adjustable* in a wide range from the ultimate conjunction to the ultimate disjunction. No fixed andness/orness.
- Using or not using annihilators for both simultaneity (conjunction) and substitutability (disjunction) for idempotent logic aggregators must be freely selectable. No fixed annihilators for idempotent conjunction/disjunction.
- The use of idempotent or nonidempotent conjunctive or disjunctive aggregators must also be freely selectable. Mathematical models of logic aggregators must support this property.

These elementary logic properties have been in plain sight to all humans for thousands of years. Yet, there is no classical mathematical logic theory that supports *all* these observable necessary properties.

[3] In this book we use the term "natural human reasoning" (or ordinary commonsense reasoning) to describe intuitive human logical reasoning in the context of everyday decision-making that is used to select one of several possible actions/activities. Such reasoning is an observable natural phenomenon which can be measured and approximately quantitatively described using graded logic models.

Observability and Explainability of Natural Human Logic

Decision making in a food store is a typical simple example of the use of natural human logic in the real life. It shows that human percepts are graded, and that the intensity of a graded percept is naturally interpreted as a degree of truth of the statement that claims the highest intensity of percept (e.g., the complete satisfaction of specific requirements). If a percept P (measured in any specific units) belongs to the interval $P \in [0, P_{max}]$, then the degree of truth t of the statement "P has the highest value" is $t = \frac{P}{P_{max}} \in [0, 1]$. Generally (but very rarely), $P \in [P_{min}, P_{max}]$, $P_{min} > 0$, and then $t = \frac{P - P_{min}}{P_{max} - P_{min}} \in [0, 1]$. So, the graded logic variables are naturally present in decision making and the logic processing of such variables needs graded logic operations.

The simple example of buying apricots in a food store is sufficient to show typical human graded percepts, continuum-valued logic variables, the need for noncommutativity of logic operations, the use of annihilators, and the need for continuous adjustment of the degree of simultaneity (and substitutability). It is easy to note many similar commonsense logic problems that are all observable and explainable. Let us show three such useful examples.

As the first example, let us consider job offers. A job offer can be evaluated using two important components: the monetary compensation (M), and the job attractiveness (J). So, the suitability of such job offers (S) is a function of M and J. Decision-makers naturally require the simultaneity of high values of both M and J, i.e., some kind of conjunction of logic variables M and J. The degree of satisfaction with the offered monetary compensation can be expressed in the range [0, 100%]. The same range holds for the job attractiveness. So, it is easy to observe and explain how decision-makers select basic properties of function $S = F(M, J)$. For example, the decision-maker can easily answer the question whether in his/her opinion the cases $M = 100\%, J = 50\%$ and $M = 50\%, J = 100\%$ are equivalent or not equivalent. The positive answer identifies commutative function $F(M, J)$, and the negative answer shows noncommutativity. In the case of noncommutativity, the decision-maker can easily answer the next question: is $S(100, 50)$ more desirable than $S(50, 100)$? The positive answer shows that the monetary compensation is more important than the attractiveness of the job. The next step would be to determine the weights that adequately express the difference in importance.

As a second similar example, let us consider a homebuyer who is interested in simultaneous satisfaction of two important requirements: the quality of home ($Q \in [0, 100\%]$) and the suitability of home location ($L \in [0, 100\%]$). So, the overall suitability $S(Q, L) \in [0, 100\%]$ is some form of conjunction, similarly to the job offer example. The homebuyer can specify the properties of the aggregator $S(Q, L)$ according to the following dialogue (Q-A-C = question–answer-conclusion):

Q1: What is more important for you: the quality of a home, or the suitability of home location?
A1: They are equally important.
C1: The aggregator should be commutative.

Q2: If Q=50% and L=90% would your overall satisfaction be 70% or below 70% or above 70%?
A2: It would be below 70%.
C2: If $S(Q, L) < (Q + L)/2$ then the aggregator is conjunctive.

Q3: Do you feel that $S(75, 75) = 75$?
A3: Yes, that is correct.
C3: The necessary aggregator is idempotent, i.e., a mean.

Q4: Would you accept a home where either $Q = 100, L = 0$, or $Q = 0, L = 100$?
A4: Of course, I would reject such homes ($S = 0$). I need both good Q and good L.
C4: The homebuyer wants a hard conjunction that supports the annihilator 0.

Q5: How about the case $Q = L = 15\%$?
A5: I would convincingly reject such a home ($S = 0$) and every other similar combination below 25%.
C5: If $S < min(Q, L)$ then that is a hyperconjunction: the simultaneity is now stronger than the pure conjunction.

Q6: What would happen in the almost perfect case $Q = L = 95\%$?
A6: I would be completely satisfied ($S = 100\%$); it is almost impossible to find such a good combination and everything above 90% in both Q and L would be perfectly acceptable for me.
C6: If $S > max(Q, L)$ then that is a hyperdisjunction: the substitutability is now stronger than the pure disjunction.

This dialogue can be based on numeric interpretation of suitability (as a percentage of satisfied requirements), but it might also be based on verbalized rating scales (e.g., lowest, very low, low, ..., high, very high, highest) for respondents who prefer strictly verbalized communication. In all cases it is easy to conclude that all homebuyers can answer the above questions, and their answers are sufficient to select an appropriate graded logic aggregator. In addition, this example illustrates the explainability of human natural logical reasoning: decision-makers know their goals and can explain their evaluation criteria.

As the third example, let us consider a research organization that wants to hire a researcher that simultaneously needs good expertise in two different areas (e.g., such as computational biology, physical chemistry, environmental economics, geophysics, and others). A computational biology candidate was interviewed in the area of biology, and his/her competence was evaluated as $B = 70\%$[4]. Then, the same candidate was interviewed in the area of computer science, and his/her competence was

[4] Each evaluation of competence of a specific candidate can be interpreted in three basic ways: (1) as a degree of truth of the assertion "the candidate achieved the highest degree of competence in a specific area," or (2) as the degree of membership of the evaluated candidate in a fuzzy set of ideal, perfectly competent candidates, and (3) as the probability that the candidate can solve a problem that requires the highest degree of competence. In many cases of decision modeling, these interpretations are equivalent [DUJ17]. Of course, in the context of Graded Logic, we use degrees of truth from interval [0,1]. However, for simplicity in communications with stakeholders this interval is frequently expressed as [0, 100%] and interpreted as a percentage of satisfied requirements.

evaluated to be $C = 70\%$. The search committee must suggest the overall suitability of this candidate $S(B, C) = S(70, 70)$. Their reasoning can be probabilistic: the degree of overall competence can be interpreted as the probability of successful problem solving, that needs simultaneous competence in two independent disciplines. Consequently, the likelihood of successful interdisciplinary research results should be the product $S(BC) = BC$, and the resulting suitability of this candidate would be $S \approx 50\%$. This is a case where the conjunctive aggregator is justifiably selected to be the nonidempotent product t-norm. This logic aggregator is called hyperconjunction, because the degree of simultaneity is higher than in the case of the pure conjunction, $S(B, C) < \min(B, C)$.

The presented examples show the observability, modelability, and explainability of necessary logic properties that graded logic operators must support. It is rather easy to see the need for continuum-valued truth, continuum-valued simultaneity/substitutability, noncommutativity, idempotence-selectability, and annihilator-selectability. If we want a human-centric logic which assumes the presence of known stakeholder/decision-maker, and provides mathematical models of valid human reasoning, then these properties are the main prerequisites for such a logic. The goal of Graded Logic is to satisfy these prerequisites.

Limitations of Classical Propositional Logic and Limitations of Graded Logic

All logic systems have their limited scope. So, it is important to know the boundaries and limitations of each of them. That holds for both classical propositional logic and Graded Logic. The above list of necessary properties of human-centric logic models used in decision making shows the basic limitations of the classical propositional logic (as well as most non-classical propositional logics [PRI09]):

(1) *No known stakeholder/decision-maker* (no stakeholder's goals and no semantic identity of logic arguments – all arguments are anonymous real numbers and logic is defined as an abstract formalized axiomatic deductive system)
(2) *Bivalent (or multivalent) truth and fixed forms of simultaneity/substitutability* (ignoring the continuous nature of natural human percepts[5]).
(3) *Commutativity of arguments* (ignoring different importance of arguments, caused by ignoring the existence of stakeholder and stakeholder's goals and requirements).
(4) *Fixed strength of conjunction/disjunction* (no graded simultaneity/substitutability).
(5) *Fixed annihilators for conjunction/disjunction* (ignoring the fact that in natural human reasoning the presence of annihilators is selectable).

The primary goal of Graded Logic is to eliminate all these limitations, while fully supporting the classical logic as a special case. However, GL has its inherent limitations. First, in this book, the goal of GL is to provide the infrastructure for a graded

[5] For example, in [GIN68] a continuum-valued logic is not presented as a model of human reasoning, but used as a tool for solving computer engineering problems (design of function generators).

propositional calculus, and nothing beyond that area. Second, GL is not a formalized axiomatic deductive system, with all consequences resulting from this fact. Primarily, there are no formal proofs of theorems that would be deductively connected to axioms. So, we must explain why we need GL and what can be considered a valid reasoning in this logic.

GL develops mathematical models that are used to approximate the reality of natural human reasoning. The natural human reasoning is an observable reality, but too complex and only partially understood. Thus, if logic is a study of valid human reasoning, there are only two approaches to logic: (1) mathematical approach which accepts reality as long as the reality can be described as a formalized axiomatic deductive system; unfortunately, that is equivalent to ignoring reality, and (2) empirical human-centric approach which tries to bring logic as close to reality as possible, based on logic models that approximate the observable, measurable, and explainable natural human commonsense reasoning; however, that creates problems of accepting empirical reasoning and explainability as a substitute for formal proofs of logic rules and properties. Classical and non-classical logics [BOO47, DEM47, BOO54, PRI09, STE22] are examples of mathematical approach, which is firmly based on formalized axiomatic deductive systems. Graded Logic is an example of empirical approach, where all GL concepts and necessary properties are supported primarily by experiments and observations of phenomena that are visible, measurable, and explainable in human reasoning and decision making.

Engineers and scientists easily subscribe to the empirical human-centric approach. On the other hand, this approach is very different from the approach used by mathematicians, logicians and philosophers. The empirical approach is based on the concept that a high applicability and a good approximation of measured reality are sufficient proofs of the validity of logic models.

Let us exemplify the empirical GL approach by the following five empirical concepts:

1. *Small errors can be neglected.* Graded conjunction and graded disjunction, modeled using power means, do not support associativity and distributivity. According to [DUJ75a], average errors of associativity are below 1%, and average errors of distributivity are below 2%. Such small errors are perfectly acceptable because they do not affect the applicability of GL propositional calculus models. Therefore, in GL, the useful properties of associativity and distributivity do not hold in theory, but they hold in practice and can be used without hesitation.
2. *The necessary precision of logic models depends on the precision of arguments.* Logic aggregators in GL are modeled using the segmented interpolative method (presented in Chap. 15). The precision of logic aggregators is limited by the imprecision of estimated parameters, the discretization of continuous parameters, and the imprecision of numeric approximations. However, the corresponding variations of the logic model results are regularly just a few percent [DUJ18]. Arguments of such models are degrees of truth specified by human decision-makers, and their precision can be 10–15% [MIL56]. Obviously, if the precision of a logic model is significantly better than the precision of arguments, then the

model is sufficiently precise and applicable. Efforts to enhance the precision of such models are possible (and often available) but not essential.

3. *Reasons that convincingly support a conjecture can substitute a formal proof*. The central result of GL is the Graded Logic Conjecture (Chap. 11), that claims that all formulas of the graded propositional calculus can be composed using 10 fundamental functions that are necessary and sufficient. Obviously, that is a conjecture and not a theorem: there is no strict formal proof that 10 fundamental functions are necessary and sufficient. However, the empirical evidence that supports the necessary status is abundant: there are many convincingly documented observations of the human use of each of the 10 functions. On the other hand, there is no known case where human propositional reasoning includes some other function, and there is no known case where the available functions are insufficient to build a desired propositional calculus model. These are all reasons that support the Graded Logic Conjecture in a convincing and sufficient way, which (in the real world) can be used as an acceptable substitute for a formal proof.
4. *The validity of a special case supports the validity of a general concept*. All binary logic models are defined in the vertices of the unit hypercube $\{0, 1\}^n$. As opposed to that, GL models are defined inside the whole hypercube $[0, 1]^n \supset \{0, 1\}^n$. In the vertices of the unit hypercube, GL is identical to the classical Boolean logic. Thus, GL is a general continuum-valued logic which is a generalization of the classical binary logic.
5. *The practical validity of useful properties is preserved in the vicinity of cases where they strictly hold*. For example, all t-norms have annihilators and neutral elements, and satisfy commutativity, monotonicity, and associativity. The popular product t-norm $z = xy$ supports all these properties. If we make the product-power function $z = (xy)^{1+\delta}$, or a weighted product-power function $z = x^{1+\delta}y^{1-\delta}$, then for $\delta \neq 0$ these functions do not have all basic t-norm properties[6] that strictly hold for $\delta = 0$. However, the useful properties disappear gradually, and for small values of δ they are preserved to the extent that can be fully acceptable in practical approximate models.

The Reasons for a Logic Decision Problem Hidden in Plain Sight

The apricot story shows the type of logical reasoning situation that happens to all humans, in all grocery stores, and in all other shopping activities. So, it is completely legitimate to ask what is the propositional calculus that properly describes the ultimately elementary commonsense logical reasoning necessary for buying apricots? After all, this is arguably the most frequent logic problem on the planet, which is hidden in plain sight of all logicians, as well as all buyers who enter any store, including those on the Internet. Yet, such problems are overlooked in the literature on mathematical logic. Why? Let us try to offer five answers to this question.

---

[6] Design and use of weighted triangular norms and conorms is extensively studied by R.R. Yager [YAG04] (see Chapter 15)

First, tradition. Most human percepts are graded, i.e., continuum-valued. However, it took 24 centuries (from Aristotle [384–322 BC] to Jan Łukasiewicz [1878–1956]) to stop using bivalent approach to logic. The point of departure was the three-valued propositional calculus, introduced by Łukasiewicz in 1917–1920. The significant interest in continuum-valued propositional logics and their applications started only after the Zadeh's introduction of fuzzy sets in 1965 [ZAD65]. That inspired other continuum-valued logic systems (e.g., [GIN68, DUJ73b, DUJ75a, RAD00]).

Second, the complexity, variability, and irregularity of human experience and natural commonsense reasoning is not compatible with strict axiomatic approach used for building formal mathematical theories. Therefore, traditional formal logic never puts the natural human reasoning and a specific human thinker (stakeholder/decision-maker) in the epicenter of interest. The inevitable result of this approach is the absence of *semantic components*: if there is no stakeholder, there cannot be the concept of importance, because importance is based on specific stakeholder's goals, interests, and requirements. So, all degrees of truth must be equally important, regardless of the ample empirical evidence that, in human reasoning, this approach is inadequate. Indeed, it is easy to see that most human actions are motivated by specific percepts of importance and the "first things first" approach. Another consequence of not having an explicitly present stakeholder is that classic formal logic is isolated from the area of natural human decision-making. So, there is no purpose and goals that can only be defined by the stakeholder. Thus, if there is no purposeful system (in the sense of Ackoff/Emery [ACK72]), then there is no decision problem of finding the best way to achieve an aim, and no need for the logic of natural human reasoning.

Third, traditional logic models use both conjunction (simultaneity) and disjunction (substitutability) as fixed (not strength-parameterized) operators. However, in natural human reasoning it is evident that decision-makers adjust the desired strength or magnitude of simultaneity and substitutability [DUJ74a]. Therefore, if we want to model natural human reasoning, then both simultaneity and substitutability must be graded, i.e., parameterized using explicitly visible strength adjustment parameters.

Fourth, traditional logic models of conjunction and disjunction use fixed annihilators. However, humans use selectable annihilators: sometimes we use simultaneity and substitutability that support annihilators, and sometimes we use simultaneity and substitutability that do not support annihilators. Human-centric graded propositional logic must have all these properties, and they were introduced in 1973 [DUJ73b, DUJ73c] in response to demand generated by decision-making problems in industrial evaluation of early mainframe computers and minicomputer systems [SHA69, KEN73, KAN70, JOS64, JOS68, JOS71, MCQ78, DUJ75b].

Fifth, it is possible to develop mathematical models of natural human reasoning and stakeholder-based decision making, and that is the primary goal of Graded Logic. However, such models are only approximations of observable and measurable natural phenomena, and the theoretical purity of basic mathematical properties of such logic models is regularly modest. For example, the fundamental algebraic properties of commutativity, associativity and distributivity for basic models of graded conjunction and graded disjunction hold correctly only in special cases of these functions, but generally they hold with acceptable small errors [DUJ75a]. Therefore, in Graded Logic, the cost of desired functionality and applicability can be the reduction of formal mathematical purity.

Applicability of Graded Logic in Decision Making

It is important to note that many decision problems with very different degrees of cost and significance can be logically equivalent and yield similar propositional calculus models. The apricot buying story is a toy decision problem. Of course, Graded Logic is not a theoretical armamentarium for solving toy decision problems. Graded Logic is primarily oriented towards real life and industrial decision-making applications where propositional calculus formulas can have 100–600 input arguments. The number of variables is a useful measure of complexity, but both the logic complexity and the significance of the decision-making problem can be visible even with the smallest number of inputs.

In the apricot buying example we assumed the use of a logic model of adjustable simultaneity that computes the overall suitability based on simultaneous high quality and high affordability of apricots. The same logic problem emerges in much more significant situations. For example, a car buyer simultaneously wants a valuable and affordable car. A homebuyer family simultaneously needs a quality home and a convenient home location, as the main components of home suitability. In the next step, the family simultaneously requires a high home suitability and an affordable home price. In industry, an electrical car company that decides whether to move the car production from its current location to another state must carefully evaluate whether the suitability of the new location and the overall cost of the whole operation are simultaneously acceptable. A government of a state that prepares for a war may decide to buy fighter jets (or a nuclear submarine) from a friendly superpower, and again, the buyer needs to simultaneously have the high suitability of the desired weapons and the affordability of that project. All these problems have very different significance and/or price, but the same logic structure.

All decision problems and solutions generated by graded logic models must be consistent with human intuitive judgement and decision making. This form of common-sense justification of logic models and their results holds equally for all degrees of problem significance, from apricots to fighter jets and nuclear submarines. Expectedly, human intuitive decision making is naturally limited to problems with a few input arguments, and the goal of graded logic models is to extend human reasoning far beyond the limits of ordinary intuitive decision problem solving. Fortunately, the common-sense justification can be applied to complex propositional calculus models, in a sequence of small steps, separately for each aggregator.

Even with only two variables the graded logic problem can become more complex. For example, when a rich buyer buys apricots or a car, the high quality of apricots or a car can be mandatory, and the affordability can be desired, but otherwise optional. So, now we need a logic function that is a model of asymmetric simultaneity: one input is mandatory and must support the annihilator 0 and the other input is optional and must not support the annihilator 0. How to make such a function? We discuss such problems in Chap. 16.

Absolute Evaluation Versus Relative Comparison in Decision Making

Graded Logic is related to decision making, and various problems of evaluation. Thus, it is useful to briefly introduce basic concepts of evaluation. The story of buying apricots for preserves can be expanded if there are prunes in a box located next to the apricots. Unavoidably, the buyer will also be exposed to the comparison of apricots and prunes since both are appropriate for preserves. This simple situation illustrates two fundamental evaluation principles. First, the comparison of two options *cannot* be based only on quality, or only on price; the comparison must be based on the overall suitability S, and most buyers think in that way. Second, the fact that suitability of one fruit is higher than the suitability of the other fruit does not mean that the decision-maker will buy the better option; the necessary condition is that the overall suitability must be above the desired minimum suitability threshold value S_{min}, and nothing will be purchased if this necessary condition is not satisfied. Thus, evaluation decisions cannot be based solely on pairwise comparisons. These facts hold equally for simple intuitive evaluations in a grocery store, and for complex and expensive industrial and other problems of evaluation, comparison, and selection of the best alternative.

At this early stage, while we are still thinking about logic used to buy apricots and prunes, it is useful to contrast and confront two different approaches to some frequent decision-making problems: *absolute evaluation* (AE) versus *relative comparison* (RC). When it comes to selecting the most suitable among several alternatives, both AE and RC offer solutions, and that is where the AE and RC approach might seem to be equivalent and competitive. There are various methods in each of these approaches [ISH10, HEN15], and some authors passionately promote one or the other. Utility theory [FIS64, FIS70, HEN15, KEE76, KEE92, KLE78, KLE80, DYE05] and scoring techniques [MIL66, MIL70, SCH69, SCH70, NIH13, DUJ18] exemplify the AE methods. AHP [SAA10, SAA77, SAA80, SAA90, SAA01, SAA06, LIB03], pairwise comparison techniques [FUL10, KOC14, KOC16], and outranking methods [FIG05, BRA05, BOU00, BEL02, TRI00] exemplify the RC approach.

The goal of AE is to evaluate a single object/candidate from the standpoint of the extent to which the object satisfies a set of stakeholder's requirements. The result of evaluation is an indicator of the overall value of each candidate. If there are multiple objects/alternatives, each object can be separately evaluated, and the ranking of alternatives is based on the comparison of overall degrees of suitability of evaluated objects. Therefore, AE logic decision methods can be used to solve both the problems of individual evaluation of a single candidate, and the problems of comparison by value of multiple candidates. The AE approach is based on clearly specified goals and requirements of a specific stakeholder/decision-maker. That is not a limitation, because if there are no goals and requirements of a stakeholder, then there is no logic problem of evaluation and comparison by value.

The RC approach basically reduces to a relative comparison and ranking of at least two alternatives without consideration of their individual values. In other words, the process of comparison and ranking is interpreted as the result of a zero-sum game. This is a typical situation in sport competitions and political elections. In all races, the winner is the fastest competitor. Gold, silver, and bronze medals are given to the three best competitors, regardless of their results. In political elections we can have several candidates, and the winner is the candidate who gets the highest number of votes. Of course, winning an election is not a convincing proof of the quality of the winner. It is only the proof of success in a zero-sum game where competitors try to get the highest number of votes, and each vote can be given to only one competitor. This problem can be considered completely different from the problem of evaluation based on value. Ranking by value and ranking by score in a zero-sum game are two different problems.

In [SAA10] there is an interesting example of presidential elections where there are two candidates: B and C. Candidate B is evaluated (by opinion polls) as a better candidate and predicted to win the election. Then, a third candidate, P, enters the competition. P is similar to B, so they split the votes, and the candidate C wins. Unfortunately, in [SAA10] this result is misinterpreted as a proof that the ranking by value of individual candidates is not possible and creates wrong results, because the ranking always depends on all competitors. Of course, voting is not the solution to an evaluation problem, and evaluation is not a substitute for a voting-based relative comparison. The candidates B, C, and P play a zero-sum game where they split N votes, and nothing else. Each vote that one candidate wins is the vote that all other candidates lose, and all pairwise comparisons can be reduced to the model of a zero-sum game. In addition, all votes only reflect the expected benefits for voters, and that is different from the "objective values" of candidates.

A simple test for classifying this type of decision methods is the question "can you evaluate a single candidate?" The affirmative answer shows AE decision methods which can be used both for the evaluation of a single candidate, and for the comparison of multiple candidates. As opposed to that, RC methods cannot be used for evaluation and can only produce the relative ranking of two or more candidates. Therefore, AE and RC are two different problems, and the value of a candidate is not equivalent to the result of a zero-sum game. However, both AE and RC include logic problems and need specific logic support.

If a value is interpreted as a graded human percept, then it can be expressed as a degree of truth. Thus, both the evaluation of alternatives, and the ranking by value, are logic problems because their solutions are based on formulas of a graded propositional calculus. Unsurprisingly, Graded Logic can be used in a variety of logic problems, including those that are not decision making. However, the primary goal of Graded Logic development is to create a human-centric logic infrastructure, and the necessary properties of such logic can be easily and naturally derived from observing logic decision problems of evaluation and ranking by value. Therefore, that approach pervades all the chapters of this book.

The Fundamental Problem of Evaluating a Single Alternative

If we define decision making as a problem where a specific decision-maker identifies m alternatives with intention to choose the best alternative, then it is important to know what happens if $m = 1$. Some authors ignore the existence of that problem. In this book, however, the problem of single alternative is considered the fundamental problem of decision making for two reasons: (1) it is a very frequent problem in human experience, and (2) the solution of the problem for $m = 1$ automatically solves the problems where $m > 1$. More precisely, in all cases where that is possible, the comparison of m alternatives (or candidates) can (and should) be reduced to m evaluations of a single alternative (or candidate). This school of thought eliminates the need for $m(m-1)/2$ pairwise comparisons, or outranking methods.

Let us first exemplify decision problems where $m = 1$. The most important decision in most human lives is marriage, and there is a single candidate that must be carefully evaluated. Indeed, the question is how suitable a single candidate is, and not who is the best (i.e. the least unsuitable) among several candidates. Generally, the best among several candidates/options, selected using pairwise comparison methods, can still be insufficiently suitable and justifiably rejected.

In most cases of buying a used car, or a used boat, a decision-maker has an opportunity to buy or not to buy a specific car or boat that unexpectedly became available. Thus, the decision-maker must perform a fast evaluation of a single object and make the accept/reject decision.

Similarly, an unemployed worker can get a single job offer and the question is whether the offer is sufficiently good to be accepted. On the other hand, a company can have a single candidate applying for an open position, and it is necessary to evaluate the competence of the single candidate and then either to accept or to reject the candidate.

In real estate a realtor offers to homebuyer a single home which must be individually evaluated and if it is accepted the search process terminates. If the home is rejected, then the search process continues. The same evaluation process is used when deciding whether to rent a specific residence for a family.

In the case of car driving exam (and many similar exams) there is a single candidate and an evaluator who must evaluate candidate's skills. There is a minimum skill level that the candidate must show, and if the candidate's skills are above that threshold value, the candidate gets a driver license. A doctoral thesis defense is also a single candidate evaluation and decision problem.

The presented examples expose the evaluation process of a single alternative/candidate as a fundamental component of human-centric decision making. In all cases, the decision is based on the overall suitability, which is a continuum-valued degree of satisfaction of decision-maker's requirements. The decision-maker also specifies a necessary minimum suitability which is the acceptability threshold value. If a single candidate evaluation process is available, then the comparison of multiple candidates is automatically solved by comparing the results of evaluation of individual candidates.

Graded Logic in the Goal-Oriented Decision Making

We assume that stakeholders/decision-makers are humans: either individuals or various human organizations. Such stakeholders have goals they want to achieve, and goal attainment is the primary motivation for stakeholder's activity. Before such activities there is a decision process which can be summarized in the following segmented form:

The stakeholder	=	a university graduate program
Stakeholder's goal	=	find exceptionally good candidates for the graduate program
Available options	=	5 candidates applying for admission in the graduate program
Suitability criterion	=	undergraduate GPA (average grade) of each candidate
Best option available	=	the best candidate in this group has GPA=3.3
Acceptance decision	=	no one is accepted because the university decided to use the minimum acceptance threshold GPA=3.5.

Fig. 1.1 A simple example of an evaluation-based stepwise decision process

- What are the goals I want to achieve?
- Based on my goals, what are my options?
- What is the suitability criterion for evaluation and comparison of options?
- What is the suitability of each option for reaching my goals?
- What is the best option, and why?
- Am I going to reject or accept (and realize) the best option?

An example of this stepwise decision process is shown in Fig. 1.1. The stakeholder's decision making is essentially an evaluation problem. The stakeholder must specify justifiable goals, identify the available goal attainment options, carefully evaluate the suitability of each option, compare them, and accept/reject the best (most suitable) option. Thus, the goal attainment depends on the correct decision making, and the decision making requires evaluation. Furthermore, each evaluation process requires a graded logic criterion model. Suitability and satisfaction are graded percepts, and the basic logic operations of simultaneity and substitutability are also a matter of degree. Graded Logic is an indispensable part of each evaluation criterion for evaluation of the overall suitability of all available options (one or more).

This short summary of the purpose and role of Graded Logic shows that a natural approach to the study of logic must be human-centric, i.e., primarily focused on the goal-oriented reasoning of a specific stakeholder. This approach yields the following central GL properties:

- *Observability*: GL is provably present and used in natural human commonsense reasoning.
- *Measurability*: GL properties can be derived from experimental measurements of human decision-making.
- *Modelability*: GL is amenable to quantitative description using mathematical models.
- *Completeness*: GL must be based on logic operations that are necessary and sufficient.
- *Usability*: GL is developed with the goal of successful use in decision engineering.
- *Explainability and confidence*: GL should be suitable for providing verbal justifications of mathematical models and their results in a way that supports confident decision-making for each specific stakeholder.

In conclusion, if we develop methodology for understanding and modeling logic processes of stakeholder's goal-oriented decision making, then we can use that knowledge to develop effective decision support methods and software systems and apply them in both human decision-making projects and automatic AI systems.

References

ACK72 Ackoff, R. C. and F. E. Emery. 1972. *On Purposeful Systems*. Aldine-Atherton, Chicago.

BEL02 Belton, V., and T.J. Stewart 2002. *Multiple Criteria Decision Analysis: An Integrated Approach*. Kluwer Academic Publishers.

BOO47 Boole, G. 1947. *The Mathematical Analysis of Logic*. Cambridge: MacMillan, Barclay & MacMillan.

BOO54 Boole, G. 1853. *An Investigation of the Laws of Thought, on Which Are Founded the Mathematical Theories of Logic and Probabilities*. Macmillan.

BOU00 Bouyssou, D., T. Merchant, M. Pirlot, P. Perny, A. Tsoukias, and P.Vincke, *Evaluation and Decision Models: A Critical Perspective*. Kluwer.

BRA05 Brans, J-P., and B. Mareschal. 2005. PROMETHEE Methods. In *Multiple Criteria Decision Analysis* ed. J. Figueira, S. Greco, and M. Ehrgott, 163–196, Springer.

BUN84 Bunn, D. W. 1984. *Applied Decision Analysis*. McGraw-Hill Book Company.

DEM47 De Morgan, A. 1847. *Formal Logic or the Calculus of Inference Necessary and Probable*. London: Taylor and Walton.

DUJ18 Dujmović, J. (2018). *Soft Computing Evaluation Logic*. J. Wiley.

DUJ73b Dujmović, J. 1973. A generalization of some functions in continuous mathematical logic—Evaluation function and its applications, (In Serbo-Croatian). In *Proceedings of the Informatica Conference*, paper d27, Bled, Yugoslavia.

DUJ73c Dujmović, J. 1973. Mixed Averaging by Levels (MAL)—A System and Computer Evaluation Method, (In Serbo-Croatian). In *Proceedings of the Informatica Conference*, paper d28, Bled, Yugoslavia.

DUJ74a Dujmović, J. (1974). Weighted conjunctive and disjunctive means and their application in system evaluation. *Journal of the University of Belgrade, EE Department., Series Mathematics and Physics* 483, 147–158.

DUJ75a Dujmović, J. 1975. Extended continuous logic and the theory of complex criteria. *Journal of the University of Belgrade, EE Department., Series Mathematics and Physics*, 537, 197–216.

DUJ75b Dujmović, J. 1975. Evaluation of digital computers using the system evaluation method MAL, (In Serbo-Croatian). In *Proceedings of the Informatica Conference 1975, paper 6.9*, Bled, Yugoslavia.

DUJ17 Dujmović, J. 2017. Relationships Between Fuzziness, Partial Truth and Probability in the Case of Repetitive Events. In *Uncertainty Modeling* ed. V. Kreinovich, 61-69, Springer.

DYE05 Dyer, J.S. 2005. MAUT—Multiattribute Utility Theory. In *Multiple Criteria Decision Analysis* ed. J. Figueira, S. Greco, and M. Ehrgott, 265–296, Springer.

EVA17 Evans, J., and B.T. St. 2017. *Thinking and Reasoning*. Oxford University Press. Oxford: United Kingdom.

FIG05 Figueira, J., V. Mousseau, and B. Roy. 2005. ELECTRE Methods. In *Multiple Criteria Decision Analysis* ed. J. Figueira, S. Greco, and M. Ehrgott, 133–162, Springer.

FIS64 Fishburn, P. C. 1964. *Decision and Value Theory*. J. Wiley.

FIS70 Fishburn, P. C. 1970. *Utility Theory for Decision-Making*. J. Wiley.

FUL10 Fülop, J., W.W. Koczkodaj, and S.J. Szarek. 2010. *A Different Perspective on a Scale for Pairwise Comparisons*. Transactions on CCI I, edited by N.T. Nguyen and R. Kowalczyk. LNCS 6220, 71–84, Springer-Verlag.

GIN68 Ginzburg, S.A. 1968. *Математическая непрерывная логика и изображение функции (Mathematical continuous logic and representation of functions)*, Energiya, Moskva.

HEN15 Hensher, D.A., J.M. Rose, and W.H. 2015. *Greene, Applied Choice Analysis*. Cambridge University Press.

JOS64 Joslin, E.O., and M.J. Mullin. 1964. Cost-value technique for evaluation of computer system proposals. In *Proceedings-Spring Joint Computer Conference*, 367–381.

JOS68 Joslin, E.O. 1968. *Computer Selection*. Addison-Wesley. (An augmented edition was published in 1977 by The Technology Press Inc., Box 125, Fairfax Station, Virginia 22039).

JOS71 Joslin, E.O. 1971. *Analysis, Design and Selection of Computer Systems. Computer Reading Series*. College Reading Ins.

KAH11 Kahneman, D. 2011. *Thinking, Fast and Slow*. New York: Farrar, Strauss and Giroux.

KAN70 Kanter, J. 1970. *Management Guide to Computer System Selection and Use*. Prentice-Hall.

KEE76 Keeney R.L., and H. Raiffa. 1976. *Decisions with Multiple Objectives/Preferences and Value Tradeoffs*. J. Wiley.

KEE92 Keeney R.L. 1992. *Value-Focused Thinking*. Harvard University Press.

KEN73 Kenney, D.P. 1973. *Minicomputers: Low-Cost Computer Power for Management*. AMACOM.

KLE78 Kleijnen, J.P.C. 1978. *Scoring Methods, Multiple Criteria, and Utility Analysis. Research Memorandum, Tilburg University, Holland, Department of Economics* (Published also in Sigmetrics Performance Evaluation Review).

KLE80 Kleijnen, J. P. C. 1980. *Computers and Profits*. Addison-Wesley.

KOC14 Koczkodaj W.W., K. Kulakowski, and A. Ligeza. 2014. On the quality evaluation of scientific entities in Poland supported by consistency-driven pairwise comparison method. *Scientometrics* 99, 911–926.

KOC16 Koczkodaj W.W., L. Mikhailov, G. Redlarski, M. Soltys, J. Szybowski, G. Tamazian, E. Wajch, and K.K.F. Yuen. 2016. Important facts and observations about pairwise comparisons. *Fundamenta Informaticae* https://doi.org/10.3233/FI-2012-0000.

LIB03 Liberatore, M.J, and R.L. Nydick, *Decision Technology, Modeling, Software, and Applications*. Wiley.

MCQ78 McQuaker, R.J.1978. *Computer Choice*. North-Holland, Amsterdam.

MIL56 Miller, G.A. 1956. The magical number seven, plus or minus two: Some limits on our capacity for processing information. *The Psychological Review*, 63, 81–97.

MIL66 Miller, J.R., III. 1966. *The Assessment of Worth: A Systematic Procedure and its Experimental Validation*. Ph.D. Dissertation, M.I.T.

MIL70 Miller, J.R., III. 1970. *Professional Decision-Making*. Praeger, New York.

NIH13 *NIH, Scoring System and Procedure*. Last reviewed on March 25, 2013. http://grants.nih.gov/grants/peer/guidelines_general/scoring_system_and_procedure.pdf.

PRI09 Priest, G. 2009. *An Introduction to Non-classical Logic*, 2nd ed. Reprinted with corrections. Cambridge University Press.

RAD00 Radojević, D.G. 2000. [0,1]-valued logic: a natural generalization of Boolean logic. *Yugoslav Journal of Operations Research* 10 (2), 185–216.

SAA01 Saaty, T.L. 2001. *Fundamentals of Decision-Making and Priority Theory*, 2nd ed. RWS Publications, Pittsburgh.

SAA06 Saaty, T.L., and L.G. Vargas. 2006. *Decision-Making with the Analytic Network Process*. Springer.

SAA10 Saaty, T.L. 2010. *Principia Mathematica Decernendi*. Pittsburgh: RWS Publications.

SAA77 Saaty, T.L. 1977. A Scaling Method of Priorities in Hierarchical Structures. *Journal of Mathematical Psychology* 15, 234–281.

SAA80 Saaty, T.L. 1980. *The Analytic Hierarchy Process*. McGraw-Hill.

SAA90 Saaty, T.L. 1990. *Decision-Making for Leaders*. RWS Publications, Pittsburgh.

SCH69 Scharf, T. 1969. Weighted Ranking by Levels. *IAG Journal* 2(2), 7–18.

SCH70 Scharf, T. 1979. Weighted Ranking by Levels Computer Evaluation Method—One Year of Experience. *IAG Journal, 3*(3), 71–91.

SHA69 Sharpe, W.F. 1969. *The Economics of Computers. A RAND Corporation Research Study*. Columbia University Press, New York.

STE22 Steinbach, B., and C. Posthoff. 2022. *Logic Functions and Equations. Fundamentals and Applications using the XBOOLE-Monitor*. Springer.

TRI00 Triantaphyllou, E. 2000. Multi-*Criteria Decision-Making Methods: A Comparative Study*. Kluwer Academic Publishers.

YAG04 Yager, R.R. 2004. Weighted triangular norms using generating functions. *International Journal of Intelligent Systems*, 19, 217–231.

ZAD65 Zadeh, L.A. 1965. Fuzzy Sets. *Information and Control* 8(3), 338–353.

Chapter 2
Human-Centric Approach to Propositional Logic: Commonsense Reasoning and Decision Making

The main goal of Chap. 1 was to show that it is necessary to clearly differentiate two approaches to propositional logic: the pragmatic *human-centric approach* and the theoretic *formal axiomatic approach*. In this chapter we discuss various properties of the human-centric approach in the case of commonsense reasoning and decision making and show main advantages of this approach. We also show that in the area of propositional calculus GL is a generalization of fuzzy logic.

Human commonsense propositional logical reasoning is observable, measurable, modelable, and explainable: it is a part of physical reality. In the real physical world, logic does not exist outside of human mental activities. Thus, it is possible (and necessary) to define logic in a natural human-centric way, i.e., strictly as the mathematical modeling of those human mental activities that are based on logic. Such activities occur primarily in the context of decision making. Based on that definition, logic is a part of observable *physical reality* and assumes the active presence of humans, including human interests, human reasoning, commonsense decision-making, and communication using a natural language. This approach is directly or indirectly oriented towards applications, i.e., building a mathematical infrastructure suitable for expanding justifiable logic models beyond natural limitations of human commonsense reasoning. Such models are the armamentarium of decision engineering.

On the other hand, the formal axiomatic approach creates a *virtual reality*: an abstract mathematical universe without the presence of humans. For example, the frequent assumption (definition) that basic logic operations satisfy commutativity automatically eliminates the presence of human beings in logic: the claim that all truths are equally important excludes the concept of importance, and that is never visible in human reasoning. Therefore, the formal axiomatic approach creates a virtual reality that is not focused on applications.

J. Dujmović, *Graded Logic*, Intelligent Systems Reference Library 273,
https://doi.org/10.1007/978-3-031-88558-7_2

The Properties of Human-Centric Approach

In this book we analyze the graded propositional logic mostly in the context of decision-making, because decision-making is a ubiquitous activity, easily observable in the behavior of society, organizations and individuals. Of course, the graded propositional logic can also be analyzed as a formal mathematical system without any specific context [RAD00]. It can also be analyzed in any other specific context. For example, S.A. Ginzburg (the author of one of early books on mathematical continuous logic [GIN68]) considers mathematical continuous logic as a tool for development of electronic circuits used as function generators.

The point of departure for development of human-centric logic is that logic models must quantitatively describe observable and measurable natural human reasoning in precisely defined situations where human reasoning can be explained and empirically verified. In the case of Graded Logic (GL), which we present in this book, in the center of interest is the logical reasoning involved in the decision-making process of a known specific (i.e., precisely defined) stakeholder/decision-maker (a single human, or a human organization).

First and foremost, based on an anthropocentric standpoint, we assume that logic, as an applied science, must be consistent with observable, measurable, modelable, and explainable properties of human natural commonsense reasoning. In addition, logic (and GL in particular) should serve for quantitative modeling of the observable and experimentally measurable processes of human cognition and decision-making. In the context of GL, we are interested only in human mental logic activities and not in computer programs that try to behave as decision-makers. Of course, GL provides an infrastructure for decision-support systems, and not the other way around. Therefore, in the context of the human-centric Graded Logic, we hold the following to be self-evident:

1. Decision-making is assumed to be a human mental process. All decisions reflect the interests, goals, and requirements of a specific stakeholder/decision-maker.
2. The stakeholder/decision-maker can be an individual, or a group of people (e.g., a family), or a human organization of any size and type. In the case of complex decisions, a typical stakeholder is a company (or another organization) where a team of experts identifies one or more development alternatives, and then systematically creates a justifiable quantitative logic criterion for evaluation and selection of the best alternative.
3. Decisions exist only in the context of stakeholder activities undertaken to satisfy desired goals. In other words, the purpose of decision-making is a stakeholder's goal attainment.
4. Goals, requirements, values, and necessary degrees of satisfaction exist only in the stakeholder's mind, or as a collective opinion of a group of experts responsible for specifying the goals of an organization. Consequently, they cannot be "objective." They reflect the best interests of the stakeholder, according to the stakeholder's best assessment, and they could be classified as "subjective." However, the common naïve view of subjective as bad and objective as good is neither reasonable nor applicable in this context. In addition, the quantitative

modeling of subjective percepts (e.g., importance and suitability) is possible, necessary, and useful.

5. Since the specification of goals and interests is strictly the stakeholder's responsibility, all assessments whether goals are appropriate, justifiable, rational, or irrational are beyond the scope of Graded Logic.
6. The scope of Graded Logic is the study of properties of the "mental processor" used for commonsense decision-making based on continuously adjustable degrees of truth. The same mental processor is used for processing degrees of truth related to all goals (rational, irrational, and others).
7. Regardless of the type of stakeholder/decision-maker, the process of decision-making must follow the same logic rules. In other words, all formal logic constructs must be consistent with natural human logical reasoning.
8. To achieve selected goals, a stakeholder must first identify all existing alternatives (options), and then organize decision-making as a process of selecting the most suitable alternative.
9. Relative ranking of alternatives is not a proof that the identified alternatives sufficiently contribute to the attainment of stakeholder's goals. Even the highest-ranking alternative can be unacceptable.
10. The number of alternatives/candidates/options can be one. The case of a single candidate reveals the very essence of decision-making: if the stakeholder can determine/compute the extent to which the single candidate satisfies stakeholder's goals (i.e., the candidate suitability degree), then it is possible to decide whether to accept or to reject the candidate.
11. If a stakeholder can evaluate each alternative/candidate, then the problem of comparison and selection of the best alternative is automatically solved.
12. Evaluation of alternatives is the key part of decision-making. The result of evaluation is the degree of overall suitability of the analyzed candidate/object/alternative.
13. Any degree of suitability is a human graded percept, i.e., a percept that is a matter of degree, in the range from the lowest (coded as 0) to the highest (coded as 1).
14. By its very nature, the degree of overall suitability is the degree of truth of the statement/assertion that the evaluated object/alternative completely satisfies all stakeholder's requirements. The statements that claim a full satisfaction of requirements (i.e. a high value of an object) can be classified as *value statements*.[1]
15. The overall suitability is a propositional logic function of suitability attributes that specify individual stakeholder's requirements. An increase of an input attribute suitability cannot cause a decrease of the overall suitability (the aggregation property of nondecreasing monotonicity).

---

[1] There are many statements that are not value statements. E.g., "each triangle has three sides" is a statement, but not a value statement. However, in decision-making, value statements have a dominant role.

16. Graded percepts are not anonymous real numbers: they always have semantic identity, i.e., an interpretation (and/or an explanation) related to the activities and goals of a specific stakeholder/decision-maker and the selected stakeholder's goals.
17. Evaluation is a fundamental component of decision making, and evaluation criteria specify graded logic conditions of simultaneity, substitutability, importance, and impact of logic arguments. Therefore, some form of graded propositional logic is necessarily used in the commonsense evaluation process.
18. Graded propositional logic operates with degrees of truth of value statements. Value statements do not exist in vacuum: each statement is in some way associated with a specific stakeholder/decision-maker. Therefore, graded propositional logic must account for the presence, role, and interests of a specific stakeholder/decision-maker.
19. In natural human decision-making, all statements have a role and a meaning that directly affect an adjustable degree of importance for a specific stakeholder.
20. If two or more value statements are simultaneously present and must be simultaneously highly satisfied (sufficiently true) to support stakeholder's goals, then the stakeholder intuitively selects a desired degree of simultaneity[2], which must also be an adjustable graded parameter in the corresponding propositional calculus formulas.
21. The highest degree of simultaneity is obtained using the *drastic conjunction* function which returns the value 1 (completely true) if and only if all arguments are completely true. Otherwise, it returns 0 (completely false)[3]. The highest degree of simultaneity can be verbalized as two conditions: everything is completely satisfied (1), and everything is not completely satisfied (0). Obviously, "everything is perfectly satisfied" is the highest level of simultaneity.
22. If a group of two or more statements describe properties that can substitute each other, so that each of them can (partially or completely) satisfy the group requirements, then the stakeholder selects/adjusts an appropriate desired degree of substitutability and uses it in intuitive reasoning and/or in the development of propositional calculus formulas.
23. The highest degree of substitutability is obtained using the *drastic disjunction* function which returns the value 0 (completely false) if and only if all arguments are completely false. Otherwise, it returns 1 (completely true), giving the highest substitutability power to all arguments: even a single nonzero input is sufficient to fully satisfy the drastic disjunction. Verbally, the drastic disjunction differentiates between two conditions: nothing is satisfied (0), and at least something is to some positive degree satisfied (1).

---

[2] It is important to note that in decision-making the desired property is the *simultaneity of high satisfaction* and not the equivalence (similarity) of arbitrary satisfaction degrees. E.g., $z = \min(x, y)$ gives high z only if both x and y are simultaneously high; on the other hand, $z = 1 - |x - y|$ gives a high value of z whenever $x \cong y$.

[3] In [GRA09] the drastic conjunction function is rightly named the *smallest aggregation function*.

24. If models of simultaneity (*conjunction*) describe patterns of human reasoning, then the intensity of simultaneity must be graded and adjustable. For desired high levels of simultaneity, all inputs can be mandatory, so that any zero input produces the zero output. For lower levels of simultaneity, that property is not necessary (and not acceptable), and the inputs can be optional. So, 0 plays the role of a selectable annihilator for the high-intensity operations of mental graded conjunction.
25. In human reasoning, the strength of substitutability (*disjunction*) is graded and adjustable. For high levels of substitutability, any fully satisfied input (having the satisfaction degree1) can be sufficient to result in a fully satisfied output. Thus, the degree of truth 1 plays the role of annihilator for the high-intensity operations of graded disjunction. For lower degrees of substitutability, inputs can be optional, and then the annihilator 1 is not supported and not used in (both mental and formal) graded disjunction. Generally, in logic models of simultaneity and substitutability, the annihilators 0 and 1 must be selectable: they can be either indispensable, or unacceptable.
26. In human reasoning, simultaneity and substitutability are complementary and concurrently present: more substitutability means less simultaneity and vice versa. Substitutability means the capability to compensate for the absence (or insufficient presence) of specific desired property by the sufficient presence of another desired property, and that necessarily/automatically causes a decrease of the need for their simultaneous presence.
27. In all cases, the natural human models of simultaneity and substitutability are dual. In verbal interpretation, if two properties must be (to some extent) simultaneously present, then that requirement is equivalent to the requirement that it is not acceptable that any one of them (either the first one *or* the second one) is insufficiently present. So, in a general case, formal models of simultaneity and substitutability are assumed to support duality. That also holds for any number of inputs greater than two.

Based on the above basic principles, let us again emphasize that a human-centric graded propositional calculus must be focused on modeling observable natural human logical reasoning. Consequently, it must support the following essential properties:

- Graded (continuum-valued) truth of statements.
- Graded (continuum-valued) importance of statements.
- Graded (continuum-valued) degrees of simultaneity/substitutability of the high truth of two or more statements.
- Simultaneous presence and complementary unification of logic simultaneity and substitutability (supported by the corresponding quantitative model) in the full range from the ultimate (drastic) conjunction to the ultimate (drastic) disjunction.
- Selectability of idempotency and the selectability of annihilators for idempotent models of simultaneity and substitutability.

There is not the slightest doubt that in human reasoning, we adjust continuum-valued degrees of truth, importance, simultaneity, and substitutability, as illustrated

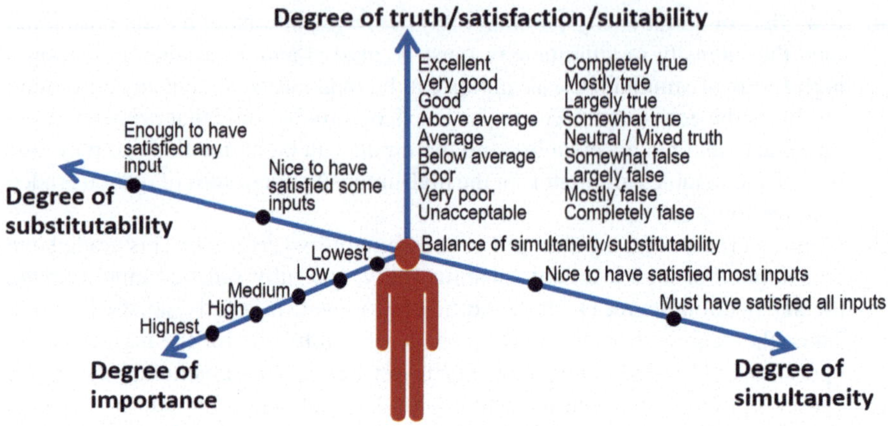

Fig. 2.1 Simultaneity, substitutability, importance, and truth/satisfaction/suitability as adjustable continuum-valued coordinates of human commonsense reasoning

in Fig. 2.1. Therefore, in human-centric logic, there should be no bivalent truth, no mandatory commutativity, no fixed andness, and no fixed orness. The main goal of Graded Logic is to provide a graded propositional calculus that supports these indispensable properties selected from observing natural human commonsense reasoning. Among possible applications of graded propositional calculus, these properties are provably necessary for decision-making. More specifically, these properties are indispensable for logical reasoning about the overall suitability of any product: from apricots and prunes to nuclear submarines.

Graded Logic as a Generalization of the Fuzzy Logic Propositional Calculus

Graded Logic was born in 1973 as a synthesis of classical Boolean logic and the theory of means [DUJ73b, DUJ73c]. Since conjunction corresponds to the minimum function and disjunction to the maximum function, it became evident that means can be interpreted as logic functions in the range between the traditional conjunction and disjunction. From that foundation, Graded Logic expanded in various directions, aiming to serve as the logic of evaluation and decision-making [DUJ18]. Graded Logic is similar to some forms of fuzzy logic.

Fuzzy logic has multiple interpretations. In its simplest form, fuzzy logic is interpreted as a propositional logic where degrees of truth belong to the interval [0,1], and logical operations are realized using t-norms. GL generalizes this form of fuzzy logic. A more precise classification, presented in [CIN23], identifies "mathematical fuzzy logic," which originated from the many-valued logic work of Łukasiewicz [GOT22] and was subsequently expanded by numerous authors in mathematics and philosophy [NGU06, PRI09, WIK25]. The introduction of fuzzy sets by Lotfi Zadeh [ZAD65] further broadened the meaning of fuzzy logic to encompass all models (and various applications) directly or indirectly related to the concept of fuzzy sets and their graded (partial) membership. This area can be referred to as "engineering fuzzy logic" because its primary motivation stems from engineering applications [SHA16, KLI95, MEN01, YAG94, ROS10, RAY15, KOS93, KAC15].

In Graded Logic, the fundamental concept is the degree of truth of a statement. In the engineering fuzzy logic, the fundamental concept is the degree of membership of an object in a fuzzy set. Suppose that an object Z must satisfy a set of requirements. Based on the actual properties of object Z, suppose that the statement "Z completely satisfies all requirements" has the degree of truth $t \in [0, 1]$. In the case of fuzzy logic, let $p \in [0, 1]$ denote the degree of membership of object Z in a fuzzy set of objects that fully satisfy all requirements. Expectedly, $p = t$ and the degree of truth and the degree of fuzzy membership are two completely equivalent ways to report the same property of object Z, which is the partial satisfaction of the given set of requirements [DUJ17]. In the area of propositional calculus, GL is a substantial generalization of fuzzy logic, expanding the concepts of conjunction, disjunction and other logic aggregation functions towards the observable properties of human reasoning. On the other hand, the term fuzzy logic is used for an area which is significantly wider than the propositional calculus and evaluation decision making [SHA16, ROS10, KLI95]. So, the graded logic and the fuzzy logic have an intersection in the area of propositional calculus. However, their union is much wider than their intersection. The primary interest of GL is the logic processing of partial truth in decision-making problems, and the primary interest of fuzzy logic is a spectrum of problems related to systems characterized by various forms of imprecision [KLI95, MEN01].

The formal theoretical approach to logic assumes that logic is constructed as a *formalized axiomatic deductive system* [RAS74, PRI09]. As an example of *formal logic* let us take the t-norm fuzzy logic, defined in [WIK22] as a logic that uses t-norms for permissible interpretations of conjunction. Then, (without justification or regret) the point of departure is the definition that the truth function of conjunction is *assumed to satisfy* commutativity, associativity, nondecreasing monotonicity, continuity, and the neutral element 1. The t-norm fuzzy logic is then declared as a "theoretical basis for approximate reasoning" but without a definition of the meaning of "approximate reasoning," without an explanation of the role and activity of human subjects involved in any form of reasoning, and without justification of the role of assumptions used as a basis of "approximate reasoning." In particular, the fundamental question why commutativity must be present in approximate reasoning remains without answer. This approach is typical for mathematical fuzzy logic, and fully unacceptable in Graded Logic.

It is important to emphasize that from a mathematical standpoint the formal axiomatic approach to fuzzy logic is fully legitimate. Basic assumptions are specified as a foundation of that logic. As long as the fundamental assumptions are consistent (free of contradictions), independent, and complete (sufficient to prove the truth of statements), their specific purposes and utilitarian values will not be questioned. On that foundation, it is possible to construct an edifice of derived properties and propositional logic models, where each component is strictly proved to be correct (consistent with all fundamental assumptions). The applicability or practicality of such an edifice is not a necessary condition and usually not a topic for discussion. It is perfectly legitimate to develop logic models that are unrelated to stakeholders, all truths can be considered equally important, the meaning of statements can be ignored, and a propositional calculus can operate in semantic vacuum, as processing of anonymous real numbers. However, if applicability is desirable or necessary, we must consider all requirements that reflect the goals and interests of stakeholders involved in logic-based reasoning.

In the area of propositional calculus, the Graded Logic is in all substantial characteristics a generalization of fuzzy logic. First, the logic range of the t-norm fuzzy logic is only in the areas of hyperconjunction and hyperdisjunction, i.e., in the areas where the strength of simultaneity is greater than or equal to the pure conjunction (the minimum function), Similarly, the strength of substitutability is greater than or equal to the pure disjunction (the maximum function). Another serious limitation of fuzzy logic is that traditional t-norm and t-conorm aggregators are not andness-directed, i.e., not parameterized for the continuous adjustment of the necessary intensity of simultaneity or substitutability, and the continuous transition in the whole range of hyperconjunction/hyperdisjunction[4]. In addition, as shown in Chapter 14, the hyperconjunctive and hyperdisjunctive aggregators cannot be used for modelling simultaneity of properties, which is necessary for logic evaluation decision problems.

In contrast, the Graded Logic continuously spans the entire spectrum of logic aggregators, from ultimate (drastic) conjunction to ultimate (drastic) disjunction (see Chap. 9). It includes the area of hyperconjunction and employs t-norms, such as $\min(x_1, \ldots, x_n)$ or $\prod_{i=1}^{n} x_i$, as models of very high simultaneity. Additionally, it uses t-conorms as models of very high substitutability. Beyond this, graded logic seamlessly extends into the area of weighted means (averaging operators), encompassing the idempotent versions of graded conjunction and graded disjunction.

Graded Logic functions consistently use weighted variables to implement the necessary variations of importance of individual inputs. Of course, all weights can be equal and in such a case, the graded logic in the areas of hyperconjunction and hyperdisjunction can have same properties as the fuzzy logic. Graded Logic satisfies nondecreasing monotonicity and continuity in the whole range of the adjustable strength of simultaneity and substitutability.

Outside the logic range of fuzzy logic, the Graded Logic satisfies associativity with errors that are negligible in applications [DUJ75a]. Similar to the extended neutral elements described in [GRA09], an idempotent graded logic function $F_n(x_1, \ldots, x_n)$ has n neutral concordance values $e_1, \ldots, e_n$, which are determined as the solutions to the equations: $F_n(x_1, \ldots, x_{k-1}, e_k, x_{k+1}, \ldots, x_n) = e_k = F_{n-1}(x_1, \ldots, x_{k-1}, x_{k+1}, \ldots, x_n), \;\; k = 1, \ldots, n$.

---

[4] Andness-directed aggregation for averaging operators was introduced by Larsen [LAR03]. Based on the concept of global andness [DUJ74a], the global andness directedness in the whole range from drastic conjunction to drastic disjunction is implemented using interpolative aggregators [DUJ14, DUJ18, DUJ21b, DUJ21c]. Global andness-directed t-norms and t-conorms were introduced by Torra [TOR22].

References

CIN23 Cintula, Petr, Christian G. Fermüller, and Carles Noguera, "Fuzzy Logic", *The Stanford Encyclopedia of Philosophy* (Summer 2023 Edition), ed. Edward N. Zalta & Uri Nodelman. https://plato.stanford.edu/archives/sum2023/entries/logic-fuzzy/.

DUJ14 Dujmović, J. 2014. Interpolative GCD aggregators. In *Proceedings of the 2014 IEEE World Congress on Computational Intelligence*, Beijing, China July 6-11. https://doi.org/10.1109/FUZZ-IEEE.2014.6891841, 1778-1785.

DUJ17 Dujmović, J. 2017. Relationships between fuzziness, partial truth and probability in the case of repetitive events. In *Uncertainty Modeling*, ed. V. Kreinovich, 61–69, Springer.

DUJ18 Dujmović, J. 2018 *Soft Computing Evaluation Logic*. J. Wiley, 2018.

DUJ21b Dujmović, J. and V. Torra. 2022. Aggregation functions in decision engineering: ten necessary properties and parameter-directedness. In *Proceedings of the 3rd International Conference on Intelligent and Fuzzy Systems, INFUS 2021 conference*, Izmir, Turkey. In C. Kahraman et al. (Eds.): *Intelligent and Fuzzy Techniques for Emerging Conditions and Digital Transformation*, Vo. 1, Springer, LNNS 307, 173–181.

DUJ21c Dujmović, J. and V. Torra. 2021. Properties and comparison of andness-characterized aggregators. *International Journal of Intelligent Systems*, 36 (3), 1366–1385.

DUJ73b Dujmović, J. 1973. A generalization of some functions in continuous mathematical logic—Evaluation function and its applications, (In Serbo-Croatian). In *Proceedings of the Informatica Conference, paper d27*, Bled, Yugoslavia.

DUJ73c Dujmović, J. 1973. Mixed averaging by levels (MAL)—A system and computer evaluation method, (In Serbo-Croatian). In *Proceedings of the Informatica Conference, paper d28*, Bled, Yugoslavia.

DUJ74a Dujmović, J. 1974. Weighted conjunctive and disjunctive means and their application in system evaluation. *Journal of the University of Belgrade, EE Department, Series Mathematics and Physics* 483, 147–158.

DUJ75a Dujmović, J. 1975. Extended continuous logic and the theory of complex criteria. *Journal of the University of Belgrade, EE Department, Series Mathematics and Physics, 537*, 197–216.

GIN68 Ginzburg, S.A. 1968. *Математическая непрерывная логика и изображение функции (Mathematical continuous logic and representation of functions)*. Energiya, Moskva.

GOT22 Gottwald, Siegfried. 2022. Many-Valued Logic, *The Stanford Encyclopedia of Philosophy* (Summer 2022 Edition), ed. Edward N. Zalta https://plato.stanford.edu/archives/sum2022/entries/logic-manyvalued/.

GRA09 Grabisch, M., J-L. Marichal, R. Mesiar, and E. Pap. 2009. *Aggregation Functions*. Cambridge University Press, Cambridge.

KAC15 Kacprzyk, J. and W. Pedrycz (Eds.). 2015. *Springer Handbook of Computational Intelligence*. Springer.

KLI95 Klir, G.J. and B. Yuan. 1995. *Fuzzy Sets and Fuzzy Logic*. Prentice-Hall.

KOS93 Kosko, B. 1993. *Fuzzy Thinking*. New York, Hyperion.

LAR03 Larsen H.L. 2003. Efficient Andness-directed importance weighted averaging operators. *International Journal of Uncertainty, Fuzziness and Knowledge-Based Systems* 12 (Suppl.) 67–82.

MEN01 Mendel, J.M. 2001. *Uncertain Rule-Based Fuzzy Logic Systems: Introduction and New Directions*. Prentice-Hall.

NGU06 Nguyen, H.T. and E.A. Walker. 2006. *A First Course in Fuzzy Logic*, 3rd edn. Chapman and Hall/CRC.

PRI09 Priest, G. 2009. An *Introduction to Non-Classical Logic*. 2nd edn reprinted with corrections. Cambridge University Press, 2009.

RAD00 Radojević, D.G. 2000. [0,1]-valued logic: A natural generalization of Boolean logic. *Yugoslav Journal of Operations Research, 10* (2), 185–216.

RAS74 Rasiowa, H. 1974. *An Algebraic Approach to Non-classical Logics*. PWN and North-Holland.

RAY15 Ray, K.S. 2015. *Soft Computing and Its Applications.* Apple Academic Press.
ROS10 Ross, T.J. 2010. *Fuzzy Logic With Engineering Applications*, 3d edn. J. Wiley.
SHA16 Shahbazova, S.N., (Ed.) 2016. *Fuzzy Logic and its Applications.* Selected Papers by Lotfi A. Zadeh. World Conference on Soft Computing.
TOR22 Torra, V. 2022. Andness Directedness for t-Norms and t-Conorms. *Mathematics, 10*, 1598. https://doi.org/10.3390/math10091598.
WIK22 Wikipedia. 2022. *The T-norm Fuzzy Logics.* https://en.wikipedia.org/wiki/T-norm_fuzzy_logics.
WIK25 Wikipedia, Fuzzy Logic. https://en.wikipedia.org/wiki/Fuzzy_logic. (Retrieved 1/18/2025).
YAG94 Yager, R.R. and D.P. Filev. 1994. *Essentials of Fuzzy Modeling and Control.* Wiley.
ZAD65 Zadeh, L.A. 1965. Fuzzy Sets. *Information and Control 8*(3), 338–353.

Chapter 3
Postulates of Graded Logic as a Fully Continuum-Valued Propositional Logic

Human reasoning and decision making are based on natural human logic, which is observable, measurable, modellable, and explainable. We believe that mathematical logic models should be derived from observable and measurable natural human logic. Consequently, the goal of Graded Logic is to model observable properties of natural human logical reasoning, and the contents of this book are selected and organized with intention to achieve this human-centric goal. Consequently, Graded Logic is not a formalized axiomatic deductive system. Instead of a set of axioms, GL is built on top of a set of human-centric postulates which play the role equivalent to axioms. The goal of this chapter is to present and justify ten basic GL postulates.

The Concept of Fully Continuum-Valued Propositional Logic

Classical and non-classical logics are usually classified according to the number of truth values that variables can take. There are three categories:

(1) Bivalent logics where variables take values from the set {0, 1}.
(2) Many-valued logics, where variables take values from the set of any number of members, primarily 3 or more.
(3) Continuum-valued logics where variables take values form the interval [0, 1].

Based on the human-centric approach, in this book we are interested only in continuum-valued logics because Graded Logic belongs to that category. However, our interpretation of human-centrism assumes more than just continuum-valued logic variables. As discussed in Chap. 2 (Fig. 2.1), the idea of human-centric continuum-valued Graded Logic is that logic models must describe operations with the following four fundamental logic percepts of a specific stakeholder/decision-maker:

- Truth
- Simultaneity/conjunction
- Substitutability/disjunction
- Importance

J. Dujmović, *Graded Logic*, Intelligent Systems Reference Library 273,
https://doi.org/10.1007/978-3-031-88558-7_3

The unique distinctive property of Graded Logic is that it is *fully continuum-valued*: all four logic percepts are graded, i.e., they belong to their respective intervals. We use the term "graded percept" to denote continuum-valued, explicitly visible, and fully adjustable percept. Non-classical continuum-valued logics, like fuzzy logics [PRI09, ROS10, KLI95, NGU06], are *partially continuum-valued*, in the sense that only truth is continuum-valued.

The most important graded percept in Graded Logic is the graded truth which always has specific "semantic identity" (it has a specific interpretation, role, and importance for a specific stakeholder/decision-maker, i.e., it is not an anonymous real number). The non-classical logics interpret continuum-valued truth as an anonymous real number from the unit interval [0, 1] without any other interpretation/identity [PRI09]. Contrastingly, Graded Logic uses the human-centric concept of "graded truth" which means that truth is both continuum-valued and assigned to a specific statement of a known stakeholder/decision-maker. So, each degree of truth has a specific interpretation and meaning. In addition, each graded truth is always paired with a *graded importance* which reflects goals and requirements of specific stakeholder/decision-maker. The concept of *semantic identity* implies that each graded truth has attributes that include *stakeholder*, *importance*, and *specific statement*. This concept is based on the interpretation of semantics as consideration of meaning, in a traditional idiomatic sense.

The need for a human-centric approach to logic might be considered self-evident. Indeed, it is difficult to use the results of logic models that are inconsistent with common sense and everyday experience in reasoning and decision-making. On the other hand, it is perfectly legitimate to create mathematical logic theories that are not human-centric and not developed with intention to be applicable. In addition, mathematical logic models need not necessarily be related to decision-making. Consequently, let us first clearly define the scope of Graded Logic, to help create realistic expectations.

Our primary domain of discourse is the propositional logic used in decision-making and in decision-support systems. Indeed, human mental activities are very often focused on solving decision-making problems: identification of alternatives, followed by evaluation, comparison, and selection of the most suitable alternative. This type of decision making is based on human natural propositional logic which combines percepts of satisfaction of requirements, resulting in the percepts of overall suitability of each alternative. Human natural propositional logic uses basic operations that are models of simultaneity, substitutability and complementing. These operations are combined to create aggregated propositional logic structures. Graded Logic is developed with the intention to quantitatively describe logic operations observable in natural human logical reasoning. Therefore, the point of departure in the development of Graded Logic is to show fundamental properties of natural human logic, observable in mental evaluation and decision-making.

Ten Postulates of Graded Logic

Human natural logic is based on a set of observable fundamental properties. These properties are used as postulates that are necessary for building Graded Logic.

The basic human-centric properties must be unconditionally upheld and mathematically implemented if we want to create quantitative logic infrastructure consistent with natural human logic. Based on observable properties of human commonsense reasoning, the necessary postulates of Graded Logic include the following requirements:

1. The truth of statements must be continuum-valued (graded in the range [0, 1]).
2. The importance of statements must be continuum-valued (graded in the range]0, 1[).
3. The simultaneity (conjunction) of statements must be continuum-valued (graded conjunction) up to the drastic conjunction, support nondecreasing monotonicity in each variable, and idempotency in extreme/boundary points (0, . . . , 0) and (1, . . . , 1).
4. The substitutability (disjunction) of statements must be continuum-valued (graded disjunction) up to the drastic disjunction, support nondecreasing monotonicity in each variable, and idempotency in extreme/boundary points (0, . . . , 0) and (1, . . . , 1).
5. The simultaneity and substitutability must be simultaneously present, unified and complementary (an increase of simultaneity must cause a corresponding decrease of substitutability and vice versa).
6. Logic neutrality must be available as a balance of simultaneity and substitutability.
7. The idempotency of logic aggregators must be selectable (either included or excluded).
8. The annihilator support for idempotent simultaneity must be selectable (either included or excluded).
9. The annihilator support for idempotent substitutability must be selectable (either included or excluded).
10. The simultaneity and substitutability models must be dual in their whole range.

These human-centric concepts are derived from observing elementary human logical reasoning. Following is a set of examples that illustrate the presence and use of the ten postulates:

1. Steve is an excellent student, but not the best in class. Thus, the statement "Steve is the best student in the class" is not completely true (it is 90% true if Steve is in the top 10% of students).
2. I need a job that is well paid and attractive. In my current situation the salary is more important than the job attractiveness.
3. A senior citizen without a car needs an apartment close to a food store and public transportation. These mandatory requirements must be simultaneously satisfied. The simultaneous proximity to parks, restaurants and public libraries is also desirable, but these are optional requirements, and the intensity of simultaneity requirement is much lower than in the case of food stores and public transportation.

4. A patient disability depends on motor symptoms (restricted mobility) and sensory symptoms (pain, numbness, tingling, itching). A high intensity of either motor or sensory symptoms is sufficient to cause the high disability reflecting a high degree of substitutability of symptoms. For the same patient tingling/itching is defined as tingling/itching either in feet or in hands. This substitutability may have a significantly lover intensity than in the case of overall motor/sensory symptoms.
5. If we need a high simultaneity, that means that all inputs must be simultaneously highly satisfied. Consequently, there is no possibility that one input can substitute another. In a symmetric case of high substitutability any input can to some extent substitute any other input and there is no need for their simultaneous presence. Therefore, more simultaneity means less substitutability and vice versa, making simultaneity and substitutability both complementary, and simultaneously present in human natural logical reasoning.
6. In all schools it is desirable that students have simultaneously high grades in all courses, and that is the simultaneity requirement. At the same time, each lower grade can be compensated by any other high grade, and that is the substitutability requirement. Since both requirements are equally present (balanced) the resulting aggregator is logic neutrality, and the average grade (GPA) is computed using the arithmetic mean. The logic neutrality is the most frequent aggregator in human natural logical reasoning.
7. The grade of student in some course is the degree of truth of statement "student completely satisfies all requirements of the course." If the student has in all courses the same grade x, then the average grade must also be x, showing that the aggregator is idempotent. On the other hand, the degrees of truth can also reflect the likelihood (probability) of success in a specific activity. In such a case the likelihood of simultaneous success in two independent such activities x and y (like the cross-country skiing and rifle shooting in biathlon) yields the probabilistic aggregation $s = xy$ and the idempotency should not be supported. The use of idempotency is selectable.
8. A family needs a residence that is close to an elementary school, day care, and two job locations for parents. These four requirements must be simultaneously satisfied, and if any of these requirements is not satisfied (0), the residence will be rejected (0 is the annihilator and in this case, it must be supported). In the senior citizen example, it is desirable to have the simultaneous proximity to parks, restaurants, and public libraries. However, these are optional requirements and if some of them are missing (0), that should not cause rejection of residence, and the annihilator 0 should not be supported. The support of annihilator is selectable by the stakeholder/decision-maker.
9. In a gated community the parking problem can be solved using either a home garage or a reserved parking space close to home. A homeowner is fully satisfied (suitability = 1) with each of these options (this form of substitutability supports the annihilator 1). The same homeowner would like to have proximity of amenities like a swimming pool, or a movie theater, or basketball or volleyball courts. Any of these amenities is desirable and can partially (but not

completely) satisfy the homeowner. Therefore, the amenities are substitutable, but not to the extent that one of them can substitute all others. So, the annihilator 1 is intentionally not supported.

10. In human logical reasoning, the requirement that conditions A, B, C, etc. must be simultaneously satisfied is equivalent to the requirement that it is not acceptable that any of the conditions (A or B or C or …) is not satisfied. Similarly, the requirement that at least one of condition A or B or C, etc. must be satisfied is equivalent to the requirement that it is unacceptable that the conditions A and B and C, etc. are simultaneously not satisfied. These equivalences are the laws of De Morgan duality of simultaneity and substitutability. Thus, De Morgan's duality provides two equivalent ways to express the same logic relationship.

The Graded Logic postulates are analyzed in detail in the next chapters of this book, and used to develop a graded propositional calculus, directly applicable in the development of decision-making models and decision-support systems. At this point, it is useful to offer only a brief and approximate explanation of these fundamental human-centric concepts.

(1) Decision-making is based on evaluation and comparison of alternatives (candidates), and evaluation is based on degrees of satisfaction of justifiable requirements. Human percepts of satisfaction of requirements are always graded. Consequently, the truth of statements that assert the full satisfaction of requirements is also always graded: the degree of truth varies in the range from false (0) to true (1) and includes all values in the range $I = [0, 1]$. The degree of truth is a continuously adjustable graded percept, and Graded Logic must be a continuum-valued logic. Bivalent or multivalent values of degrees of truth can exist as special cases, or approximations of natural continuously adjustable percepts. If the intensity of a graded percept P varies in the range $P_{min} \le P \le P_{max}$ (regularly, $P_{min} = 0$), then the degree of truth t of statement "P has the maximum value" is $t = (P - P_{min})/(P_{max} - P_{min})$, and $t \in I$.

(2) In each decision problem we can identify a stakeholder/decision-maker. The stakeholder can be an individual or an organization that has desired goals. The stakeholder/decision-maker specifies requirements that various attributes must satisfy to achieve desired goals. So, the percepts of satisfaction of requirements reflect the goals and interests of the stakeholder. Each attribute and the corresponding requirement always have specific importance for the stakeholder/decision-maker. Consequently, in a general case, all propositions and value statements have continuously adjustable importance degrees in the range]0, 1[(see Fig. 2.1). It is reasonable to assume that importance cannot be 0 because the inputs without importance should be omitted. The ultimate importance 1 is also questionable, because we assume that the importance degrees are normalized so that their collective summary value is 1; in such a case, the input that has importance 1 would eliminate all other inputs in a group, and that is not acceptable. Graded propositional calculus must provide ways for expressing the adjustable importance of all individual statements.

(3) The simultaneity of desired high degrees of truth is modeled using a conjunction which also must be graded. Indeed, in natural decision-making, decision-makers need to continuously adjust the desired degree of rewarding the simultaneity of high truth of inputs, or the degree of penalizing the presence of low degrees of truth. *Graded conjunction* (denoted Δ), in the case of equal weights (commutativity), is a logic aggregator located between the drastic conjunction $\lfloor\prod_{i=1}^{n} x_i\rfloor$ and the arithmetic mean: $\lfloor\prod_{i=1}^{n} x_i\rfloor \le x_1\Delta\ldots\Delta x_n \le \frac{x_1+\cdots+x_n}{n}, \quad x_1 \in I, \quad i = 1,\ldots,n, \; n > 1$.
The floor function $\lfloor t\rfloor$ returns the largest integer that is less than or equal to t. Therefore, the drastic conjunction returns the value $\lfloor\prod_{i=1}^{n} x_i\rfloor = 1$ if and only if $x_1 = \ldots = x_n = 1$. In all other cases, $\lfloor\prod_{i=1}^{n} x_i\rfloor = 0$. Obviously, the drastic conjunction is the model of the ultimate intensity of simultaneity.
The graded conjunction Δ has the adjustable intensity (called *andness*) that decision-maker selects to position the graded conjunction in desired location inside the interval between the drastic conjunction and the arithmetic mean. Each argument of the graded conjunction function is a degree of truth that positively contributes to the aggregated output value. Thus, the graded conjunction function must support nondecreasing monotonicity in each argument:

$$\forall t \in [0, 1 - x_i] \Rightarrow x_1\Delta\ldots\Delta(x_i + t)\Delta\ldots\Delta x_n \ge x_1\Delta\ldots\Delta x_i\Delta\ldots\Delta x_n,\; i = 1,\ldots, n.$$

Each argument can increase up to its maximum value 1. If all arguments have the maximum value 1 then the goal of simultaneous achieving high degrees of truth is fully satisfied, and the conjunction of n arguments becomes true. That yields the following necessary boundary conditions:

$$1\Delta\ldots\Delta 1 = 1, \quad 0\Delta\ldots\Delta 0 = 0.$$

In the range of conjunctive aggregators, we also have the pure conjunction $x_1 \wedge \ldots \wedge x_n = \min(x_1, \ldots x_n)$ that satisfies the following condition:

$$\left\lfloor\prod_{i=1}^{n} x_i\right\rfloor \le x_1 \wedge \cdots \wedge x_n \le \frac{x_1 + \cdots + x_n}{n}.$$

In the range $x_1 \wedge \cdots \wedge x_n \le x_1\Delta\ldots\Delta x_n \le \frac{x_1+\cdots+x_n}{n}$ the conjunctive aggregator $x_1\Delta\cdots\Delta x_n$ is a mean, and it is idempotent. So, the graded conjunction is equal to the arithmetic mean only in the case where all arguments have the same value (that is the reason for using "$\le$" and not "$<$"). Similarly, the range $\lfloor\prod_{i=1}^{n} x_i\rfloor \le x_1\Delta\cdots\Delta x_n < x_1 \wedge\cdots\wedge x_n$ includes nonidempotent aggregators. Therefore, the graded conjunction must provide the selectability of idempotent and nonidempotent properties.

(4) The substitutability of inputs is modeled using graded disjunction (denoted ∇) which enables the decision-maker to continuously adjust the capability

of a single input to compensate for the impact of low inputs. In the case of equal weights, the graded disjunction is a logic aggregator located between the arithmetic mean and the drastic disjunction function $1 - \lfloor \prod_{i=1}^{n} (1 - x_i) \rfloor$. Therefore,

$$\frac{x_1 + \cdots + x_n}{n} \le x_1 \nabla \cdots \nabla x_n \le 1 - \left\lfloor \prod_{i=1}^{n} (1 - x_i) \right\rfloor, \; x_1 \in I, \; i = 1, \ldots, n.$$

The drastic disjunction function is defined as De Morgan dual of the drastic conjunction, and therefore it is a model of the ultimate intensity of substitutability. The drastic disjunction returns the value 0 only if $x_1 = \cdots = x_n = 0$ and in all other cases (for the smallest positive value of a single argument) the drastic disjunction returns the value 1.
The graded disjunction ∇ has the adjustable intensity (called *orness*) that the decision-maker selects to position the graded disjunction in desired location inside the interval between the arithmetic mean and the drastic disjunction. As in the case of simultaneity, each argument of the graded disjunction function positively contributes to the aggregated output value, and the graded disjunction function supports the nondecreasing monotonicity in each argument:

$$\forall t \in [0, 1 - x_i] \Rightarrow x_1 \nabla \cdots \nabla (x_i + t) \nabla \cdots \nabla x_n \ge x_1 \nabla \cdots \nabla x_i \nabla \cdots \nabla x_n, \; i = 1, \ldots., n.$$

Each argument can decrease up to its minimum value 0. If all arguments have the minimum value 0 then the goal of having at least one positive input degree of truth is fully unsatisfied, and the disjunction of n arguments becomes false. That yields the following necessary boundary conditions:

$$0 \nabla \cdots \nabla 0 = 0, \quad 1 \nabla \cdots \nabla 1 = 1.$$

In the range of disjunctive aggregators, we also have the pure disjunction $x_1 \vee \cdots \vee x_n = \max(x_1, \ldots x_n)$ that satisfies the following condition:

$$\frac{x_1 + \cdots + x_n}{n} \le x_1 \vee \cdots \vee x_n \le 1 - \left\lfloor \prod_{i=1}^{n} (1 - x_i) \right\rfloor.$$

In the range $\frac{x_1 + \cdots + x_n}{n} \le x_1 \nabla \cdots \nabla x_n \le x_1 \vee \cdots \vee x_n$ the disjunctive aggregator $x_1 \nabla \cdots \nabla x_n$ is a mean, and it is idempotent. Similarly, the range $x_1 \vee \cdots \vee x_n < x_1 \nabla \cdots \nabla x_n \le 1 - \lfloor \prod_{i=1}^{n} (1 - x_i) \rfloor$ includes nonidempotent aggregators. Therefore, similarly to the graded conjunction, the graded disjunction must provide the selectability of idempotent and nonidempotent properties.

(5) In natural human reasoning, both simultaneity and substitutability are simultaneously present. In the case of conjunctive aggregators, it is desirable or necessary to simultaneously have high truth of inputs, but at the same time it is possible to

partially compensate for the impact of low inputs using the opposite compensatory impact of high inputs. If the desired simultaneity is high, then all input degrees of truth are requested to be simultaneously high. In such a case, it is not acceptable to request high substitutability where a single input can significantly compensate for the low values of the remaining inputs. Thus, more simultaneity means less substitutability. Similarly, if the desired substitutability is high, then a single high input is sufficient to compensate for the low values of remaining inputs. In such a case, we cannot request that all inputs should be simultaneously high. Therefore, simultaneity and substitutability are complementary: more simultaneity means less substitutability and vice versa. The complementarity of simultaneity and substitutability makes graded conjunction and graded disjunction complementary special cases of the most general basic graded logic function, called the *graded conjunction/disjunction* (GCD, denoted $\diamondsuit$):

$$\left\lfloor \prod_{i=1}^{n} x_i \right\rfloor \le x_1 \diamondsuit \cdots \diamondsuit x_n \le 1 - \left\lfloor \prod_{i=1}^{n} (1 - x_i) \right\rfloor, \quad x_1 \in I, \quad i = 1, \ldots, n.$$

To satisfy this condition, GCD must support a continuous transition from the drastic conjunction to the drastic disjunction. Considering that GCD includes both conjunctive and disjunctive properties, it follows that the boundary conditions of GCD are $0\diamondsuit\cdots\diamondsuit 0 = 0$ and $1\diamondsuit\cdots\diamondsuit 1 = 1$. In addition, GCD must provide the selectability of idempotent and nonidempotent properties.

(6) The presence of simultaneity and substitutability can be balanced. The resulting operator is called the *logic neutrality*, and it must be implemented as the arithmetic mean. In the special case of two variables, $(x + y)/2 = 0.5(x \wedge y) + 0.5(x \vee y)$, i.e., the arithmetic mean can be interpreted as "50% of conjunction and 50% of disjunction." Generally, the balanced presence of simultaneity and substitutability yields the arithmetic mean.[1] Since simultaneity is denoted using symbol Δ and substitutability is denoted using symbol ∇, the logic neutrality can be symbolically denoted by combining these symbols, as follows:

$$x_1 \ominus x_2 \ominus \cdots \ominus x_n = \frac{x_1 + x_2 + \cdots + x_n}{n}.$$

Unsurprisingly, the logic neutrality supports the nondecreasing monotonicity in each argument:

$$\forall t \in [0, 1 - x_i] \Rightarrow x_1 \ominus \cdots \ominus (x_i + t) \ominus \cdots \ominus x_n \ge x_1 \ominus \cdots \ominus x_i \ominus \cdots \ominus x_n, \quad i = 1, \ldots, n.$$

[1] It is important to note that the reduction to the arithmetic mean is mandatory. There are aggregators that do not satisfy that condition. One of them is the gamma aggregator [ZYS79, ZIM80], which in the case of two variables is $z(x, y; \gamma) = (xy)^{1-\gamma}(1 - (1 - x)(1 - y))^{\gamma}$. So, $z(0, y; \gamma) = z(x, 0; \gamma) = 0$, $z(1, 1; \gamma) = 1$, $z(x, y; 0) = xy$, $z(x, y; 1) = 1 - (1 - x)(1 - y)$, $z(x, y; 0.5) = (xy(x + y - xy))^{1/2}$, $z(a, a; 0.5) = \left(2a^3 - a^4\right)^{1/2} \ne a > 0$. Therefore, the gamma aggregator has either conjunctive or disjunctive properties, but it is not idempotent, and it cannot be reduced to the arithmetic mean.

Generally, $x_1\Delta \dots \Delta x_n < x_1 \Diamond \cdots \Diamond x_n < x_1\nabla \dots \nabla x_n$, except in boundary points $(0, \dots, 0)$ and $(1, \dots, 1)$, and, for idempotent aggregators, except in points where we have idempotency $x\Delta \dots \Delta x = x \Diamond \cdots \Diamond x = x\nabla \dots \nabla x = x$. Because of the central location of logic neutrality, this aggregator can be considered the centroid of all GCD aggregators.

(7) Basic logic aggregators that are models of simultaneity and substitutability can be ether idempotent or nonidempotent. Idempotent aggregators satisfy conditions $\forall x \in [0, 1] \Rightarrow x\Delta \cdots \Delta x = x$ and $\forall x \in [0, 1] \Rightarrow x\nabla \dots \nabla x = x$. Idempotent aggregators are means because each mean M by definition satisfies $\min(x_1, \dots, x_n) \leq \mathrm{M}(x_1, \dots, x_n) \leq \max(x_1, \dots, x_n)$, and by inserting $x_1 = \cdots = x_n = x$ it follows $x \leq M(x, \dots, x) \leq x$, i.e., $M(x, \dots, x) = x$. The desired simultaneity can be stronger than in the case of pure conjunction. In such cases $x\Delta \cdots \Delta x < x$ and such conjunctive aggregators are nonidempotent. Likewise, the desired substitutability can be stronger than in the case of pure disjunction. In such cases $x\nabla \cdots \nabla x > x$ and such disjunctive aggregators are nonidempotent.

(8) The percept of satisfaction of a requirement is equivalent to the degree of truth of the value statement that asserts the complete satisfaction of the requirement. At the high degree of simultaneity, the satisfaction of all inputs can become mandatory: if any input is 0, the output must also be 0. So, the value 0 is the annihilator for graded conjunction. If the graded conjunction supports the annihilator 0, it is called *hard* and denoted $\overline{\Delta}$; so, for each positive t we have $t\overline{\Delta}0 = 0$. At lower degrees of simultaneity, decision-makers can consider that high input degrees of truth are *desired* but *not mandatory*. Then, zero inputs will not automatically produce the zero output, i.e., the zero inputs can be partially compensated by those inputs that have positive values. In such cases, the annihilator 0 must not be supported, and such a form of graded conjunction is called *soft* and denoted $\underline{\Delta}$. Thus, $\forall t > 0 \Rightarrow t\underline{\Delta}0 > 0$, and in Graded Logic, decision-makers must have the possibility to select or reject the annihilator 0 for conjunctive functions that can be either hard or soft.

(9) In the symmetric case of substitutability, the situation is similar to the case of simultaneity. At the high degree of substitutability, the full satisfaction (1) of a single input can become sufficient to cause the full satisfaction (1) of the output. In this case, the graded substitutability is *hard* ($\overline{\nabla}$), and the truth value 1 is the annihilator for the operation of substitutability: $\forall t \geq 0 \Rightarrow t\overline{\nabla}1 = 1$. Of course, at lower levels of substitutability, decision-makers can decide that inputs can significantly compensate each other, but not to the extent that a single input is sufficient to fully compensate for the absence of other inputs. Such graded disjunction is *soft* ($\underline{\nabla}$) and $\forall t < 1 \Rightarrow t\underline{\nabla}1 < 1$. That creates the need for selectability of annihilators: in GL models, the annihilators of graded conjunction and graded disjunction can be either *indispensable* or *unacceptable*.

(10) Graded logic must support De Morgan's duality of basic conjunctive and disjunctive functions. In verbal interpretation, if a decision-maker requires that all inputs must simultaneously have high truth values, this requirement is equivalent to the requirement that it is not acceptable that any of inputs has a low input value. This rule can be written as follows: x_1 *and* x_2 *and* ... *and* $x_n = 1 - (1 - x_1)$ *or* $(1 - x_2)$ *or* ... *or* $(1 - x_n)$. Also, if a single highly true input can satisfy a requirement, then this criterion is equivalent to the requirement that all inputs should not be simultaneously low: x_1 *or* x_2 *or* ... *or* $x_n = 1 - (1 - x_1)$ *and* $(1 - x_2)$ *and* ... *and* $(1 - x_n)$. Thus, De Morgan's duality can be interpreted as two equivalent ways to express the same propositional logic relationship.

Our task is now to build Graded Logic that is based on the above ten principles. It is not difficult to see that the classic bivalent Boolean logic is a special case of Graded Logic, obtained taking special cases of the ten fundamental properties:

1. Truth can take only the extreme values 0 or 1.
2. All statements have the same importance, yielding the commutativity of inputs.
3. Simultaneity has a single value $x_1 \wedge x_2 \wedge \cdots \wedge x_n = min(x_1, x_2, \ldots, x_n)$.
4. Substitutability has a single value $x_1 \vee x_2 \vee \cdots \vee x_n = max(x_1, x_2, \ldots, x_n)$.
5. Only two extreme cases are used: full conjunction (without disjunctive properties) or full disjunction (without conjunctive properties). No transition between these extremes.
6. Logic neutrality is not used.
7. Logic aggregators are strictly idempotent. Nonidempotent aggregators are not supported.
8. Annihilator 0 is always used for conjunction ($\forall x \in \{0, 1\} \Rightarrow 0 \wedge x = 0$).
9. Annihilator 1 is always used for disjunction ($\forall x \in \{0, 1\} \Rightarrow 1 \vee x = 1$).
10. De Morgan duality holds for selected full conjunction and full disjunction functions: $x_1 \wedge \cdots \wedge x_n = \overline{\overline{x_1} \vee \cdots \vee \overline{x_n}}$, $\quad x_1 \vee \cdots \vee x_n = \overline{\overline{x_1} \wedge \cdots \wedge \overline{x_n}}$.

If classical Boolean logic is a special case of Graded Logic, then Graded Logic is a seamless generalization of Boolean logic. This important relationship indicates that, similarly to Boolean logic, Graded Logic can also have applicability in areas different from its primary domain of decision making. The recent explosion of public interest in AI is based on applications where machines behave and reason like humans. Obviously, the human-centric Graded Logic facilitates moving in that direction.

Prerequisites for Applicability

Applicability is a natural motivation for the development of any theory. In other words, reduction to practice should be a natural test of the success of any theory. Graded Logic is based on these concepts. Expectedly, there are prerequisites for applicability, and it is important to keep them in mind in all phases of the development of theoretical concepts. If various applicability prerequisites could be reduced to a single word, then that word would be *simplicity*.

In the case of graded percepts and graded properties, the simplicity and the resulting applicability are achieved if all graded percepts and their parameters are *explicitly visible and easily adjustable.* Since means are functions located between two extreme cases, $min(x_1, \ldots, x_n)$ and $\max(x_1, \ldots, x_n)$, and these extreme cases are logic functions of conjunction and disjunction, it is hidden in plain sight that all means can (and should) be primarily interpreted as logic functions.[2] If means are logic functions, then the natural question is to determine the degree of conjunction (the andness α) and the degree of disjunction (the orness ω), as parameters that determine the strength of conjunction and the strength of disjunction of any mean. If we create graded logic functions using the quasi-arithmetic mean $y(\boldsymbol{x}; \boldsymbol{W}) = F^{-1}\left(\sum_{i=1}^{n} W_i F(x_i)\right)$, $\boldsymbol{x} = (x_1, \ldots, x_n)$, $\boldsymbol{W} = (W_1, \ldots, W_n)$, $0 < W_i < 1$, $i = 1, \ldots, n$, $\sum_{i=1}^{n} W_i = 1$, then the degrees of importance expressed using weights $W_1, \ldots, W_n$ are both explicitly visible and easily adjustable. However, the conjunction and disjunction degrees (α and ω) are hidden inside the structure of function F and are not visible and not easily adjustable. Even in the simple special cases, like $F(x_i) = 1/x_i$, where we get the traditional harmonic mean $y(\boldsymbol{x}; \boldsymbol{W}) = 1/\left(\sum_{i=1}^{n} W_i/x_i\right)$, the strength of simultaneity is not explicitly visible. Unfortunately, the intensity of simultaneity and substitutability are primary inputs in decision models. Thus, we must provide a logic aggregator $y(\boldsymbol{x}; \boldsymbol{W}, \alpha)$ where the user can directly insert the desired value of α. Such aggregators are called "andness-directed" and presented in Chap. 15. In subsequent chapters of this book, we present various techniques that support the applicability of GL, but the roots of these techniques must already be present in the postulates of GL.

References

KLI95 Klir, G.J., and B. Yuan. 1995. *Fuzzy Sets and Fuzzy Logic*. Prentice-Hall.

NGU06 Nguyen, H.T., and E.A. Walker. 2006. *A First Course in Fuzzy Logic*, 3rd edn, Chapman & Hall/CRC.

PRI09 Priest, G. 2009. *An Introduction to Non-Classical Logic*, 2nd edn. reprinted with corrections. Cambridge University Press.

ROS10 Ross, T.J. 2010. *Fuzzy Logic With Engineering Applications*, 3rd edn. J. Wiley.

ZIM80 Zimmermann, H.-J., and P. Zysno. 1980. Latent connectives in human decision-making. *Fuzzy Sets and Systems* 4: 37–51.

ZYS79 Zysno, P. 1979. *One Class of Operators for the Aggregation of Fuzzy Sets*. Amsterdam: EURO III Congress.

[2] Of course, this is an intentionally provocative statement (why primarily?).

Chapter 4
The Graded Percepts of Truth, Importance, Simultaneity and Substitutability

There are statements which are completely true or completely false. Assuming the Euclidean geometry, a provably false statement is "the sum of angles in a triangle is 179 degrees." Examples of provably true statements are "7 is a prime number" and "the area of a rectangle with length L and width W is LW." The statement "*x is prime*" is true for $x = 7$ and false for $x = 8$. Such statements, with easily provable bivalent truthfulness, reflect objective reality and are not related to any specific stakeholder or their interests.

There are other statements with provable bivalent truthfulness that are related to a specific stakeholder; for example, the statement "I have two sisters" can be either true or false. However, in human communication, most statements originating from a given stakeholder reflect the stakeholder's interests and can be partially true. For example, the statements "I am tired," "today is a very warm day," "you drive too fast," "I have a good job," "Max is an excellent student," "ChatGPT is always right," and "this is a very old car" are regularly only partially true. A rather large class of statements reflects a degree of satisfaction of stakeholder's requirements and can be classified as *value statements*. Such statements report the result of an evaluation, i.e., a degree of satisfaction of specific requirements (e.g., "this house completely satisfies all requirements of our family"). Of course, the requirements always reflect individual goals and interests of a specific stakeholder.

In Chap. 1, the story about buying apricots, prunes, and nuclear submarines used the concept of the degree of truth without precisely explaining its meaning. So, what is the truth and what is its degree? The search for an answer to this question directs us to all the greatest philosophers who offered deep discussions of this topic in the last 2–3 millennia. The spirit of our time is to avoid outdated profound studies, and to find answers to all questions using a smartphone and some quick AI-enhanced search. All dictionaries define truth as the property of being in accord with facts or reality. To be fully aligned with the spirit of our time we will offer the answer to that difficult question using the following trivial example: a 1-L bottle (1000 ml) contains 750 ml

J. Dujmović, *Graded Logic*, Intelligent Systems Reference Library 273,
https://doi.org/10.1007/978-3-031-88558-7_4

of wine. That is a fact or reality (the author consumed 250 ml writing this chapter). Now come the degrees of truth, defined for three statements:

(1) The degree of truth of statement "The bottle is full" is 0.75 (or 75%).
(2) The degree of truth of statement "The bottle is empty" is 0.25 (or 25%).
(3) The degree of truth of statement "The bottle contains 750 ml of wine" is 1 (or 100%).

In these examples the degree of truth is used as a fraction of fulfillment of the associated assertion. This interpretation is suitable for evaluation problems: the statement "my car satisfies 90% of my requirements" is equivalent to the statement "the degree of truth of the statement 'my car fully satisfies all my requirements' is 0.9 (or 90%)." In other words, in 10% of observed cases the car causes problems to the owner, and in 90% of cases the car fully satisfies the owner's needs. The satisfaction of 90% of requirements is an approximate estimate by the car owner, based on subjectively and imprecisely defined requirements. On the other hand, the contents of the bottle are objectively measured. In both cases the precision of measurement is the best possible, and the assertions of complete fulfillment are only partially true.

These interpretations look trivial, but they still need an explanation of the standard negation $not(x) = 1 - x$. The negation of statement "the bottle is full" is "the bottle is empty" and vice versa. Note that the negation of statement "the bottle is full" should not be the statement "the bottle is not full", because in such a case the statements "the bottle is empty" and "the bottle is not full" would be equivalent, and that is not the case.

The word "percept" denotes a mental impression that is a result of the natural human process of perception or cognition. We are primarily interested in percepts that are results of logical reasoning. Almost all human percepts are graded, and almost all graded percepts can be expressed using degrees of truth of appropriate statements. Table 4.1 shows 40 examples of graded percepts. If a graded percept p has a magnitude m in the interval $[0, M]$, then the degree of truth x of the assertion *"p has the maximum value M"* is $x = m/M$. The negation of this assertion is *"p has the minimum value 0"* and its degree of truth is $\overline{x} = \frac{M-m}{M} = 1 - x$. Since $x \in [0, 1]$ it is possible and convenient to interpret the degree of truth x as a stakeholder-specific normalized value of the graded percept p. Observations of human reasoning show that human mind operates with normalized values of graded percepts. Indeed, all graded percepts vary between a minimum value and a maximum value, and that range is intuitively sensed the same for all percepts. In Graded Logic, we process normalized graded percepts using their degrees of truth and define the degree of truth as a fraction of the maximum value of a graded percept.

Sample graded percepts, presented in Table 4.1, are classified in two groups: *observable characteristics* of objects, and *compound percepts* resulting from the process of cognition. It is easy to note that all items in Table 4.1 (as well as many others) are graded, i.e. each of them has a magnitude in the range from the lowest (numerically coded as 0) to the highest (normalized, and numerically coded as 1). For example, the concept of satisfaction is related to specific requirements that can be satisfied in the range from no satisfaction to complete satisfaction. So, satisfaction

Table. 4.1 Two basic types of graded percepts

EXAMPLES OF GRADED PERCEPTS			
Observable graded characteristics of an object		Graded compound percepts resulting from the process of cognition	
Round	Flat	Truth	Likelihood
Heavy	Short	Importance	Significance
Light	Sharp	Satisfaction	Probability
Dark	Weight	Suitability	Possibility
Inexpensive	Young	Preference	Attractiveness
Soft	Metallic	Confidence	Worth
Long	Container	Trust	Happiness
Tall	Oval	Quality	Aptness
Old	Hard	Value	Reliability
Distance	Speed	Similarity	Uncertainty

is a matter of degree, i.e., it is graded. People can feel the magnitude of satisfaction, and express that magnitude numerically, using the normalized range [0,1] or [0, 100%]. E.g., the statement "my job satisfies 85% of my requirements" is equivalent to the degree of truth $x = 0.85$ of the statement "my job completely satisfies all my requirements". Obviously, satisfaction does not have objectively measurable "physical units." If components of job satisfaction are salary and the number of hours at work per week, then these inputs are observable, objectively measurable, and transformed to normalized percepts of satisfaction with salary and satisfaction with hours at work. In the next steps various individual normalized components of satisfaction are logically combined to yield the overall normalized degree of job satisfaction.

In the context of decision-making, a particularly influential graded percept is importance, which reflects the interests of stakeholder/decision-maker in the high degree of truth of specific statement. The range of importance can be verbalized as [negligible, extreme], and expressed numerically as]0, 1]. We usually assume that zero importance means that such input is omitted from consideration, and therefore, all inputs must have a positive importance. In the context of decision-making, the percept of importance is not considered in isolation, but as a parameter used to differentiate the strength of impact of two or more degrees of truth. Thus, the degrees of truth of value statements are regularly paired with their degrees of importance. The graded percept of importance is an important adjustable component in intuitive decision-making, but sometimes it can also be defined in an objective way. For example, the grades of students can be weighted with the number of hours of lectures per week and used to compute the weighted average grade (GPA). A detailed analysis of percepts of importance is presented in Chap. 12.

In the context of decision-making, the percept of simultaneity is also graded. To satisfy their goals, decision-makers specify multiple requirements and request their simultaneously high satisfaction. The sufficient simultaneously high satisfaction is expressed as the simultaneity of high degrees of truth of logic arguments. Sufficient simultaneity is rewarded, and insufficient simultaneity is penalized. The degree of penalizing/rewarding the simultaneity of high degrees of truth is adjustable and reflects the stakeholder's graded percept of simultaneity. According to postulates of Graded Logic presented in Chap. 3, simultaneity and substitutability are complementary, and consequently substitutability is also a graded percept. A detailed analysis of graded simultaneity and substitutability is presented in Chaps. 7 and 9.

The graded percepts of truth, importance, and simultaneity/substitutability are the only graded percepts that are observable in natural human decision-making. Consequently, logic models should be developed with the goal to support graded truth, graded importance, and graded simultaneity/substitutability. We classify such logics as *fully continuum-valued*. The Graded Logic presented in this book is fully continuum-valued.

Rating Scales

In human communication, the magnitude of graded percepts is frequently expressed verbally, using discrete rating scales [DUJ18]. An example of a 5-level rating scale for satisfaction or suitability is *lowest* $\prec$ *low* $\prec$ *medium* $\prec$ *high* $\prec$ *highest*. The symbol "$\prec$" denotes the strictly increasing ordering of the linguistic labels of a rating scale.

The linguistic labels can be adjusted according to the meaning of a particular graded percept. For example, the rating scale for the degree of satisfaction can also be *not at all* $\prec$ *slightly* $\prec$ *moderately* $\prec$ *considerably* $\prec$ *completely*. The linguistic labels of rating scales are assumed to be equidistant [ROH07, MEN10] and consequently they can be expressed using numeric (integer) labels 0, 1, 2, ..., N. Therefore, the label $k \in \{0, \ldots, N\}$ corresponds to the degree of truth $x = k/N$. So, degrees of truth of 5-level scales are 0, 0.25, 0.5, 0.75, 1. These values should be interpreted approximately, considering the imprecision of human estimates, where the label *high* might easily denote a value between 0.7 and 0. 8, and not only the ideal central point 0.75.

The rating scale precision can be increased by adding more labels. An example of a 9-level rating scale of magnitude is *lowest* $\prec$ *very low* $\prec$ *low* $\prec$ *mid* $-$ *low* $\prec$ *medium* $\prec$ *mid* $-$ *high* $\prec$ *high* $\prec$ *very high* $\prec$ *highest*. Table 4.2 shows examples of 5-level and 9-level rating scales and the correspondence between linguistic labels and degrees of truth.

Rating scales are strictly increasing linear sequences of equidistant linguistically labeled normalized human graded percepts, and not the values of variables that create graded percepts. The *cardinality of rating scale* is the number of labels which determine the precision of rating scale. In GL applications, it is convenient to have a middle point of rating scales. In such cases the numeric labels are $0, \ldots, N$ and N is an even number, and the middle point is $N/2$. It is convenient to use $N = 2^m, m = 1, 2, 3, 4, \ldots$ as shown in Fig. 4.1. Generally, rating scale must satisfy strict monotonicity, linearity, and balance. In the 5-level rating scale from the lowest degree of truth to the highest degree of truth we satisfy the following properties:

Table. 4.2 Linguistic labels and corresponding degrees of truth

Linguistic label	Linguistic label	Numeric label	Degree of truth
Lowest	Not at all	0	0
Low	Slightly	1	0.25
Medium	Moderately	2	0.50
High	Considerably	3	0.75
Highest	Completely	4	1
Linguistic label	**Linguistic label**	**Numeric label**	**Degree of truth**
Lowest	Unacceptable	0	0
Very low	Very poor	1	0.125
Low	Poor	2	0.25
Mid-low	Below average	3	0.375
Medium	Average	4	0.5
Mid-high	Above average	5	0.625
High	Good	6	0.75
Very high	Very good	7	0.875
Highest	Excellent	8	1

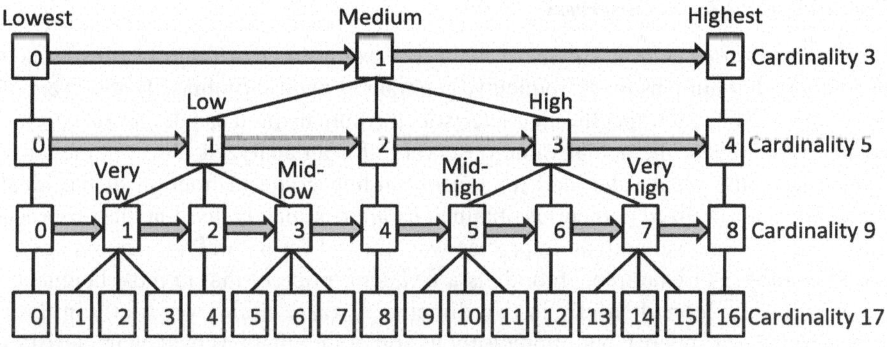

Fig. 4.1 Rating scales with cardinality $N + 1 = 3, 5, 9, 17$.

Cardinality: $N + 1 = 5$

Strict monotonicity: $0 = lowest \prec low \prec medium \prec high \prec highest = 1$ $(N = 4)$.

Linearity: $low - lowest = medium - low = high - medium = highest - high = 1/N$.

Balance: symmetry (in number, distribution and labeling) between favorable/superior options (those above the middle point) and unfavorable/inferior options (those below the middle point): $(lowest + highest)/2 = (low + high)/2 = 1/2$.

The cardinality and organization of rating scales are limited by human limitations in discernment of discrete levels of unidimensional stimuli. According to [MIL56], the most frequent cardinality is between 5 and 9. Figure 4.1 indicates that rating scales can have higher cardinality, but in such cases the scales are based on a process of stepwise refinement, where in each step we process a low cardinality subscale [DUJ18]. More details about experiments with rating scales and design of rating scales can be found in [PRE00, ROH07, MEN10, SMI13, DUJ18].

If the decision-making is interpreted as processing inputs to produce an output, then in a general case both inputs and the output are graded. Both inputs and the output have semantic identity – they have meaning and role for decision-maker. If the decision-making is based on logic operations, then the computation of output is based on a formula of graded propositional calculus. In the process of selecting the best among several alternatives, the human mind has the observable and measurable ability to process graded percepts and produce and use graded results. Based on appropriate experiments, human graded logic can be analyzed and modeled, and the resulting mathematical models can be used to develop decision-support systems.

Unbiased and Biased Observers

Observable graded characteristics of an object, exemplified in Table 4.1, are defined as percepts that humans most frequently generate as neutral (unbiased) observers of given objects. For each specific characteristic, the human mind can define an extreme (ideal) value. Then, the actual value observable for an analyzed object is mentally compared to the ideal value and the corresponding grade (a fraction of the ideal value, i.e., a normalized percept) is obtained by approximately dividing the observed value and the ideal value in an imprecise way, typical for natural human perception and reasoning. For example, "round" is a frequent property that in most languages is not precisely defined strictly in two dimensions (as a circle), or strictly in three dimensions (as a sphere), but imprecisely as something that has oval characteristics comparable to the ideal cases of circle and sphere. So, for all practical purposes, the Moon, car tires, tennis balls and all current US coins might be considered close to 100% round. On the other hand, eggs, watermelons, bottles and glasses are not 100% round. Eggs and watermelons are usually elliptical, and observers might claim that they are close to 80% round. Bottles, airplanes, submarines, cans and glasses might be 50% round if they have cross sections that are perfectly round. Some shoes and boots might be considered 30% round, same as violins and cellos. Nails, pins, flutes, and wires are maybe 10% round because their shapes are dominated by their length and not by the shape of their perfectly round cross section. Arrows, rivers, and hammers might be considered close to 0% round.

All presented objects can be observed from the standpoint of unbiased estimating to what extent they are round and to each of them human neutral observers can assign the degree of roundness which is normalized in the range from 0 to the maximum value 1 or 100%. Of course, the degree of roundness can be interpreted as the degree of truth of the assertion "the object is perfectly round." Consequently, "round" is one of graded percepts that is observer-neutral because most people share the same understanding of the notion of roundness and their estimated grades of roundness differ less than in the case of determining the grades of compound concepts that are not observer-independent. Of course, all these claims must be experimentally verified. Some verification can come from explanations provided by actual decision-makers, based on their real experiences; such results are rather credible, but their number is rather modest. More frequently, verification comes from organized experiments with human subjects, introduced in Chap. 5.

It is easy to note that some object characteristics are not observer-independent. For example, each computer has memory, and the necessary memory capacity depends on the needs of each specific observer/user. There is always a normalized graded percept of satisfaction with available memory capacity, but the maximum capacity that yields the full satisfaction can be different for each observer. So, the suitability attribute (memory capacity) can have different ranges, but its normalized graded percept of suitability (a degree of truth) is always in the standard range [0, 1].

Graded compound percepts, exemplified in Table 4.1, are defined as percepts that humans generate by mentally combining various normalized inputs in a graded logic way, i.e., using a natural mental propositional calculus. In the process of decision-making for typical job selection, presented in [DUJ18], decision-makers use 20 input attributes to determine the overall suitability of a job offer. Similarly, a car selection exemplified in [MES17] is based on 29 inputs, and a home selection presented in [DUJ18], is based on 36 inputs. Complex professional evaluation problems can include a few hundred attributes (e.g., see [SU82, SU87]). Both mental and formal decision models are based on the observable model of human reasoning presented in Chap. 6. Therefore, GL reasoning includes graded percepts shown in Table 4.1, where the most important are graded truth, importance, suitability, and satisfaction, but also various graded percepts that determine the desired strength of logic operations, such as graded simultaneity (conjunction degree), graded substitutability (disjunction degree) and partial absorption, discussed in subsequent chapters.

In the case of compound percepts shown in Table 4.1 we cannot expect that the observers are neutral and that all observers behave in a similar way. In the simple case of the apricot preserve maker, described in Chap. 1, we differentiated various types of buyers: those who think that the quality and the affordability of the product are equally important, and those who think that affordability is more important. Obviously, each buyer has individual goals and interests, and these goals and interests affect the resulting concept of suitability of product.

A similar situation occurs in the example of job selection, where the job offer suitability depends on monetary compensation and job attractiveness. Some candidates primarily seek a high monetary compensation, and the job attractiveness is a secondary requirement. Other candidates might be focused on promising properties of a new job and for them the job attractiveness is a primary concern. When the same job is offered to such different kinds of candidates their evaluation will generate different overall job suitability degrees. We can consider that such candidates are always biased, reflecting their individual goals and experiences. It is important to realize that such bias is also present in all propositional logic experiments where respondents play the role of a specific decision-maker which is defined to have interests different from respondents' personal interests and experiences.

Grading, as an organized process of creating a publicly available graded percept, is a part of many professions. All instructors in all schools try to teach students specific material and at the end of course the success of students is measured with grades that express the degree of satisfaction of course requirements. All evaluations are based on a set of requirements and the overall satisfaction or suitability is expressed using numerical or symbolic grades. In the case of school grades $x_1, \ldots, x_n$, most schools also offer an aggregated indicator GPA (grade point average) $GPA = (x_1 + \ldots + x_n)/n$ and that indicator is also graded. A more precise version of the GPA indicator considers that some courses need more effort than other courses, and in such cases, GPA is a weighted average: $GPA = w_1x_1 + \ldots + w_nx_n$. The weights are assumed to be positive and, to preserve the idempotency $GPA = w_1x + \ldots + w_nx = x$, the sum of weights must satisfy $w_1 + \ldots + w_n = 1$.

The wide use of the Internet has resulted in permanent grading of services: after most taxi rides, customers are requested to provide a graded percept (1–5 stars) of satisfaction with that experience. Patients are requested to evaluate medical services, students evaluate instructors, and managers evaluate workers. In all such activities the result is a graded percept of satisfaction in the range from a selected minimum value to a selected maximum value. The human mind has a remarkable ability to process graded percepts. Human reasoning in the context of decision-making is provably based on processing of graded percepts using a graded propositional calculus. Therefore, a human-centric approach to logic naturally yields GL as the logic of human reasoning with graded percepts.

Graded percepts and their processing to create decisions are not an exclusive human mental activity. They have deep roots in the animal world of our planet. In one of his email messages to the BISC community, Lotfi Zadeh noticed that many human percepts are shared by animals. Among physical percepts, he mentioned distance, speed, weight, size, and color. Among examples of mental percepts, he included likelihood (probability), possibility, similarity, class, and uncertainty. It is easy to verify Zadeh's observations. For example, the decision process is clearly visible when predators (e.g., lions) hunt other animals for food. Before trying to hunt a potential prey, predators estimate the graded percepts of distance, as well as the percepts of size, speed, and risk of attacking the potential prey. After aggregating all these inputs, they decide whether to hunt, or to wait for a better opportunity.

Discretization Problems

Classical logic works with extreme values of the degree of truth: the value "true" means completely true, and the value "false" means completely false. As opposed to that, inspired by natural human reasoning, GL and all other continuum-valued logics [GIN68, RAD00] operate with graded degrees of truth, including all values in the interval between completely false and completely true. Observations show that in natural human reasoning the concepts of "completely true" and "completely false" are only infrequent extreme cases. E.g., assuming that the maximum value of the average grade in university education is $GPA_{max} = 4$, let us investigate the meaning of assertion "Max is an excellent student." If Max has $GPA = 3.8$, then everybody would agree that Max is an excellent student. However, assuming that "excellent" means the highest grade, the binary logic would assign the status of excellent student only to those fantastic students who have $GPA = 4$, and in such a case the result would be that the assertion "Max is an excellent student" is completely false. Since $3.8/4 = 0.95$,, it is obviously more natural and more understandable to claim that the assertion "Max is an excellent student" is 95% true, and therefore sufficiently acceptable. In most real situations, truth has a continuous nature, and consequently most logic theories based on discrete values of truth can be considered theoretical oversimplifications, inconsistent with natural human reasoning.

The example of "excellent student" assertion is suitable for exposing other questionable, but rather frequent, forms of reasoning based on discretization. Since $GPA = 4$ is obviously inappropriate requirement for assigning the status of excellent student, we might be tempted to use another lower "reasonable threshold," e.g. $GPA = 3.75$, and then establish the criterion that the status of "excellent student" means $GPA > 3.75$, while $GPA = 3.75$ (and slightly below) would be interpreted as a "good student." Of course, this is an example of discretization, which is very frequent in "administrative reasoning." Unfortunately, most people are ready to accept (or propose) such form of interpretation for these reasons: (1) it is simple, (2) it must be correct because it is frequently used, and (3) in all natural languages humans express magnitudes of percepts using a small number of ordered linguistic labels and such experiences make discretization a "natural way" of verbalized dealing with the intensity of graded percepts.

In most cases, however, the distance between simplicity and oversimplification is very small. Mathematically, discretization means that *microscopic (negligible) differences of inputs* cause *macroscopic (significant) differences of output*. That is the undesirable (unjustifiable and/or unacceptable) property of all discontinuities, which is inconsistent with the continuous properties of natural graded percepts. So, if we accept the threshold-based reasoning, then the student with $GPA = 3.75$ is "good" and the student with $GPA = 3.7501$ is "excellent." Is this reasonable? Of course, it is not reasonable, but most people are ready to accept simplicity as a justification for imprecision or questionable correctness. Questionable effects of discretization, occurring at the border between linguistic labels, are not frequent, compared to acceptable classifications in areas apart from the discretization points. For example, the dramatic discerning of $GPA = 3.75$ and $GPA = 3.7501$ is highly questionable but obviously rather rare. On the other hand, almost everybody can agree that a student who has $GPA = 3.8$ is excellent, and a student who has $GPA = 3.2$ is good; such acceptable classifications are significantly more frequent than those problematic cases that are on opposite sides, but very close to discretization points. In a sense, the low frequency of highly questionable effects mitigates the discretization problems.

A more serious concern is that discretization regularly causes ethical problems. This form of reasoning is frequently applied in retirement rules: e.g., "no retirement benefits if service time is less than 10 years, and if it is 10 years and more, then there are retirement benefits based on a specific formula." Thus, if after 9 years and 9 months of service a worker is killed in a traffic accident, there will be nothing to help his or her family members (that happened in the author's family). Is this ethical or not ethical? Obviously, a better solution would be a continuous increase of benefits over time, but in such a case the formula would be more complex, and maybe beyond the easy reach of administrators who propose and/or approve the retirement benefit rules.

Similar situation is encountered in COVID-19 vaccination rules [DOO20, GAY20, KAT21, COV21]. The initial governmental (CDC) vaccination priority model was based on four vaccine allocation phases (subsequently refined and updated):

Phase 1a: High risk health workers, residents of long-term care facilities, first responders.
Phase 1b: Age $\geq$ 75 years, or frontline non-health care essential workers.
Phase 1c: $65 \leq \text{Age} < 75$, or people with high-risk medical conditions (age 16–64).
Phase 2: K-12 teachers, homeless people, people with disabilities.
Phase 3: Children and young adults (lower risk).
Phase 4: Everybody else.

The drawbacks of the initial governmental priority model can be summarized as follows [DUJ21a]:

1. A coarse discretization of phases (the absence of a continuous model)
2. The age threshold justification problem (if there is a provable optimum value of threshold age, it is highly unlikely that it might be a round value, 75)
3. The absence of priority differentiation inside individual phases
4. Poor justification of phase belonging in the vicinity of all borders between phases
5. Undefined duration of individual phases
6. Incompleteness (the absence of more precise non-medical criteria: patient's role and responsibility, social and ethical attributes).

These rules create an obvious discretization problem for person P1 whose age is 75 and belongs to Phase 1b, and person P2 whose age is 74.5 and belongs to Phase 1c. The difference in age (6 months) is clearly insignificant and negligible. If the Phase 1b lasts less than 6 months, then P2 cannot satisfy the age criterion before the beginning of Phase 1.c. Therefore, according to the CDC vaccination priority rules, P1 can be vaccinated first in Phase 1b, and P2 can be vaccinated last in Phase 1c, making a huge vaccination time difference between P1 and P2. Assuming that both P1 and P2 are vaccinated in the middle points of their phases, the average difference is 50% of the duration of Phase 1b + Phase 1c. The maximum difference is the duration of Phase 1b + Phase 1c. During this interval, P2 has much higher chances (compared to the very similar person P1) to be infected and become one of 6.7 million people who died worldwide from COVID-19 before 2023. According to [WOR23], in the USA, the total number of coronavirus cases on February 4, 2023 was 104,453,003 and the total number of deaths was 1,135,957. Thus, it is natural to ask how many lives are lost because of poor (imprecise) vaccination priority decisions? Nobody knows the exact answer, but everybody knows that it is not zero. In addition, it is obvious that the economic damage caused directly and indirectly by coronavirus cases is both huge and significantly reducible.

Clearly, this is primarily an ethical problem: there is no justification for why a microscopic (i.e., negligible) difference between P1 and P2 causes a macroscopic (i.e., significant) difference in survival chances. If the government selected to use a continuous vaccination priority model and to vaccinate people according to a continuously adjustable "vaccination priority number," then the microscopic difference in age of P1 and P2 would result in fully justifiable microscopic difference in vaccination priority, and P1 and P2 would have almost identical survival chances. That would eliminate easily visible ethical problems caused by the primarily political solution based on oversimplified discretization.

The COVID-19 decision example [DUJ21a] shows that vaccination decision methods must be precise, i.e., quantitative and continuous. Such methods need the aggregation of various inputs using graded logic functions. The effort in developing optimum vaccination priority decisions is equally important as efforts in vaccine development, production, distribution, and organized vaccination. Continuous models for vaccination priority solve many problems: (1) the dynamics of supply of vaccines, (2) the number of necessary medical facilities and personnel, (3) the total time and the cost for protecting the whole population, and (4) the scheduling of vaccination.

Discretization problems can be consequences of the unnecessary oversimplification of decision models. In all described cases, the discretization problems can be avoided by using continuously adjustable degrees of truth. However, there are cases where discretization is necessary and acceptable. The binary discretization is used in all situations where there are no other options other than binary pass/fail, yes/no, and accept/reject decisions. For example, there are exams where grading is based on credit/no-credit or pass/fail grades (e.g., a driver exam). In such cases the suitability of a candidate is computed as a suitability degree in a specific interval, and the passing or credit grade is assigned if the suitability degree is above a specific fixed (and justifiable) threshold. A contract for 24/7 maintenance of some sensitive equipment could be either available or not available. A contrabass player must have a car that has enough space for this large instrument. The available volume is a continuous variable, but the suitability of that volume becomes zero if the instrument cannot fit in the car.

There are various cases where we must use discrete arguments of evaluation models, and let us show a few typical examples. In the home evaluation the number of bedrooms and/or bathrooms is discrete: 1, 2, 3, etc. Computer equipment can be brand new, or refurbished, and there is nothing in between. The number of USB ports and the number of disk units are natural numbers (1, 2, 3,...). The computer memory capacity is not continuously adjustable but regularly comes in discrete steps (e.g. 4, 8, 16, 32, 64 GB).

It is important to differentiate two kinds of discretization: (1) the natural discretization based on naturally discrete arguments, and (2) the discretization motivated by simplification. The natural discretization is unavoidable and acceptable because degrees of truth are undefined between the justified discrete values. On the other hand, discretization motivated by simplification should not be acceptable without a strong justification, because it can frequently yield a harmful oversimplification. That is particularly visible in arbitrary selected threshold values of degrees of truth. Examples of discretization motivated by simplification are all low granularity rating scales, such as discrete letter grades (A, B, C, D) or discrete numeric grades used in most schools, or population clustering according to insufficiently justified age thresholds. For naturally continuous variables, such as time, geometric size, product quality, suitability, and price, it is natural to use continuously adjustable degrees of truth.

Linguistic labels are not continuous but discrete, and we use them in rating scales, and in everyday natural human communication. There is no possibility to precisely express continuous variables using linguistic labels. On the other hand, degrees of truth are real numbers and offer continuous adjustability. Therefore, there are two possibilities: either we use discretization that can cause questionable effects in discretization points, or we base our reasoning on continuously adjustable degrees of truth. Unsurprisingly, GL is focused on developing precise logic models that are based on continuously adjustable degrees of truth, simultaneity, substitutability, and importance.

References

COV21 COVID19.CA.GOV, Vaccines. 11/14/2021. https://covid19.ca.gov/vaccines/#California's-vaccination-plan.

DOO20 Dooling, K., et al. 2020. *The Advisory Committee on Immunization Practices' Updated Interim Recommendation for Allocation of COVID-19 Vaccine—United States*. https://www.cdc.gov/mmwr/volumes/69/wr/mm695152e2.htm?s_cid=mm695152e2_w.

DUJ18 Dujmović, J., 2018. *Soft Computing Evaluation Logic*. J. Wiley.

DUJ21a Dujmović, J., and D. Tomasevich. 2021. COVID-19 Vaccination priority evaluation. In Explainable AI and Other Applications of Fuzzy Techniques ed Julia Rayz, Victor Raskin, Scott Dick, and Vladik Kreinovich. (Proceedings of the North American Fuzzy Information Processing Society, NAFIPS 2021) Springer 2021.

GAY20 Gayle, H., W. Foege, L. Brown, and B. Kahn (Editors). 2020. *Framework for Equitable Allocation of COVID-19 Vaccine*. http://nap.edu/25917. Washington, DC: The National Academies Press.

GIN68 Ginzburg, S.A., Математическая непрерывная логика и изображение функции (Mathematical continuous logic and representation of functions), Energiya, Moskva, 1968.

KAT21 Kates, J., J. Tolbert, and J. Michaud. 2021. The COVID-19 "Vaccination Line": An Update on State Prioritization Plans, Jan 11, 2021. https://www.kff.org/coronavirus-COVID-19/issue-brief/the-COVID-19-vaccination-line-an-update-on-state-prioritization-plans/

MEN10 Mendel, J.M., and D. Wu. 2010. *Perceptual Computing*. IEEE Press and J. Wiley.

MES17 Messer, K.D., and W.L. Allen, III. 2017. *The science of strategic conservation: Protecting more with less*. Cambridge University Press, 2017.

MIL56 Miller, G.A. 1956. The magical number seven, plus or minus two: Some limits on our capacity for processing information. *The Psychological Review,* 63: 81–97.

PRE00 Preston, C.C., and A.M Colman. 2000. Optimal number of response categories in rating scales: Reliability, validity, discriminating power, and respondent preferences. *Acta Psychologica* 104: 1–15.

RAD00 Radojević, D.G. 2000. [0,1]-valued logic: A natural generalization of Boolean logic. *Yugoslav Journal of Operations Research* 10(2), 185–216.

ROH07 Rohrmann, B. 2007. *Verbal Qualifiers for Rating Scales: Sociolinguistic and Psychometric Data. Project Report*, Australia: University of Melbourne. http://rohrmannresearch.net/pdfs/rohrmann-vqs-report.pdf.

SMI13 Smith, W.D. 2013. *Rating Scale Research Relevant to Score Voting*. http://rangevoting.org/RateScaleResearch.html.

SU82 Su, S.Y.W., D.S. Batory, J. Dujmović, et al. 1982. *A DMS Cost/Benefit Decision Model: Cost and Preference Parameters. National Bureau of Standards.* Washington D.C., No. GCR 82- 373. NTIS No. PB 82-169566 (393 pages).

SU87 Su, S.Y.W., J. Dujmović, D.S. Batory, S.B. Navathe, and R. Elnicki. 1987. A cost-benefit decision model: Analysis, comparison, and selection of data management systems. *ACM Transactions on Database Systems* 12(3), 472–520.

WOR23 Worldometer. 2023. COVID-19 coronavirus pandemic. https://www.worldometers.info/coronavirus/country/us/.

Chapter 5
Experimental Measurements of Basic Graded Percepts

Observations of human reasoning indicate that graded percepts are generated, normalized, memorized, and processed by the human mind. Graded percepts normalized in the range [0, 1] play a fundamental role in decision-making. Of course, the only way to justify claims that the human mind generates and uses normalized graded percepts is to perform experiments and measurements that can verify the expected properties. In this chapter we present basic concepts of developing experiments with graded percepts, and sample results of typical experimental measurements.

Decision-making experiments with human subjects are delicate and deserve attention. There are six important components that affect such experiments:

(1) The role of respondents
(2) Design of narrative used for eliciting responses
(3) Preparation of respondents.
(4) Motivation and effort of respondents.
(5) Filtering of measurements and the selection of measured results.
(6) Processing selected results.

Below we discuss problems related to these six components.

(1) The role of respondent. In most surveys the respondents are requested to provide their personal opinions formalized as answers to specific questions. Generally, such answers reflect social, economic, ethnic and educational background of respondents. In addition, the answers may also depend on age, gender, and specific experiences of each respondent. In other words, the respondents represent themselves, with all specific differences that humans can have. That is exactly what we try to avoid in the experiments that investigate human commonsense logical reasoning.

In the context of human-centric approach to logic, we assume that human commonsense logical reasoning is a general human mental property, independent of social, ethnic, and economic conditions. Indeed, the conjunction, disjunction, and negation do not depend on the wealth, age, or ethnicity of specific stakeholder/decision-maker. Therefore, the logic experiments must be designed in a way that is

J. Dujmović, *Graded Logic*, Intelligent Systems Reference Library 273,
https://doi.org/10.1007/978-3-031-88558-7_5

independent of respondents' social and economic status. That is possible and easy if we study the simultaneity of specific observable properties of evaluated objects (see Chap. 14). However, if we analyze the simultaneity of respondent's goals (e.g., the evaluation of job offers, where a decision-maker requires the simultaneity of the goal to get a good monetary compensation and the goal to have an attractive job) then, depending on the age, education, and economic status, each respondent can have different requirements, resulting in a different degree of simultaneity and different importance weights of input components. That is certainly not a respondent-neutral situation, and in such cases, it is necessary to equalize respondents by asking them to play the role of specific clearly defined stakeholder and not to represent themselves based on their current social and economic status. Generally, in each experiment that investigates human commonsense logical reasoning, the first step is to define the role of respondent, and there are two possibilities: (1) the respondents represent themselves, and (2) all respondents play the role of the same specific stakeholder/decision-maker.

(2) Design of narrative. All experiments with human subjects are based on a written request that the respondents read before and/or during experiments. Such a narrative must be as short and precise as possible. The main problem is that the participants in experiments are not exposed to real decision situations, but in most cases, they are requested to *play the role of a specific decision-maker*. For example, if respondents are requested to play the role of an apricot-buyer who makes preserves, then there will be a significant difference between a respondent who routinely makes preserves each year, and a respondent who never made preserves and thinks that glass jars with marmalade in a local food store can only be an industrial product. Some people have difficulties playing the role of anybody except themselves: impecunious people can face difficulties to play the role of wealthy people and vice versa. Therefore, the narrative of decision-making experiments must be as convincing as possible, to ensure that all respondents properly play the same role. For that reason, it is necessary to properly prepare the respondents for experiments.

(3) Preparation of respondents. The preparation of respondents is always useful but not always possible. We differentiate four kinds of respondents: (P0) *unprepared*, (P1) *underprepared*, (P2) *prepared,* and (P3) *overprepared.*

Unprepared respondents are people who had no contact with organizers of experiments and have no clear understanding of the purpose and significance of experiments. Sometimes, these are volunteers who participate over the Internet in a recreational way and with minimum motivation. Their preparation is reduced to a short and frequently incomplete narrative provided together with questions. Unprepared respondents read the narrative partially, or superficially. They answer questions without clear understanding of the nature and goals of the experiment, and without awareness of the responsibility for the quality of responses.

Underprepared respondents are respondents who had minimum contact with the organizers of experiments. Such respondents might be informed about the goal of experiment but don't have individual interest in the experiment and its results. Sometimes such respondents are participating in experiments long time after the preparation session, when the experiment guidelines have faded out. Sometimes, these are

Table. 5.1 Classification of respondents and types of experiments

Types of respondents based on combination of preparation and motivation		Respondent preparation degree			
		Unprepared (P0)	Underprepared (P1)	Prepared (P2)	Overprepared (P3)
Respondent motivation and effort	Low (M1)	P0M1	P1M1	P2M1	P3M1
	Medium (M2)	P0M2	P1M2	P2M2	P3M2
	High (M3)	P0M3	P1M3	P2M3	P3M3

students who are encouraged to participate anonymously as volunteers. Having that activity very low on their priority list, they participate only after several reminders.

Prepared respondents are respondents who had regular personal contact with the organizers of experiments, where they are informed about the type, significance, and goals of experiments. Typically, they are given precise guidelines and examples of questions and answers. Such respondents participate in experiments shortly after the preparation, and sometimes collectively, as students who take a decision-making experiment in a classroom and submit results at the end of class.

Overprepared respondents are respondents who are not only fully prepared for participating in the experiments, but usually fully aware of the expected or desired results of experiments. These are people very close to the organizers of experiments, or the organizers and related researchers themselves. Unsurprisingly, such participants can be biased and in majority of cases should be disqualified to take part in experiments. For example, the author of this book was never a participant in experiments reported in the book.

(4) Motivation and effort of respondents. Motivation and effort of respondents are important factors contributing to the credibility of results. According to Table 5.1, we differentiate three levels of motivation: low, medium, and high.

Low motivation is a serious problem that is sometimes characterized by respondents entering random numbers instead of carefully selected answers. Another manifestation of low motivation is ultimate imprecision (entering 0 instead of all low values, or 1 instead of all high values), violating the rules of commutativity (if commutativity is supported), violating the use of annihilators, or replacing all logic operations with a trivial arithmetic mean. The problem with low motivation is that it creates worthless answers that damage the whole experiment and must be automatically (algorithmically) detected and disqualified. Low motivation occurs primarily in experiments where respondents are anonymous, and their primary concern is minimizing the time of participation in experiments. In some experiment, the low motivation respondents can be detected using special attention check questions.

Medium motivation and effort are the most frequent case that assumes respondents who understand the goals of experiment and try to contribute to results with careful selection of answers, in a way consistent with experiment guidelines. Medium motivation is typical for respondents who are not anonymous, e.g., respondents who are either paid for participation, or unpaid students replying in the context of an official coursework assignment.

Table. 5.2 Experiments with two groups of P1M2 respondents

Group	Number of respondents	Object	Round R_1; R_2 [%]	Hard H_1 [%]	Soft S_2 [%]	σ_{round} [%]	σ_{hard} [%]	σ_{soft} [%]
1	59	Tomato	77.9	20.5		17.9	19.1	
		Avocado	57.1	38.0		21.7	24.5	
		Brick	6.2	94.9		13.8	13.6	
		Tennis ball	95.3	52.6		17.2	23.9	
		Pancake	73.3	9.1		24.5	13.2	
2	42	Tomato	79.2		74.1	18.7		23.5
		Avocado	58.2		55.6	25.7		25.2
		Brick	3.0		4.1	9.5		16.2
		Tennis ball	95.4		47.4	15.8		24.5
		Pancake	77.8		85.5	27.5		17.0

High motivation and effort of respondents is typical for participants in experiments who are researchers in decision-making areas (e.g., graduate students). Highly motivated participants are usually individuals recruited by organizers of experiments in situations where the organizers are primarily interested in high quality results, regardless that their number can be small.

Table 5.1 shows twelve possible combinations of motivation and preparation. Of course, all combinations are not equally likely and equally desirable. The shaded combinations are highly unlikely. On the other hand, the most frequent and acceptable combinations are P1M2 and P2M2. These combinations create useful results. A sample of such results is presented in Table 5.2.

(5) Filtering of measurements and the selection of measured results. In ideal situations all collected results should be usable for making analytic models of commonsense logical reasoning. In real situations, however, that is possible only for highly motivated and sufficiently prepared participants. In cases of insufficiently motivated and/or insufficiently prepared participants, it is easy to find responses that convincingly disqualify some of collected results. The filtering of measurements is the process of excluding disqualified responses.

Disqualifying responses differ in cases of estimating graded characteristics of objects, and in cases of analyzing goal-based mental logic aggregation. Typical examples of such disqualifying responses include the following:

- Answers consisting of random numbers (e.g., a mix of results that are insufficiently correlated to expected "ideal" values, i.e., correlated below some low threshold value).
- Obvious meaningless results (e.g., claiming that a coin is soft, or the car tire is not round).
- Disjunctive (or even hyperdisjunctive) responses in strictly conjunctive aggregation experiments (accepting or even rewarding a single input in cases where requested relationship is the simultaneous satisfaction of two or more inputs).
- Responses based on a formula (in aggregation experiments, by far the most frequent violation of experiment rules consists of using for all aggregations the

arithmetic mean formula instead of supplying the requested intuitive aggregated percept).

- Violating the request for commutativity, i.e., equal importance of inputs (different ordering of commutative inputs give significantly different aggregated results).
- Violating requests for annihilators in strongly conjunctive aggregation problems (e.g., giving positive answers for zero values of inputs that, according to experiment rules, must be satisfied).
- Disrespectful/meaningless responses (e.g., giving the same response, like 0 or 1, to all questions).
- Failing specific number of *attention check* (AC) questions (some experiments can include several AC questions requesting respondents to provide a specific answer; correct answers verify that the respondents are alert and carefully read the requests of the experiment). Failing AC questions usually indicates a low motivation.

The disqualifying results can be algorithmically detected and eliminated by a filtering program before the processing of experimental measurements. The decision to disqualify and exclude some of collected responses is justified by balancing the goal to get trustworthy inputs for modeling human reasoning, and the goal to avoid a biased analysis that would be obtained by eliminating all inputs inconsistent with desired results. Of course, the purpose of these experiments is *not* to investigate the responses of a population which often has questionable trustworthiness, but to analyze the reasoning of trustworthy participants. Consequently, the elimination of disqualifying responses is fully legitimate, but this process is neither trivial nor riskless, and must be motivated only by the goal to maximize the credibility of conclusions derived by processing the selected responses.

(6) Processing of selected results. An experiment with *unfiltered responses* obtained from two different groups of underprepared respondents who have medium motivation (unpaid students of Computer Science) is presented in Table 5.2 (see also [DUJ22] and Chap. 14), The goal of this experiment was to analyze mental graded percepts for object that can be round and hard, or round and soft. The first group of 59 respondents was asked to report to what extent selected objects are simultaneously round and hard. The second group of 42 respondents was given the same objects and requested to report to what extent selected objects are simultaneously round and soft. The first goal of the experiment was to find whether the two independent groups have consistent graded percepts about roundness of objects. The second goal was to find whether the soft and hard characteristics are complements (negations) of each other.

For each object, let R_1 and R_2 be the mean values of the intensity of percepts of roundness obtained from the group 1 and the group 2, respectively. The absolute difference between the percepts of roundness of the two groups is $|R_1 - R_2|$, presented in Table 5.3. Then, let H_1 be the percept of the hardness of object created by the group 1, and let S_2 be the percept of the softness of object created by the group 2. In an ideal case, we expect that $H_1 + S_2 = 100\%$. The inconsistency between the percepts of hardness and the percepts of softness is $|H_1 - (100 - S_2)| = |S_2 - (100 -$

Table. 5.3 Differences between graded percepts

	Differences (errors)	
	Roundness of objects	Hard/soft complements
Object	$\|R_1 - R_2\|$ [%]	$\|H_1 + S_2 - 100\|$ [%]
Tomato	1.3	5.4
Avocado	1.1	6.4
Brick	3.2	1.0
Tennis ball	0.1	0.0
Pancake	4.5	5.4
Mean values [%]	2.04	3.64

$H_1)| = |H_1 + S_2 - 100|$.. These results are presented in Table 5.3. Following are the observations that can be derived from this experiment with graded percepts:

- The human mind creates justifiable graded percepts of characteristics of objects.
- Different groups of respondents can create very similar graded percepts of characteristics of objects (absolute differences from 2 to 4%).
- Observer-independent graded percepts are frequently present in human reasoning. Such graded percepts facilitate human reasoning and communication.
- The human mind creates consistent graded percepts of the complements of the characteristics of a given object.

Standard deviations of presented responses are moderate. High values of standard deviations can indicate the presence of outlier values and the need for filtering responses. They can also indicate the presence of controversial questions and the need to redesign the experiment, omitting or replacing those questions that create too wide a spectrum of responses. For example, the highest standard deviation of responses of the pancake questions shows the presence of respondents who are not familiar with pancakes (or avocados) as a food. Thus, the experiment developers may consider the option to eliminate such questions and replace them with objects that are provably familiar to all respondents. Food or fruit that are present only in some restricted regions or cultures, should not be used in logic experiments because we are interested in logic and not interested in the degree of familiarity with objects. Therefore, the experiments with human subjects can be designed in two phases: in the first phase the goal is to test the validity of questions, and after the questions are accepted, in the second phase, the goal is to analyze and model human logical reasoning.

Another experiment, devoted to the study of mental simultaneity models and reported in more detail in Chap. 14, was performed with 153 psychology students by David Martin Obando M.A. in the Personality & Well-Being Lab of Dr. Ryan T. Howell at San Francisco State University. The respondents were requested to play the role of a homebuyer and evaluate homes based on their quality and the suitability of location. The experiment included 20 questions that can be answered in 3 to 5 min,

Table. 5.4 Experiments with 153 psychology students

Number of participants	Number of removed responses	The reason for removal
153 (100%)	18 (12%)	Incomplete or duplicate answers
135 (88%)	74 (48%)	3 failed AC questions
61 (40%)	19 (12%)	2 failed AC questions
42 (28%)	27 (18%)	1 failed AC questions
15 (10%)	Complete answers that passed all AC questions	

expanded with 3 AC questions. This experiment provides realistic insight into the role of AC questions, shown in Table 5.4.

Obviously, the results in Table 5.4 are discouraging because they show that only 10% of participants can sufficiently focus on the experiment that takes less than 5 min. Another disappointment: 12% of useless incomplete answers. On the other hand, failed AC questions are not the proof of unacceptable answers for the analysis of simultaneity. Similarly, the complete answers that pass all AC tests are not a proof that the answers are useful. It is necessary to develop specific acceptance tests for participants' answers and that filtering problem is discussed in Chap. 14.

Reference

DUJ22 Dujmović, J., and D. Tomasevich. 2022. Experimental Analysis and Modeling of Human Conjunctive Logic Aggregation. In *2022 IEEE International Conference on Fuzzy Systems (FUZZ-IEEE)*, 1–8, Padua, Italy. https://doi.org/10.1109/FUZZ-IEEE55066.2022.9882665.

Chapter 6
A Basic Model of Human Decision-Making

Identifying and understanding the stakeholder/decision-maker is the indispensable first step in all human-centric propositional logic problems. In the case of decision-making, we assume that the stakeholder is an individual (or an organization) interested in achieving specific goals. The goals are usually derived from current activities that need improvement. Stakeholder's goals attainment is the primary motivation for decision-making, which includes the following main steps:

(1) Specification of desired goals.
(2) Identification of alternatives that support goal attainment.
(3) Development of criterion for evaluation and comparison of alternatives.
(4) Evaluation of alternatives and selection of the most suitable alternative/action.
(5) The decision whether to realize or to postpone the most suitable alternative/action.

In human activities there are countless decision problems. Table 6.1 shows nontrivial examples of different stakeholders, different alternatives (evaluated objects), and typical decision problems that include graded propositional logic. For each object/alternative, the result of the process of decision-making is the percept of overall satisfaction of stakeholder's requirement. That resulting percept of satisfaction/suitability is graded and can be interpreted as the degree of truth of the statement "the selected alternative completely satisfies all stakeholder's requirements."

Generally, each stakeholder is assumed to bear the consequences of selected most suitable alternative/action. Thus, the stakeholder/decision-maker is always interested to use a correct (best possible) intuitive or formal criterion for determining the degree of satisfaction of desired goals and requirements. An observable model of human intuitive decision-making is presented in Fig. 6.1. This model includes a sequence of several mental activities that can be easily identified when we observe the process of intuitive decision-making. In Chap. 1 we presented a trivial problem of decision-making when buying apricots. Now, we are going to explain the process of intuitive decision-making, and the development of the human decision-making model shown

J. Dujmović, *Graded Logic*, Intelligent Systems Reference Library 273,
https://doi.org/10.1007/978-3-031-88558-7_6

Table. 6.1 Sample stakeholders and decision problemsa

Stakeholder	Object/alternative	Decision problem (selection of the best alternative)
Car buyer	Car	Select the most suitable car
Company	Job applicant	Evaluate the suitability of applicant
Student	University	Select the most suitable university
University	Professor of Math	Compare candidates and select the best professor
Engineer	Laptop computer	Select the most suitable laptop computer
Patient	Kidney transplant	Determine the kidney transplant priority score
Patient	COVID vaccine	Determine the COVID vaccination priority score
City	Airport location	Find the most suitable location for a new airport
Family	Home	Select the most suitable family home
Job applicant	Job offers	Compare job offers and find the best offer
Company	Computer system	Select the most suitable computer system
Hotel company	Hotel location	Find the most suitable location for a new hotel
Military	Fighter jet	Find the best new fighter jet
State	Agricultural land	Find a suitability map for growing corn in the state
Patient	Disability	Find a disability degree caused by specific disease
Software engineer	Search engine	Compare performance of search engines
Senior citizen	Retirement home	Find the best retirement home in a given area
Teenager	Smart phone	Find the best smart phone for social networking
President	Vice president	Compare candidates for vice presidents
Family	Vacation place	Select an attractive and affordable vacation place

in Fig. 6.1, using a slightly more complex example. The stakeholder/decision-maker will be an engineer, and the corresponding decision problem will be the selection of a new laptop computer. The laptop is necessary to help achieve stakeholder's professional goals. Of course, the selection of a new laptop is a very frequent decision problem, familiar to many other people.

Suppose that the stakeholder (an engineer) identified m objects (e.g., 4 laptops manufactured by Dell, HP, Apple, and Lenovo) that are alternatives in the decision-making process. In a special case, m can be 1. Consequently, the goal of evaluation is to find the overall degree of suitability of each of m candidates.

To derive the overall degree of suitability, the first step is to find suitability attributes, i.e., all inputs that significantly affect the overall suitability of evaluated objects. Of course, there are also various attributes that do not affect the overall suitability or affect it in a negligible way; such attributes should not be used as suitability attributes. For example, the color of a laptop computer is an attribute, but it is usually neutral, or easily selectable, and for most decision-makers it will not significantly affect the laptop suitability. On the other hand, the size of memory is a suitability attribute because it significantly affects the capability and suitability of any computer.

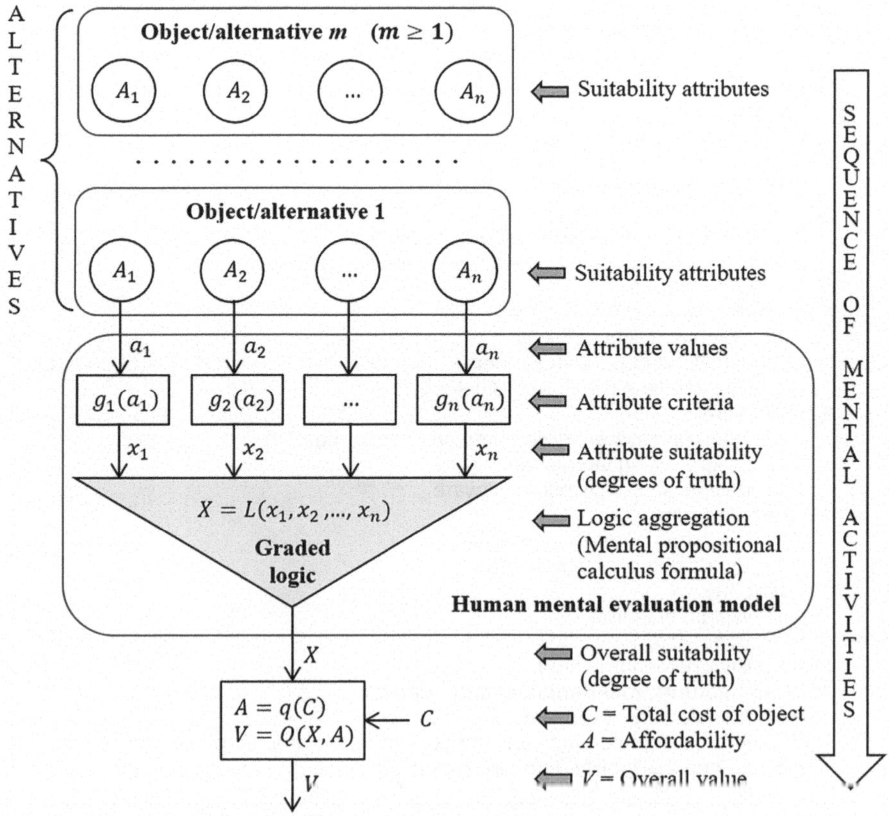

Fig. 6.1 A basic observable model of human intuitive decision-making

A natural way humans use to develop a list of necessary and sufficient suitability attributes is to work in simple steps looking systematically for components of each attribute. Our stakeholder (an engineer) knows that laptops are produced for various application areas: work, education, gaming, family life and entertainment. So, the desired laptops will be selected from the group of machines for work, and in such a case the stakeholder's initial idea is that the laptop suitability depends on the following five significant components:

1 Laptop
 11 Hardware
 12 Software
 13 Performance
 14 Usability
 15 Manufacturer, maintenance, and support

Attribute tree for project Laptop

- **1 Laptop**
  - **11 Hardware**
    - 111 Processor (clock rate, cache size, number of cores)
    - 112 Memory capacity (RAM)
    - 113 Storage capacity (SSD or HDD)
    - 114 Input/output devices
      - 1141 Keyboard quality and organization
      - 1142 Graphics card and display (resolution and size)
      - 1143 Camera
      - 1144 Audio and speakers
    - 115 Communication devices
      - 1151 Ports (type and number)
      - 1152 Wireless (Wi-Fi and Bluetooth)
  - **12 Software**
    - 121 Operating system (Win home/pro, Linux, macOS)
    - 122 Programming languages
    - 123 Utilities and application software
  - **13 Performance**
    - 131 Published performance results
    - 132 User's personal benchmarks
  - **14 Usability**
    - 141 Weight of laptop
    - 142 Battery life
    - 143 Battery recharge time
  - **15 Manufacturer, maintenance, and support**
    - 151 Manufacturer's warranty
    - 152 HW/SW maintenance and support (manufacturer, employer, and others)
    - 153 Laptop availability (time on market)

Fig. 6.2 The suitability attribute tree structure

So, the graded percept of the overall laptop suitability depends on graded percepts of suitability of hardware, software, performance, usability, and manufacturer, hardware/software maintenance, and support. It is obvious that each of these components is compound and can be decomposed into simpler attributes. This hierarchical decomposition process yields the suitability attribute tree structure shown in Fig. 6.2. The suitability attributes are the leaves of the suitability attribute tree. Therefore, in this example, the stakeholder created $n = 20$ suitability attributes shown in Fig. 6.3. These are the attributes denoted $A_1, \ldots, A_n$ in the first step of mental activities shown in Fig. 6.1.

It is crucial to realize that each suitability attribute has its distinctive name, role, function, units of measurement, and specific importance for the stakeholder/decision-maker. These components are collectively denoted as a *semantic identity* of each attribute. Using these attributes in the natural mental evaluation process is impossible unless their semantic identity is well understood and permanently considered. Different identities are easily visible in units of measurement. For example, the capacity of memory and storage is measured in gigabytes, the weight of laptop is

Suitability attributes for project Laptop

1. Processor (clock rate, cache size, number of cores)
2. Memory capacity (RAM)
3. Storage capacity (SSD or HDD)
4. Keyboard quality and organization
5. Graphics card and display (resolution and size)
6. Camera
7. Audio and speakers
8. Ports (type and number)
9. Wireless (Wi-Fi and Bluetooth)
10. Operating system (Win home/pro, Linux, macOS)
11. Programming languages
12. Utilities and application software
13. Published performance results
14. User's personal benchmarks
15. Weight of laptop
16. Battery life
17. Battery recharge time
18. Manufacturer's warranty
19. HW/SW maintenance and support (manufacturer, employer, and others)
20. Laptop availability (time on market)

Fig. 6.3 Suitability attributes for evaluation of laptop computer

measured in pounds or kilograms, battery life is measured in hours, processor clock rate in gigahertz, the laptop time on market is measured in months. Some attributes, like the number of USB ports, have no unit of measurement, but certainly have a clear semantic identity. Some attributes, like the quality and organization of keyboard, are themselves compound. E.g., the keyboard is characterized by mechanical quality of keys, functionality of keys, distribution of keys, backlit support, and physical convenience for the stakeholder, that collectively give the specific semantic identity of this suitability attribute.

The list of suitability attributes shown in Fig. 6.3 could not be easily developed unless the stakeholder/decision-maker has a sufficient understanding of the objects of evaluation: their components and principles of functioning and use. In addition, it is difficult to develop the lists of suitability attributes unless the stakeholder uses the systematic hierarchical decomposition process. In such a way it is possible to develop lists that are complete and realistically reflect the goals and requirements of the stakeholder.

Let $A_1, \ldots, A_n$ be the set of n suitability attributes selected by a decision-maker. In the intuitive decision-making process, the number of suitability attributes (denoted n in Fig. 6.1) is usually very limited, and 20 attributes might be considered an upper value used in intuitive decision problems, such as selection of laptops, cars, schools, homes, job offers, retirement homes, etc. The list of attributes shown in Fig. 6.3 clearly reflects the requirements of the selected stakeholder (an engineer). Other possible stakeholders (business executives, students, medical personnel, gamers, family users, etc.) might develop completely different lists of suitability attributes.

From GL standpoint, it is very important to note the obvious fact that all attributes don't have equal importance for the stakeholder, showing that commutativity of inputs in propositional calculus operations is not an acceptable property. The assumed stakeholder is an engineer who might develop professional software using selected programming languages, and in such a case the availability of necessary programming languages is an important requirement. Similarly, engineering projects might need high performance, and the assumed stakeholder could use some of her/his software as important user's performance benchmarks. For such a user the high quality of audio and high power of speakers might have rather low importance. On the other hand, for stakeholders who are not software developers, programming languages might be completely excluded from the list of suitability attributes, and (assuming primary interest in games and movies) the high quality and power of audio equipment might be a mandatory requirement.

According to Fig. 6.1, there are m alternatives, and each alternative has n observable and measurable attributes $A_1, \ldots, A_n$ that affect the overall suitability degree of the analyzed alternative. For each evaluated object (laptop) these suitability attributes have specific values. In Fig. 6.1 they are denoted $a_1, \ldots, a_n$ and used as inputs in the human mental evaluation model. For example, if A_3 denotes the storage capacity (#3 in Fig. 6.3), then its value can be $a_3 = 256GB$. For each value of selected suitability attribute the stakeholder performs intuitive evaluation using a suitable intuitive attribute criterion. The intuitive criterion can be approximately described as shown in Fig. 6.4: the stakeholder wants to reject laptops that offer 128 GB or less of storage. The stakeholder is completely satisfied if the storage has 512 GB or more. For the storage capacity of 256 GB, the graded suitability is 70% (i.e., the degree of truth of the total satisfaction statement "256 GB of storage completely satisfies all my requirements" is 0.7).

Attribute criteria are functions that map the value of attribute a_k to the corresponding normalized suitability degree x_k. In other words, the attribute criterion is a function $g_k : \mathbb{R} \to [0, 1]$, where a_k is a real number and $x_k = g_k(a_k), k \in \{1, \ldots, n\}$ is the suitability degree. In our terminology, *suitability* is a normalized percept of satisfaction, i.e., the degree of truth of the corresponding total satisfaction statement.

The attribute criterion for storage capacity in Fig. 6.4 is an increasing function of the available capacity. Unsurprisingly, there is a wide variety of various intuitive attribute criteria used by human stakeholders/decision-makers, and three more characteristic cases are included in Fig. 6.4. The criterion for weight (#15 in Fig. 6.3) is a decreasing function. It reflects rather strong preference for very light laptops: ideally, the stakeholder would like a laptop with weight up to 2 pounds and would reject laptops above 4 pounds.

The criterion for the time on market (#20 in Fig. 6.3) shows a cautious stakeholder who would not accept a laptop that is too new and not proved as a successful product. So, the most acceptable laptops are those that have been on the market already for 6 months and received positive reviews. However, the stakeholder wants to have a

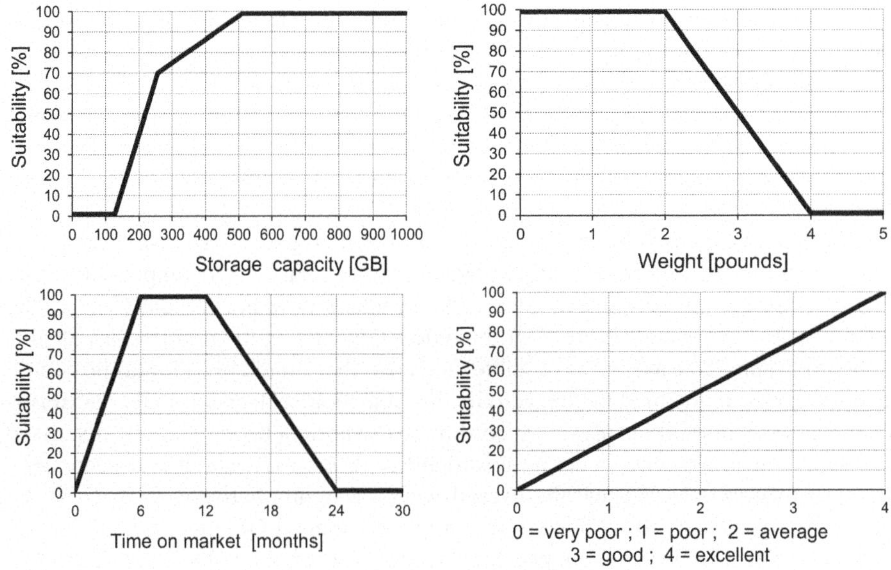

Fig. 6.4 Sample attribute criteria

new product and as shown in Fig. 6.4, after 12 months the suitability of laptops starts decreasing to reach zero after two years.

If the attribute values are measurable (as the memory capacity, weight, and the time on market) the decision-maker can establish a justifiable mental grading function that assigns suitability to specific values of a_k. There are situations, however, where a_k is not directly measurable. For example, utilities and application software (#12 in Fig. 6.3) include a variety of programs of various origins and various quality. The decision-maker can analyze such software, and the results of analysis express as a linguistic label on a rating scale. If the utility and application programs can be evaluated as very poor, poor, average, good, or excellent, then the decision-maker is directly assessing the attribute suitability x_k using a simple linear criterion shown in Fig. 6.4.

According to Fig. 6.1, in the process of natural intuitive decision-making, the first two significant mental activities are the development of the list of suitability attributes, and the attribute-by-attribute evaluation of selected object/alternative. At the end of this process the decision-maker knows the suitability degrees $x_1, \ldots, x_n$, which are all of them GL variables from interval [0, 1]. If the decision-maker knows how good all components of an object/alternative are, now is the time to answer the crucial question "how good is the evaluated object as a whole?". The overall suitability $X \in [0, 1]$ is also a GL variable, interpreted as the degree of truth of the value statement "evaluated object completely satisfies all stakeholder's requirements." As shown in Fig. 6.1, X is computed as a graded propositional logic function $X = L(x_1, x_2, \ldots, x_n)$, using human mental propositional calculus. Therefore, now we face the problem of identifying GL functions that are used within the human mental propositional calculus, and the rest of this book is contributing answers to this fundamental question.

The propositional logic function $L : [0, 1]^n \rightarrow [0, 1]$ is not the only GL function used within the human intuitive decision-making process shown in Fig. 6.1. Like laptop computers, many objects have the cost C that critically affects the final decision about the overall value of an object. We assume that the cost reflects all the aggregated expenses for buying an object or realizing an alternative. E.g., in the case of laptops, the cost reflects all hardware, software, and other components of the laptop. Consequently, the stakeholder/decision-maker pays the amount C to get the suitability X. The overall value V of the object depends on the relationship of suitability (that should be as high as possible) and cost (that should be as low as possible). In other words, in the final mental activity, the stakeholder/decision-maker performs a mental cost/suitability analysis. According to the apricot buyer example discussed in Chap. 1, it is possible to define affordability $A = q(C)$ which is a decreasing (mental) function of C. Then, the overall value of an alternative is $V = Q(X, A)$, where $Q : [0, 1]^2 \rightarrow [0, 1]$ denotes a conjunctive mental GL function. Of course, there are many decision-making problems where cost is not involved. E.g., a student can evaluate and compare the websites of two potential universities and there is no cost involved in such an analysis. In this situation we simply have the overall value $V = X$.

To summarize this chapter, we have shown that humans use GL as a fundamental component of a mental (intuitive) decision-making process. This process includes the evaluation of objects and follows the general decision-making model shown in Fig. 6.1. This model is proved by observability of human reasoning, as well as by many successful industrial decision projects based on methodology derived from the presented model. Human mental processor can create, memorize, and use graded percepts and process them by combining GL functions and reasoning according to formulas of graded propositional calculus. To study GL, we must first understand the basic mental logic functions that model simultaneity (conjunction) and substitutability (disjunction). This is done in the next chapter.

Chapter 7
Graded Simultaneity and Graded Substitutability: Complementary and Unified

Classical logic is not graded. It is bivalent: all assertions can be either completely true (numerically coded 1) or completely false (numerically coded 0) [BOO47, DEM47, BOO54, STE22]. So, logic variables are binary, $x_i \in \{0, 1\}, i = 1, \ldots, n, n > 0$, and logic functions create binary results $y = f(x_1, \ldots, x_n) \in \{0, 1\}$. There are 2^n vertices in the n-dimensional hypercube, and in each vertex the logic function $f : \{0, 1\}^n \rightarrow \{0, 1\}$ can have the value 0 or 1. Thus, there are 2^{2^n} different logic functions of n variables (16 functions of 2 variables, 256 functions of 3 variables, etc.). The classical logic model of simultaneity is conjunction, the model of substitutability is disjunction, and the third necessary basic logic function is complementing/negation, defined as follows:

$$y_c = x_1 \wedge x_2 \wedge \ldots \wedge x_n = \min(x_1, \ldots, x_n) \quad \text{(conjunction, logic connective AND)}$$

$$y_d = x_1 \vee x_2 \vee \ldots \vee x_n = \max(x_1, \ldots, x_n) \quad \text{(disjunction, logic connective OR)}$$

$$y = \overline{x} = not(x) = 1 - x \quad \text{(complementing/negation).}$$

Bivalent conjunction and disjunction are commutative ($x \wedge y = y \wedge x$; $x \vee y = y \vee x$), associative ($(x \wedge y) \wedge z = x \wedge (y \wedge z)$; $(x \vee y) \vee z = x \vee (y \vee z)$), distributive ($x \wedge (y \vee z) = (x \wedge y) \vee (x \wedge z)$; $x \vee (y \wedge z) = (x \vee y) \wedge (x \vee z)$), and idempotent ($x \wedge \ldots \wedge x = x$; $x \vee \ldots \vee x = x$). Using these three basic functions it is possible to create all other bivalent logic functions. In the case of two variables there are 16 functions $z = f(x, y)$, where 6 functions are trivial reductions to constants or single variables: $z = 0, z = 1, z = \overline{x}, z = \overline{y}, z = x$ (the total absorption of y), and $z = y$ (the total absorption of x). The remaining 10 functions include two variables as follows (I = idempotent, C = commutative):

$z = x \wedge y$ (conjunction, and) [I, C]

J. Dujmović, *Graded Logic*, Intelligent Systems Reference Library 273,
https://doi.org/10.1007/978-3-031-88558-7_7

$z = x \vee y$ (disjunction, or) [I, C]

$z = x|y = \bar{x} \vee \bar{y} = \overline{x \wedge y}$ (negated and, nand) [C]

$z = x \downarrow y = \bar{x} \wedge \bar{y} = \overline{x \vee y}$ (negated or, nor) [C]

$z = x \rightarrow y = \overline{x \wedge \bar{y}} = \bar{x} \vee y$ (implication, x implies y)

$z = y \rightarrow x = \overline{y \wedge \bar{x}} = \bar{y} \vee x$ (implication, y implies x)

$z = x \wedge \bar{y} = \overline{x \rightarrow y}$ (abjunction)

$z = \bar{x} \wedge y = \overline{y \rightarrow x}$ (abjunction)

$z = x \oplus y = (x \wedge \bar{y}) \vee (\bar{x} \wedge y) = \overline{x \sim y}$ (exclusive or) [C]

$z = x \sim y = (\bar{x} \wedge \bar{y}) \vee (x \wedge y) = \overline{x \oplus y}$ (equivalence) [C]

In these functions, variables $x = 1$ or $y = 1$ denote the complete satisfaction of requirements specified by x and y. Similarly, variables $x = 0$ or $y = 0$ denote the complete nonsatisfaction of requirements specified by x and y. Consequently, the basic functions of two variables can be verbally interpreted as follows:

Conjunction: it is necessary to simultaneously provide the satisfaction of all inputs

Disjunction: it is sufficient to provide satisfaction of one or more inputs

Nand: it is not necessary to provide satisfaction of all inputs; it is sufficient to have one or more inputs that are not satisfied

Nor: it is not sufficient to provide satisfaction of one or more inputs; it is necessary that all inputs are not satisfied

Implication: if x implies y, then it is not possible/acceptable that x is satisfied, and y is not satisfied

Abjunction: one input is desirable, and it must be satisfied, and the other input is undesirable and it must not be satisfied

Exclusive or: a requirement for the difference of inputs; the condition $x \neq y$.

Equivalence: a requirement for the equality of inputs; the condition $x = y$.

Conjunction and disjunction are related by De Morgan duality $x \wedge y = \overline{\bar{x} \vee \bar{y}}$, and $x \vee y = \overline{\bar{x} \wedge \bar{y}}$. Assuming $x \in \{0, 1\}$ and $y \in \{0, 1\}$, the basic functions of two variables can also be written using arithmetic operations of addition, subtraction, and multiplication, as in the following examples:

$\bar{x} = 1 - x$
$x \wedge y = xy$
$x \vee y = x + y - xy = x + \bar{x}y = y + x\bar{y}$
$x \oplus y = x + y - 2xy = x(1 - y) + y(1 - x) = x\bar{y} + y\bar{x} = |x - y|$
$x \sim y = 1 - x - y + 2xy = 1 - |x - y|$
$x \rightarrow y = 1 - x + xy = 1 - x(1 - y)$

Using distributivity (or simply checking for $x = 0$ and $x = 1$) it is easy to verify the following two absorption relations:

$$z = x \wedge (x \vee y) = (x \vee 0) \wedge (x \vee y) = x \vee (0 \wedge y) = x \vee 0 = x$$
$$z = x \vee (x \wedge y) = (x \wedge 1) \vee (x \wedge y) = x \wedge (1 \vee y) = x \wedge 1 = x$$

In both cases the primary variable x *totally absorbs* the impact of the secondary variable y. In the first (conjunctive) case, the satisfaction of x is necessary. In the second (disjunctive) case the satisfaction of x is sufficient to satisfy the condition. The GL versions of these two relations are very important because they permit the conjunctive or disjunctive partial (graded) absorption of the secondary variable. In conjunctive case that makes a mandatory-optional condition, where x is a mandatory variable (one that must be satisfied) and y is an optional variable (desired but not necessary); see Chap. 16. In disjunctive case that makes a sufficient-optional condition, where x is a sufficient variable (one that is enough to satisfy) and y is an optional variable (contributing but not sufficient); see Chap. 17. Of course, such conditions are very frequently encountered in human mental criteria.

Similarly to two values of truth, classical logic uses a single form of conjunction (the minimum function) and a single form of disjunction (the maximum function). As opposed to that, in observable natural human reasoning, both simultaneity and substitutability are graded: decision-makers select a desired intensity of simultaneity or substitutability and use them in decision-making. The human-centric concepts and models of graded conjunction and graded disjunction were introduced in [DUJ73b, DUJ73c, DUJ74a, DUJ74b] and then used as a foundation for the development of GL.

The fundamental GL functions are the graded (or partial) conjunction and the graded (or partial) disjunction. These graded functions use a conjunction degree as a quantifier for continuous adjustment of the intensity of simultaneity, and a disjunction degree for continuous adjustment of the intensity of substitutability. More precisely, these fundamental GL functions are based on the following crucial concepts that are observable and measurable in natural (intuitive) human reasoning:

- Both simultaneity and substitutability are graded: they have continuously adjustable intensity.
- Simultaneity and substitutability are complementary (increasing simultaneity decreases substitutability and vice versa) with continuously adjustable ratio between conjunctive and disjunctive properties.
- Both simultaneity and substitutability are compensatory with continuously adjustable strength of compensatory properties. In most cases a decrement of an input argument can be compensated by an increment of another input argument, yielding unchanged output aggregated value; exception are the minimum and maximum functions, as well as those cases where a decrement is so big that it cannot be compensated.
- Simultaneity and substitutability have selectable use of annihilators 0 (for graded conjunction) and 1 (for graded disjunction).
- Simultaneity and substitutability have selectable use of idempotency; both idempotent and nonidempotent logic aggregators must be supported.

- Generally, both simultaneity and substitutability are noncommutative aggregators because input arguments are degrees of truth that generally have different degrees of importance. Commutativity is present as a special case.
- Human natural propositional logic combines formal logic components (degree of simultaneity or substitutability) and semantic components (meaning and importance of arguments). Thus, GL operators must support these properties.

In subsequent sections and chapters, we systematically present these fundamental concepts as observable and explainable properties of simultaneity and substitutability in natural human reasoning. After that, we can focus on mathematical aspects of modeling graded conjunction and graded disjunction.

Graded Simultaneity (Graded Conjunction) and Graded Substitutability (Graded Disjunction)

In natural human reasoning, simultaneity and substitutability have adjustable intensity. The intensity of simultaneity is visible if we look at the penalty that the stakeholder/decision-maker applies in cases where the inputs are not simultaneously highly satisfied. Similarly, the intensity of substitutability is visible if we look at the reward that the stakeholder/decision-maker applies in cases where a single input is satisfied. The intensity of graded conjunction and graded disjunction can also be visible in the way decision-makers compensate for deficiencies of some inputs using excessess of other inputs.

The essential goal of simultaneity is the simultaneous *high satisfaction* of all inputs. Note that simultaneity is different from equivalence. In equivalence the inputs must be equal or similar, regardless of their value. In the case of simultaneity, the inputs must be equal or similar, and in addition, they must be high, i.e., as true as possible. Therefore, the lowest input is one that is the least supportive of the requirement for simultaneous high satisfaction (high degree of truth). The low inputs are the most harmful for our goal, and that is easily observable in human reasoning.

In the simplest case of two different variables, the low variable is more influential than the high variable, and the output value is closer to the minimum input value. In Fig. 7.1 we combine a high input and a low input, and identify three characteristic intensities of simultaneity: medium, high, and highest. If the percept of the intensity of simultaneity has a medium value (which is the middle point between the lowest and the highest possible intensity of simultaneity) then the output value is in the middle between the high and the low input values. As the intensity of simultaneity increases, the necessary penalty increases, and the output value approaches the low input value. At the high intensity of simultaneity, the output degree of truth equals the low input value. The reasoning behind this form of simultaneity is that a chain is only as strong as its weakest link. In graded propositional calculus, this form of reasoning also reflects situations where the success of a team depends on its weakest member.

In the case of very high (strong) simultaneity the output value can be below the value of the lowest input. Typical examples of such simultaneity are all cases where we have a sequence (or a group) of operations (independent events) and input degrees

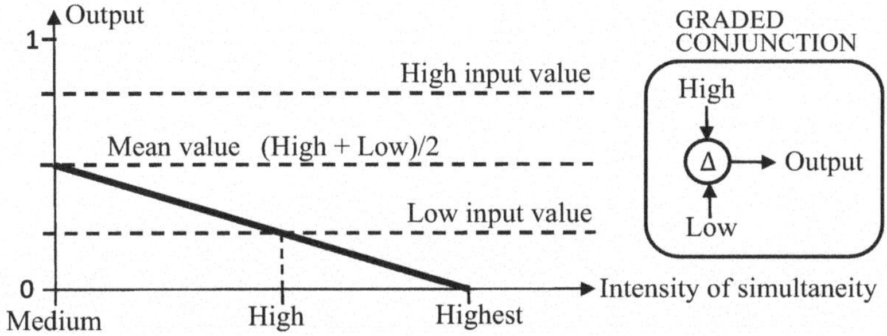

Fig. 7.1 The output of graded conjunction as a function of the intensity of simultaneity

of truth reflect the likelihood of successful completion of individual operations. So, the degree of truth of statement asserting a successful execution of the whole sequence is the same as the probability of occurrence of all independent events, which is the product of probabilities of individual operations/events. Of course, the product of degrees of truth, in cases where at least two degrees are less than 1, is less than the value of the lowest input, as shown in Fig. 7.1. While this probabilistic interpretation is easily understandable and frequently used, decision-makers can also use various intensities of simultaneity that are above or below this characteristic value. In the case of a very high simultaneity, the output value can be in the range from 0 to the lowest input value.

What is the ultimate (highest) intensity of simultaneity? Obviously, the highest simultaneity of high degrees of truth is the unique case where all inputs have the highest value 1. In this case we accept only those candidates that are perfect is each input argument. All those candidates that are not perfect in one or more inputs are rejected, i.e., for them the degree of truth is 0. Such a logic operation is called *drastic conjunction*: its output is 0 in all situations, except if the extreme case where all inputs are 1. According to Fig. 7.1, the output value 0 is achieved for the highest intensity of simultaneity even if all inputs are positive, but at least one of them is less than 1.

The medium, high, and the highest intensities of simultaneity shown in Fig. 7.1 are characteristic special cases. However, human graded percepts are continuously adjustable, and this also holds for the percept of simultaneity that can be continuously selected it the whole range from the lowest to the highest. Let us now show a few numeric examples of the effects of various degrees of simultaneity in the simultaneity operations. In our first and simplest example in Chap. 1, the buyer of apricots (or any other product) simultaneously wants a high quality and a high affordability (low price). The nature of simultaneity is visible in equivalent combinations of graded quality (Q) and graded affordability (A). So, the overall suitability is $S = Q\Delta A$, where the symbol Δ denotes the GL operation of graded conjunction. We assume that the graded conjunction is a parameterized operator, i.e., it includes a parameter that the decision-maker uses to adjust a desired intensity of simultaneity and the corresponding desired degree of penalizing inconvenient combinations of inputs.

Table. 7.1 Effects of increasing intensity of simultaneity

α	Operation	Graded conjunction $S = Q \Delta A$	Q=0.8 A=0.5	Q=0.8 A=0.2	Q=0.5 A=0.32	Q=1 A=0.16	Q=0.8 A=0.8
0.5	Arithmetic mean	$S = (Q + A)/2$	0.65	0.5	0.41	0.58	0.8
0.56	Heronian mean	$S = (Q + \sqrt{QA} + A)/3$	0.644	0.457	0.407	0.52	0.8
0.67	Geometric mean	$S = \sqrt{QA}$	0.632	0.4	0.4	0.4	0.8
0.77	Harmonic mean	$S = 2QA/(Q + A)$	0.615	0.32	0.39	0.276	0.8
1.00	Pure conjunction	$S = min(Q, A)$	0.5	0.2	0.32	0.16	0.8
1.25	Product (t-norm)	$S = QA$	0.4	0.16	0.16	0.16	0.64
2	Drastic conjunction	$S = \lfloor QA \rfloor$	0	0	0	0	0

This is exemplified in Table. 7.1, where we use several analytic models to show the effects of increasing degree of simultaneity and the possibility to select analytic models that support desired properties of the graded conjunction.

In Table. 7.1 each row (except the first row) has the intensity of simultaneity that is greater than the intensity in the previous row. It is convenient to have an indicator of the intensity of simultaneity and we use the conjunction degree, or andness, denoted α [DUJ74a]. This indicator will be studied in Chaps. 8–10; in the context of Table. 7.1 it is only important to note that its value is increasing in each row, showing the increasing intensity of simultaneity. Consequently, the aggregated values of suitability in each column are monotonically decreasing, showing increasing penalizing of low input values. Unsurprisingly, the values in the column Q = 0.8, A = 0.2 are less than the corresponding values in the column Q = 0.8, A = 0.5, penalizing rather strictly the inconvenient value A = 0.2 (an exception is the drastic conjunction where all presented Q, A combinations are drastically penalized).

The row of the geometric mean and the row of the product operator show that in these cases the combinations (Q = 0.8, A = 0.2), (Q = 0.5, A = 0.32), and (Q = 1, A = 0.16) are considered equivalent, yielding the same value of suitability. The penalty caused by the low input values is clearly visible: the first row of Table. 7.1 shows the mean values, and all other rows show smaller values indicating the increasingly strong impact of low values. An exception is the last column where we have equal values Q = A = 0.8; if the decision-maker decides to use S = 0.8, then this property is called idempotency, and it is a fundamental characteristic for all means. Indeed, by definition, any mean $M(x_1, \ldots, x_n)$ is a function that returns the value between the minimum value and the maximum value:

$$min(x_1, \ldots, x_n) \leq M(x_1, \ldots, x_n) \leq max(x_1, \ldots, x_n).$$

By inserting $x_1 = \ldots = x_n = x$ we get $x = \min(x, \ldots, x) \leq M(x, \ldots, x) \leq \max(x, \ldots, x) = x$, and consequently $M(x, \ldots, x) = x$.

If a decision-maker wants to have $M(x, \ldots, x) < x$ then s/he must not use means but functions that are more conjunctive than means. Among such functions are t-norms [BEL07, BEL16, GRA09], and one of them is the product used in Table. 7.1.

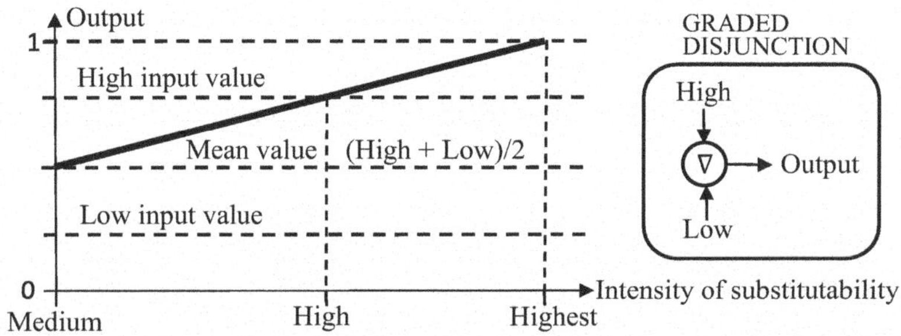

Fig. 7.2 The output of graded disjunction as a function of the intensity of substitutability

Let us reuse the previously discussed computational biology example: if a leading researcher in computational biology needs both a strong competence in biology (B) and a strong competence in computing (C), then her/his suitability (S) for successful research in such an area depends on a very strong simultaneity of B and C. That can be defined as a product $S = B\Delta C = BC$. So, if a candidate for such a position has $B = C = 0.8$, then her/his overall suitability will only be $S = 0.64$.

A typical example of graded substitutability is visible in many medical conditions where patient's disability (D) depends on motor symptoms (M) and sensory symptoms (S) [DUJ08a]. The motor symptoms are inability to move (e.g., walk/run/keep balance), and the sensory symptoms are most frequently pain and/or numbness/itching. A single component (either M or S) can be sufficient to cause a degree of disability $D = M\nabla S$, where ∇ denotes the operation of graded disjunction. The disability is predominantly affected by the higher of the two input values (M or S). If the motor symptoms are significantly higher than the sensory symptoms, then M predominantly affects the value of D. Similar situation holds for S, assuming that S is significantly higher than M.

The analysis of graded substitutability is very similar to the analysis of graded simultaneity. That is visible from De Morgan duality and in GL we usually define the graded disjunction directly as De Morgan dual of graded conjunction: $M\nabla S = 1 - (1 - M)\Delta(1 - S)$. This definition directly maps the properties of graded conjunction to dual properties of graded disjunction. That is illustrated in Fig. 7.2 where we show the impact of the intensity of substitutability on the resulting output value (e.g., the patient disability). We again have three characteristic degrees of substitutability, denoted medium, high, and highest. As the desired degree of substitutability increases, so does the impact of the value $\max(M, S)$ on the output value D. The high degree of substitutability yields the pure disjunction $M\nabla S = \max(M, S)$.

The highest degree of substitutability yields the extreme operation called *drastic disjunction*, where the output value is 0 in only one case, where all inputs are 0. In all other cases the output value is 1. Acceptable are all candidates who have the slightest positive degree of truth in any of inputs. In this case of ultimate permissiveness, rejected are only those candidates who fail (without exception) each of input requirements. All other candidates are considered fully acceptable, characterized by the output value 1.

Complementary and Unified Properties of Simultaneity and Substitutability Models

Simultaneity and substitutability are complementary requirements, and graded conjunction and graded disjunction are complementary operations that satisfy De Morgan duality:

$$x\Delta y + (1-x)\nabla(1-y) = 1, \quad x\nabla y + (1-x)\Delta(1-y) = 1.$$

De Morgan's laws show that the operations Δ and ∇ are not independent. For example, if $x\Delta y = min(x, y)$ then

$$\begin{aligned} x\nabla y = 1 - (1-x)\Delta(1-y) &= 1 - \min\bigl[(1-x), (1-y)\bigr] \\ &= 1 - (1 - \max(x, y)) = \max(x, y). \end{aligned}$$

So, $\min(x, y)$ and $\max(x, y)$ are dual operations. Similarly, if $x\Delta y = xy$, then

$$x\nabla y = 1 - (1-x)(1-y) = 1 - (1 - x - y + xy) = x + y - xy.$$

In this case, xy and $x + y - xy$ are dual operations. It is worth noting that if $x\Delta y = (x + y)/2$ then we have

$$x\nabla y = 1 - \frac{(1-x) + (1-y)}{2} = \frac{x+y}{2}.$$

Therefore, in this case $x\Delta y = x\nabla y$ and the arithmetic mean is an autodual function, i.e., a function that is dual to itself with respect to a specific duality transformation (in this case De Morgan's law). This function is the central point (a centroid) between all logic aggregators.

If the stakeholder decides to use a significant strength of substitutability, then a single input (or a small subset of inputs) can highly satisfy stakeholder's requirements. Obviously, that is opposite to simultaneity requirement where we simultaneously need high satisfaction of *all* inputs. Similarly, if we use a significant strength of simultaneity, then *all* inputs are needed to simultaneously contribute to the output degree of truth, and that is opposite from the substitutability requirement where a single input can significantly contribute to the output.

When the strength of simultaneity decreases, individual inputs become more significant contributors to the output degree of truth, and they become more capable to compensate low values of remaining inputs. Therefore, by decreasing the strength of simultaneity, we automatically increase the strength of substitutability, and by increasing the strength of simultaneity, we automatically decrease the strength of substitutability. Figure 7.3 illustrates this concept. The complementary properties of simultaneity (graded conjunction) and substitutability (graded disjunction) are

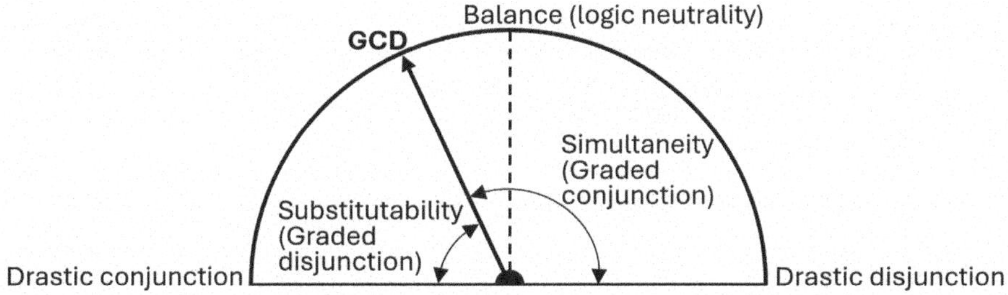

Fig. 7.3 Unified conjunctive/disjunctive aggregator GCD with adjustable complementary simultaneity and substitutability

simultaneously present. Consequently, the simultaneity and substitutability can be interpreted as two complementary components of a single *unified logic aggregator*, called *Graded Conjunction/Disjunction (GCD)*, studied in Chap. 8. The continuous and complementary adjustment of the ratio of simultaneity and substitutability is an observable and explainable commonsense mental process. In some cases, we use simultaneity as a dominant component of GCD. In other cases, we use substitutability as a dominant component of GCD. Rather frequently, simultaneity and substitutability are balanced. Such an operator is a logic neutrality, and its analytic form is the arithmetic mean.

References

BEL07. Beliakov, G., A. Pradera, and T. Calvo. 2007. *Aggregation Functions: A Guide for Practitioners*. Springer.

BEL16. Beliakov, G., H. Bustince Sola, and T. Calvo Sanchez. 2016. A Practical Guide to Averaging Functions. *Studies in Fuzziness and Soft Computing*, 329. Springer.

BOO47 Boole, G. 1847. *The Mathematical Analysis of Logic*. Cambridge: MacMillan, Barclay & MacMillan.

BOO54 Boole, G. 1853. *An Investigation of the Laws of Thought, on Which are Founded the Mathematical Theories of Logic and Probabilities*. Macmillan.

DEM47 De Morgan, A. 1847. *Formal Logic or the Calculus of Inference Necessary and Probable*. London: Taylor and Walton.

DUJ08a. Dujmović, J., J.W. Ralph, and L.J. Dorfman. 2008. Evaluation of Disease Severity and Patient Disability Using the LSP Method. In *Proceedings of the 12th Information Processing and Management of Uncertainty International Conference (IPMU 2008)*ed. L. Magdalena, M. Ojeda-Aciego, J.L. Verdegay, 1398–1405, Torremolinos (Malaga).

DUJ73b. Dujmović, J. 1973. A Generalization of Some Functions in Continuous Mathematical Logic—Evaluation Function and its Applications, (In Serbo-Croatian). *Proceedings of the Informatica Conference*, paper d27, Bled, Yugoslavia.

DUJ73c. Dujmović, J. 1973. Mixed Averaging by Levels (MAL)—A System and Computer Evaluation Method, (In Serbo-Croatian). *Proceedings of the Informatica Conference*, paper d28, Bled, Yugoslavia.

DUJ74a. Dujmović, J. 1974. Weighted Conjunctive and Disjunctive Means and their Application in System Evaluation. *Journal of the University of Belgrade, EE Department, Series Mathematics and Physics* (483): 147–158.

DUJ74b. Dujmović, J. 1974. New Results in the Development of the Mixed Averaging by Levels Method for System Evaluation, (In Serbo-Croatian). *Proceedings of the Informatica Conference 1974*, paper 4.36, Bled, Yugoslavia.

GRA09. Grabisch, M., J.-L. Marichal, R. Mesiar, and E. Pap. 2009. *Aggregation Functions.* Cambridge: Cambridge University Press.

STE22. Steinbach, B., and C. Posthoff. 2022. *Logic Functions and Equations. Fundamentals and Applications using the XBOOLE-Monitor*. Springer.

Chapter 8
A First Look at GCD: The Concept of Graded Conjunction/Disjunction

Classical bivalent logic defines conjunction and disjunction as independent and separate basic logic operations. In GL, the situation is different. Graded conjunction (the model of simultaneity) and graded disjunction (the model of substitutability) are not independent and different operations, but adjustable and complementary special cases of a single fundamental GL function that is called the *Graded Conjunction/Disjunction* (GCD).[1] GCD unites conjunctive and disjunctive properties in the single fundamental parameterized operation, where the intensity of conjunction or disjunction is adjusted by selecting the value of a single parameter. This is the consequence of the human-centric approach to GL. It is very important to recognize that simultaneity and substitutability are concurrently present and permanently balanced in natural human reasoning. Consequently, GCD is a model of human reasoning that permits logical fusion of input degrees of truth with adjustable balance between conjunctive and disjunctive properties. More simultaneity yields less substitutability and vice versa, as shown in Fig. 7.3.

The simplest way to illustrate the concurrent presence of simultaneity and substitutability in human reasoning is to observe any logic aggregation that results in using the arithmetic mean. E.g., the GPA of students is usually defined as the arithmetic mean of grades in all courses taken in the school. The arithmetic mean is a logic function that keeps in equilibrium two opposite requirements: (1) *the conjunctive*

---

[1] During 50 years of using GL, there were many changes and updates of terminology. GCD was introduced in 1973 in [DUJ73b] under the names "evaluation function" and "disjunctive/conjunctive function." Then the name was changed to Generalized Conjunction/Disjunction (GCD) in [DUJ07a], and in the context of GL the abbreviation GCD is interpreted as Graded Conjunction/Disjunction. Graded Logic was initially called Extended Continuous Logic in [DUJ75a], then Weighted Compensative Logic in [DUJ15a], and the GL was introduced in [DUJ18]. The evaluation method based on GL aggregation functions was introduced in 1973 under the name MAL (Mixed Averaging by Levels) [DUJ73c, DUJ74b] and then renamed LSP (Logic Scoring of Preference) in [DUJ87]. Disjunction degree and conjunction degree are the terms introduced in [DUJ73b, DUJ74a]. The terms "andness" and "orness" were introduced by R. R. Yager in 1988 for the conjunction degree and the disjunction degree in the context of OWA aggregator [YAG88].

J. Dujmović, *Graded Logic*, Intelligent Systems Reference Library 273,
https://doi.org/10.1007/978-3-031-88558-7_8

requirement is that students should simultaneously excel in all courses (simultaneity of high grades), and (2) *the disjunctive requirement* is that each low grade can be compensated by any other high grade; that is clearly a disjunctive property (substitutability of grades). These two requirements are equally present, and that is visible if the simple arithmetic mean of two variables is written as a combination of 50% of conjunction and 50% of disjunction:

$$z = \frac{x + y}{2} = 0.5(x \wedge y) + 0.5(x \vee y).$$

Therefore, the arithmetic mean has the status of a logic function that is *logically neutral* because of perfect balance between conjunctive and disjunctive properties.

The fundamental concept of GCD, and the human-centric root of the propositional calculus in GL, are based on provable observation that *in natural human reasoning simultaneity and substitutability coexist in simultaneous, adjustable, and complementary way.* To show that idea, we can use a weighting factor α and create the following example of a simple additive form of GCD:

$$z = \alpha(x \wedge y) + (1 - \alpha)(x \vee y), \;\; 0 \leq \alpha \leq 1.$$

In the range $0.5 < \alpha \leq 1$, we can increase conjunctive properties of GCD and decrease the disjunctive properties of this function; in the range $0 \leq \alpha < 0.5$ we can amplify disjunctive properties and attenuate conjunctive properties. The parameter α enables a continuous transition from the pure conjunction (high simultaneity) to the pure disjunction (high substitutability). This transition enables continuous and complementary adjustment of logic properties of GCD. Since humans continuously adjust the intensity of simultaneity and substitutability in intuitive evaluation criteria, it follows that GCD can serve as a model of human logic operations. If the intensity of simultaneity is greater than the intensity of substitutability, then in GL we consider that such GCD is *conjunctive*. Similarly, if the intensity of substitutability is greater than the intensity of simultaneity, then such GCD is *disjunctive*. Otherwise, if the intensities of simultaneity and substitutability are the same, then we have equilibrium and the GCD is *neutral*. If the parameter α denotes the intensity of simultaneity, it represents the conjunction degree, and its complement $\omega = 1 - \alpha$ represents the disjunction degree, as discussed in the Chap. 9. The conjunction degree is also called *andness*, and the disjunction degree is also called *orness*.

In GL, if $\alpha > \omega$ (i.e., if $\alpha > 1 - \alpha$ and $\alpha > 1/2, \omega < 1/2$) then the GCD is *conjunctive* and used as a mathematical model of graded simultaneity. Similarly, if $\alpha < \omega$ and $\alpha < 1/2, \omega > 1/2$, then the GCD is *disjunctive* and used as a mathematical model of graded substitutability. If the intensity of simultaneity is equal to the intensity of substitutability, then we get GCD that is logically neutral (i.e., in logic equilibrium, equally conjunctive and disjunctive). That special case of GCD is the arithmetic mean, where $\alpha = \omega = 1/2$.

Mathematical literature [BEL07, GRA09, BEL16] does not interpret conjunctive and disjunctive functions in the context of logic where it is necessary to have a continuous parameterized transition from the highest simultaneity to the highest substitutability. In GL interpretation, one half of that path is going through the territory of conjunctive aggregators and after the logic neutrality (the arithmetic mean), another half of the path is going through the territory of disjunctive aggregators. However, mathematical literature considers conjunctive only those aggregation functions that satisfy $f(x_1, \ldots, x_n) \le \min(x_1, \ldots, x_n)$ (typically t-norms), and disjunctive only those aggregation functions that satisfy $f(x_1, \ldots, x_n) \ge \max(x_1, \ldots, x_n)$ (typically t-conorms). Functions that satisfy $\min(x_1, \ldots, x_n) \le f(x_1, \ldots, x_n) \le \max(x_1, \ldots, x_n)$ are classified as averaging aggregation functions (because they are means), without logic properties. In GL, we interpret means as logic functions and seamlessly connect the territories of t-norms, means, and t-conorms in a continuum of logic aggregators that are either predominantly conjunctive or predominantly disjunctive, and having a neutral aggregator (the arithmetic mean) as a central point (centroid of all logic aggregators). Without this interpretation it is impossible to model natural human reasoning and to create and use GL.

For a general GCD function we use the symbol $\Diamond$, which is a combination of symbols $\wedge$ and $\vee$. The output values of GCD, and the continuous transition from the highest conjunction to the highest disjunction, are shown in Fig. 8.1 and Table 8.1, where we aggregate two values denoted *high* (H) and *low* (L), assuming $0 < L < H < 1$. The output value $H \Diamond L \in [0, 1]$ is a linear increasing function of the intensity of substitutability (or a linear decreasing function of the intensity of simultaneity). For the lowest degree of substitutability (and the highest degree of simultaneity) we have the output $H \Diamond L = 0$; this special case of the GCD function is called the *drastic conjunction*. Similarly, for the lowest degree of simultaneity (and the highest degree of substitutability) we have the output value$H \Diamond L = 1$; this special case of the GCD function is called the *drastic disjunction*. Other characteristic properties are obtained for low substitutability (and high simultaneity), where the GCD function becomes the pure conjunction and $H \Diamond L = \min(H, L) = L$. Likewise, we have low simultaneity (and high substitutability), where the GCD function becomes the pure disjunction and $H \Diamond L = \max(H, L) = H$. In the central point, we have the logic neutrality (or logic equilibrium) obtained if GCD becomes the arithmetic mean, and $H \Diamond L = (H + L)/2$.

On the path from the drastic conjunction to the drastic disjunction we have the following four characteristic ranges that depend on the intensity of simultaneity/substitutability:

- Conjunctive functions:
  - Nonidempotent hyperconjunction: $H \Diamond L < \min(H, L)$; $x \Diamond x < x, 0 < x < 1$
  - Idempotent conjunction: $\min(H, L) \le H \Diamond L < (H + L)/2$; $x \Diamond x = x \in [0, 1]$
- Neutrality: $H \Diamond L = (H + L)/2$ (idempotent, $x \Diamond x = x \in [0, 1]$).
- Disjunctive functions:
  - Idempotent disjunction: $(H + L)/2 < H \Diamond L \le \max(H, L)$; $x \Diamond x = x \in [0, 1]$
  - Nonidempotent hyperdisjunction: $\max(H, L) < H \Diamond L$; $x \Diamond x > x, 0 < x < 1$.

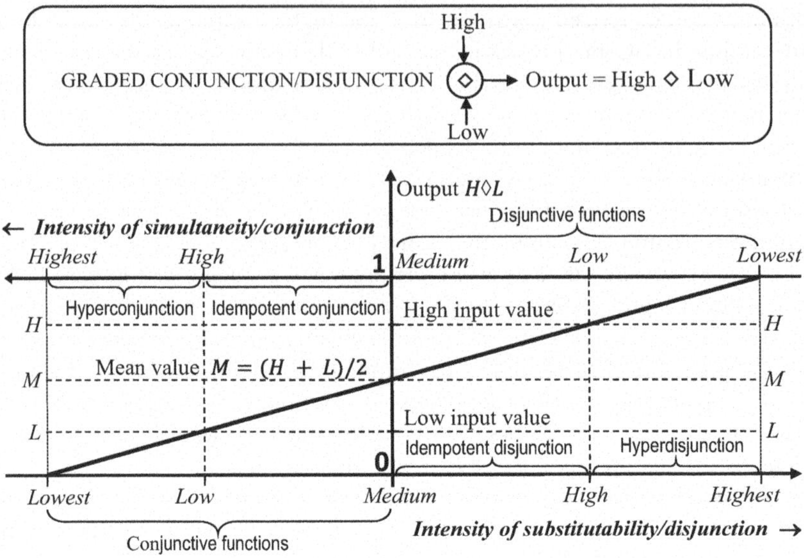

Fig. 8.1 The output value of GCD as a function of the intensity of simultaneity and the intensity of substitutability

Table 8.1 Characteristic complementary values of simultaneity and substitutability

Substitutability	Complementing	Simultaneity	Operator	Model: $z = x \diamond y$
Lowest	**Simultaneity**	Highest	Drastic conjunction	$z = \lfloor xy \rfloor$
Low		High	Pure conjunction	$z = \min(x, y)$
Medium		Medium	Arithmetic mean	$z = (x + y)/2$
High		Low	Pure disjunction	$z = \max(x, y)$
Highest	**Substitutability**	Lowest	Drastic disjunction	$z = 1 - \lfloor (1 - x)(1 - y) \rfloor$

The complementing of conjunctive and disjunctive properties (simultaneity and substitutability) is presented in Table 8.1. We also included analytic models of five special cases of GCD that are the border aggregators of the four characteristic ranges of the GCD function of two variables. Hyperconjunctive aggregators are special cases of GCD located in the space between the drastic conjunction and the pure conjunction. Idempotent conjunctive aggregators are found between the pure conjunction and the arithmetic mean. The location of idempotent disjunctive aggregators is between the arithmetic mean and the pure disjunction. Finally, the hyperdisjunctive versions of GCD are found between the pure disjunction and the drastic disjunction. Later, in Chap. 11, we are going to separately study each of these four characteristic types of

GCD functions and present analytic models of GCD with all necessary detail. At this point it is only necessary to develop informal intuitive feeling for properties of GCD. In Table 8.1 it is important to see that simultaneity and substitutability complement each other: when one of them increases, the other one decreases and vice versa.

Compensatory Properties of GCD

GCD is a *compensatory operator*: it has the property that the deficiency of an input (degree of truth) can be compensated by the excess of another input. This property is observable in natural human reasoning about the value of an object, and that is always necessary in decision-making. If an evaluated object has multiple components that contribute to the percept of the overall value, then the contributions are expressed using degrees of truth of corresponding statements (e.g., "the attribute x fully satisfies stakeholder's requirements"). Then, in most cases, a deficiency of any component can be compensated by excess of one or more of other components (e.g., a low grade in specific course can be compensated by a high grade in another course).

In the trivial example $z = \alpha(x \wedge y) + (1-\alpha)(x \vee y)$, assuming $x < y$, the deficiency p of argument x can be compensated by using the excess q of argument y as follows: $z = \alpha x + (1-\alpha)y = \alpha(x-p) + (1-\alpha)(y+q)$, and consequently, $q = \alpha p/(1-\alpha)$.

On the other hand, the classical conjunction $y_c = x_1 \wedge x_2 \wedge \ldots \wedge x_n$ and the classical disjunction $y_d = x_1 \vee x_2 \vee \ldots \vee x_n$ are *not compensatory*. Assuming for simplicity that the arguments are sorted ($x_1 \leq x_2 \leq \ldots \leq x_n$), it follows $y_c = x_1$ and $y_d = x_n$. The non-compensatory conjunction means that the value $y_c = x_1$ cannot be compensated/increased by increasing the values of $x_2, \ldots, x_n$. Similarly, the non-compensatory disjunction means that the value $y_d = x_n$ cannot be compensated/decreased by decreasing the values of $x_1, \ldots, x_{n-1}$. Therefore, classical conjunction is not upwards compensatory, and the classical disjunction is not downwards compensatory. From the human centric point of view, these properties are not desirable.

Non-compensatory pure conjunction and pure disjunction occur in human reasoning only in rare exceptional situations. Most observations show that in human reasoning we need and frequently use both the upwards compensatory simultaneity (graded conjunction), and the downwards compensatory substitutability (graded disjunction). Therefore, we will develop GCD function that has these properties.

Let us now investigate our initial example of apricots buyer, where the suitability $S = Q\Delta A$ is defined as a graded conjunction of quality Q *and* affordability A. If $Q = Q_0$ and $A = A_0$, then the resulting suitability is $S_0 = Q_0\Delta A_0$. If the quality is decreased, then the overall suitability will also decrease: $(1-q)Q_0\Delta A_0 = (1-s)S_0$, where $q > 0$ and $s > 0$. To compensate the decrement of quality the vendor can reduce the price and increase the affordability, yielding $(1-q)Q_0\Delta(1+a)A_0 = S_0 = Q_0\Delta A_0$. The product (e.g., the apricots) is now less attractive, but its price is more attractive, and the buyer considers this combination equally suitable as the initial combination Q_0, A_0. If the graded conjunction is selected to be a product, $Q_0\Delta A_0 = Q_0A_0$, then we have $Q_0A_0(1-q)\left(1+a_p\right) = Q_0A_0$. So, $(1-q)\left(1+a_p\right) = 1$ and the necessary increment of affordability is $a_p = q/(1-q)$. For example, if the quality is decreased 20%, then $q = 0.2$ and $a_p = 0.25$. If we use the geometric mean

$Q_0 \Delta A_0 = \sqrt{Q_0 A_0}$ then from $\sqrt{Q_0 A_0(1-q)(1+a_g)} = \sqrt{Q_0 A_0}$ we get the same result $a_g = q/(1-q)$. In these examples compensation does not depend on the values of Q_0 and A_0. However, if we use the harmonic mean $Q_0 \Delta A_0 = 2Q_0A_0/(Q_0 + A_0)$, then we have compensation that depends on the values of Q_0 and A_0:

$$\frac{2Q_0A_0(1-q)(1+a_h)}{Q_0(1-q)+A_0(1+a_h)} = \frac{2Q_0A_0}{Q_0+A_0}, \quad a_h = \frac{qA_0}{Q_0(1-q)-qA_0}.$$

With exception of full and drastic conjunction and disjunction, and for arguments in]0, 1[, all versions of GCD are compensatory operators.

It is important to emphasize that in natural human reasoning we intuitively adjust the desired ratio between the intensity of simultaneity and substitutability and by adjusting that ratio we also automatically adjust the intensity of compensatory properties, and the use of annihilators. Once the ratio of simultaneity/substitutability is selected, the only other independent parameters are the degrees of importance of individual inputs. In mathematical models of GCD the intensity of simultaneity (or substitutability) is adjusted independently from the degrees of relative importance of inputs. In natural human reasoning, the intuitive feeling of the overall importance of an input is a compound percept affected by both the high degree of simultaneity/substitutability, and the relative importance of individual inputs. We discuss these topics in next chapters.

Nondecreasing Monotonicity

In Graded Logic, we assume that all arguments and the resulting value of the GCD function $y = x_1 \Diamond \ldots \Diamond x_n$ are degrees of truth. In both conjunctive and disjunctive versions of GCD, all arguments positively contribute to the resulting degrees of truth. More precisely, if $0 < x_i < 1$, $i = 1, \ldots, n$, and GCD is not a full or drastic conjunction/disjunction, then we expect that any increase of the input degree of truth causes an increase of the output degree of truth:

$$(a_1, \ldots, a_n) < (b_1, \ldots, b_n) \Rightarrow a_1 \Diamond \ldots \Diamond a_n < b_1 \Diamond \ldots \Diamond b_n.$$

In most cases, GCD is a continuous differentiable function and in such cases $\frac{\partial y}{\partial x_i} > 0$, $i = 1, \ldots, n$.

In a general case, for arbitrary values of arguments, and for all forms of conjunctive or disjunctive aggregation, we assume that GCD satisfies *nondecreasing monotonicity in each argument*. In other words, if the degree of truth of an argument increases, the resulting degree of truth cannot decrease:

$$(a_1, \ldots, a_n) \le (b_1, \ldots, b_n) \Rightarrow a_1 \Diamond \ldots \Diamond a_n \le b_1 \Diamond \ldots \Diamond b_n.$$

This general property is visible in natural human decision-making, where arguments represent degrees of satisfaction of individual suitability attributes (Fig. 6.1):

if the suitability of an input attribute increases, the resulting aggregated suitability cannot decrease (improving a component cannot degrade the system).

If a graded propositional calculus formula consists of any superposition of GCD functions, then the nondecreasing monotonicity propagates from a GCD function to another GCD function and consequently it holds for the whole propositional calculus formula. Of course, there are graded propositional calculus formulas where an increase of input can cause a decrease of output, but such formulas must contain functions that are different from GCD and use negation.

Algebraic Properties of GCD

In the case of two variables, the GCD function $z = x\Diamond y$, $x \in [0, 1]$, $y \in [0, 1]$, $z \in [0, 1]$ generates results in the range $\lfloor xy \rfloor \le x\Diamond y \le 1 - \lfloor (1-x)(1-y) \rfloor$. Therefore, GCD can be considered a generalization/extension of two important special cases: (1) bivalent logic and (2) non-classical continuum-valued logics, including the fuzzy logic based on t-norms and t-conorms. In these special cases GCD satisfies various classic algebraic properties that are not satisfied in a general case.

The classic bivalent logic is based on conjunction, disjunction, and negation, and satisfies the following properties:

$$x \in \{0, 1\}, y \in \{0, 1\}, z \in \{0, 1\} \qquad \text{(bivalency)}$$
$$x \wedge y = \min(x, y) \qquad \text{(conjunction)}$$
$$x \vee y = \max(x, y) \qquad \text{(disjunction)}$$
$$y = \overline{x} = 1 - x \qquad \text{(negation)}$$
$$x \wedge x = x \vee x = x \qquad \text{(idempotency)}$$
$$x \wedge y = y \wedge x; x \vee y = y \vee x \qquad \text{(commutativity)}$$
$$(x \wedge y) \wedge z = x \wedge (y \wedge z); (x \vee y) \vee z = x \vee (y \vee z) \qquad \text{(associativity)}$$
$$x \wedge (y \vee z) = (x \wedge y) \vee (x \wedge z); x \vee (y \wedge z) = (x \vee y) \wedge (x \vee z) \qquad \text{(distributivity)}$$
$$\overline{x \wedge y} = \overline{x} \vee \overline{y}; \overline{x \vee y} = \overline{x} \wedge \overline{y} \qquad \text{(duality)}$$
$$x \wedge (x \vee y) = x; \qquad x \vee (x \wedge y) = x \qquad \text{(absorption)}$$

These basic algebraic properties can be easily verified/proved by inserting all combinations of values $\{0, 1\}^n$ in formulas with n variables.

In the case of continuum-valued (continuous) logic, all degrees of truth have continuous values; in the case of three variables, $x \in [0, 1], y \in [0, 1], z \in [0, 1]$. However, assuming that conjunction and disjunction are defined as min and max respectively, all other algebraic properties are the same as in the bivalent case. We can prove the properties using $n!$ permutations of n sorted variables; for example, we can prove the duality and absorption properties as follows:

$$x > y \Rightarrow \overline{x \wedge y} = \overline{y}; \overline{x} \vee \overline{y} = \overline{y}; \overline{x \vee y} = \overline{x}; \overline{x} \wedge \overline{y} = \overline{x} \qquad \text{(duality)}$$
$$y > x \Rightarrow \overline{x \wedge y} = \overline{x}; \overline{x} \vee \overline{y} = \overline{x}; \overline{x \vee y} = \overline{y}; \overline{x} \wedge \overline{y} = \overline{y} \qquad \text{(duality)}$$

$$x > y \Rightarrow x \wedge (x \vee y) = x \wedge x = x; x \vee (x \wedge y) = x \vee y = x \qquad \text{(absorption of y)}$$
$$y > x \Rightarrow x \wedge (x \vee y) = x \wedge y = x; x \vee (x \wedge y) = x \vee x = x. \qquad \text{(absorption of y)}$$

In the case of associativity and $n = 3$, we have $3! = 6$ permutations:

$$x > y > z \Rightarrow (x \wedge y) \wedge z = y \wedge z = z;\ x \wedge (y \wedge z) = x \wedge z = z \quad (= \min(x, y, z))$$
$$x > z > y \Rightarrow (x \wedge y) \wedge z = y \wedge z = y; x \wedge (y \wedge z) = x \wedge y = y \quad (= \min(x, y, z))$$
$$y > x > z \Rightarrow (x \wedge y) \wedge z = x \wedge z = z; x \wedge (y \wedge z) = x \wedge z = z \quad (= \min(x, y, z))$$
$$y > z > x \Rightarrow (x \wedge y) \wedge z = x \wedge z = x; x \wedge (y \wedge z) = x \wedge z = x \quad (= \min(x, y, z))$$
$$z > x > y \Rightarrow (x \wedge y) \wedge z = y \wedge z = y; x \wedge (y \wedge z) = x \wedge y = y \quad (= \min(x, y, z))$$
$$z > y > x \Rightarrow (x \wedge y) \wedge z = x \wedge z = x; x \wedge (y \wedge z) = x \wedge y = x \quad (= \min(x, y, z)).$$

A similar approach can be used in the case of distributivity. For conjunction, we have:

$$x > y > z \Rightarrow x \wedge (y \vee z) = x \wedge y = y;\ (x \wedge y) \vee (x \wedge z) = y \vee z = y$$
$$x > z > y \Rightarrow x \wedge (y \vee z) = x \wedge z = z;\ (x \wedge y) \vee (x \wedge z) = y \vee z = z$$
$$y > x > z \Rightarrow x \wedge (y \vee z) = x \wedge y = x;\ (x \wedge y) \vee (x \wedge z) = x \vee z = x$$
$$y > z > x \Rightarrow x \wedge (y \vee z) = x \wedge y = x;\ (x \wedge y) \vee (x \wedge z) = x \vee x = x$$
$$z > x > y \Rightarrow x \wedge (y \vee z) = x \wedge z = x;\ (x \wedge y) \vee (x \wedge z) = y \vee x = x$$
$$z > y > x \Rightarrow x \wedge (y \vee z) = x \wedge z = x;\ (x \wedge y) \vee (x \wedge z) = y \vee x = x.$$

Similarly, for disjunction, we have:

$$x > y > z \Rightarrow x \vee (y \wedge z) = x \vee z = y;\ (x \vee y) \wedge (x \vee z) = x \wedge y = y$$
$$x > z > y \Rightarrow x \vee (y \wedge z) = x \vee y = x;\ (x \vee y) \wedge (x \vee z) = x \wedge x = x$$
$$y > x > z \Rightarrow x \vee (y \wedge z) = x \vee z = x;\ (x \vee y) \wedge (x \vee z) = y \wedge x = x$$
$$y > z > x \Rightarrow x \vee (y \wedge z) = x \vee z = z;\ (x \vee y) \wedge (x \vee z) = y \wedge z = z$$
$$z > x > y \Rightarrow x \vee (y \wedge z) = x \vee y = x;\ (x \vee y) \wedge (x \vee z) = x \wedge x = x$$
$$z > y > x \Rightarrow x \vee (y \wedge z) = x \vee y = y;\ (x \vee y) \wedge (x \vee z) = y \wedge z = y.$$

In the case of logic neutrality, both the graded conjunction and the graded disjunction become the arithmetic mean $y = x_1 \ominus \ldots \ominus x_n = (x_1 + \ldots + x_n)/n$. This function is commutative (because addition is commutative), autodual (satisfying $x_1 \ominus \ldots \ominus x_n + \overline{x_1} \ominus \ldots \ominus \overline{x_n} = 1$), associative, and distributive. For duality we have $\overline{y} = 1 - x_1 \ominus \ldots \ominus x_n = 1 - \frac{x_1 + \ldots + x_n}{n} = \frac{n-(x_1+\ldots+x_n)}{n} = \frac{(1-x_1)+\ldots+(1-x_n)}{n} = \frac{\overline{x_1}+\ldots+\overline{x_n}}{n} = \overline{x_1} \ominus \ldots \ominus \overline{x_n}$.

Associativity must preserve the equal weights of arguments, e.g., for n = 3 we have

$$\frac{x+y+z}{3} = \frac{(x+y)+z}{3} = \frac{2(x+y)/2+z}{3} = \frac{2}{3}\left(\frac{x+y}{2}\right) + \frac{z}{3}$$
$$= \frac{x}{3} + \frac{2}{3}\left(\frac{y+z}{2}\right) = \frac{x+(y+z)}{3}.$$

In the case of distributivity $x \wedge (y \vee z) = (x \wedge y) \vee (x \wedge z)$, if we insert the arithmetic mean in the operations of the left-hand expression, we get $\frac{x}{2} + \frac{1}{2}\frac{y+z}{2} = \frac{x}{2} + \frac{y}{4} + \frac{z}{4}$. If we do the same with the right-hand expression, we get the same expression: $\frac{1}{2}\frac{x+y}{2} + \frac{1}{2}\frac{x+z}{2} = \frac{x}{2} + \frac{y}{4} + \frac{z}{4}$.

The conclusion of the above analysis is that in general, the GCD function $y = x_1 \Diamond \ldots \Diamond x_n$ provides the mapping $[0, 1]^n \to [0, 1]$ that satisfies basic algebraic properties of commutativity, associativity, and distributivity in cases where $\Diamond = \wedge$, $\Diamond = \vee$, $\Diamond = \ominus$.

What happens in other cases? That depends on various ways GCD can be realized (see Chaps. 13 and 15). In many simple cases, the idempotent version of GCD can be implemented as a power mean:

$$x_1 \Diamond \ldots \Diamond x_n = \left(\frac{x_1^{r_n} + \ldots + x_n^{r_n}}{n}\right)^{\frac{1}{r_n}}, \quad -\infty \le r_n \le +\infty.$$

In this case, the desired logic properties are adjusted using the parameter r_n. Obviously, if $r_n = 1$ (neutrality), or $r_n = -\infty$ (the pure conjunction, min), or $r_n = +\infty$ (the pure disjunction, max), this model offers commutativity, associativity, and distributivity. If r_n is constant (not a function of n), we can also have associativity. For example, if $r_n = -1$, the power mean becomes the harmonic mean, and for $n = 4$ we have the following partial conjunction:

$$y_{14} = \frac{4}{\frac{1}{x_1} + \frac{1}{x_2} + \frac{1}{x_3} + \frac{1}{x_4}} = \frac{4x_1x_2x_3x_4}{x_2x_3x_4 + x_1x_3x_4 + x_1x_2x_4 + x_1x_2x_3}.$$

To show associativity, we can get the same result using aggregation in pairs, as follows:

$$y_{12} = \frac{2x_1x_2}{x_1+x_2}, \; y_{34} = \frac{2x_3x_4}{x_3+x_4}, \; y_{22} = \frac{2y_{12}y_{34}}{y_{12}+y_{34}} = \frac{2\frac{2x_1x_2}{x_1+x_2}\frac{2x_3x_4}{x_3+x_4}}{\frac{2x_1x_2}{x_1+x_2} + \frac{2x_3x_4}{x_3+x_4}}$$
$$= \frac{4x_1x_2x_3x_4}{x_1x_2(x_3+x_4) + x_3x_4(x_1+x_2)} = y_{14}.$$

Table 8.2 Sample differences between associativity patterns in an example with 4 variables

Type of aggregator	Strength of aggregator	r_2	r_4	Error e_4 [%]
Graded disjunction	Strong	9.5208	12.2695	2.145
	Average	3.9294	4.8250	1.663
	Weak	2.0183	2.3018	0.693
Neutrality (balance of conjunction/disjunction)		1	1	0
Graded conjunction	Weak	0.2621	0.1561	0.302
	Average	-0.7201	-0.7205	0.001
	Strong	-3.5101	-2.8233	1.813

In the next chapter we will show that if we want consistent logic properties for any number of variables, then r_n depends on n and in such a case $y_{14} \neq y_{22}$, i.e., no associativity. However, this fact does not create practical problems. Let us now see the difference $d_4 = |y_{14} - y_{22}|$ and the percent error $e_4[\%] = 100|y_{14} - y_{22}|/(0.5y_{14} + 0.5y_{22})$. To create realistic numerical results, suppose that the arguments have typical different values $x_1 = 0.5, x_2 = 0.65, x_3 = 0.8, x_4 = 0.95$. Using these inputs, we can compute the associativity errors $e_4[\%]$ shown in Table 8.2.[2]

To interpret the results shown in Table 8.2 it is necessary to emphasize that the input arguments are human-generated degrees of truth. The precision of such arguments is limited [MIL56], and the typical precision of inputs can be from 5 to 15%. The results in Table 8.2 show that the errors are small, mostly below 2%. Obviously, if the precision of a mathematical model is for an order of magnitude higher than the precision of inputs, we can conclude that the model is qualified to be used in practical problem solving. A detailed analysis of this problem can be found in [DUJ75a] where the average absolute associativity errors are shown to be typically 1% or less than 1%. In addition, the average absolute distributivity errors are shown to be below 1.5%. Consequently, human-centric models of graded conjunction and graded disjunction can be considered fully applicable from the standpoint of precision of algebraic properties of basic logic functions that model simultaneity, substitutability and compound propositional formulas based on graded conjunction and graded disjunction (i.e., all special cases of GCD).

Graded conjunctions can be stronger than the pure conjunction and in Graded Logic we call such functions *hyperconjunction*. The area of hyperconjunction is between the full and drastic conjunction. Among many functions in this territory, the most popular are triangular norms (t-norms) [FOD94]. The most frequently used representative of t-norms is the product $T(x, y) = xy,\ x \in [0, 1],\ y \in [0, 1]$. The product can be used to see general algebraic properties of t-norms: commutativity $(T(x, y) = T(y, x))$, nondecreasing monotonicity $((a, b) \leq (A, B) \Rightarrow T(a, b) \leq T(A, B))$, associativity $(T(x, T(y, z)) = T(T(x, y), z))$, neutral element $(T(x, 1) = x)$ and annihilator $(T(x, 0) = 0)$. In the case of n variables this hyperconjunctive GCD aggregator is $y = \prod_{i=1}^{n} x_i$, and it satisfies the same algebraic properties. If we decide to abandon commutativity and use the weighted aggregator $y = \prod_{i=1}^{n} x_i^{p_i}$,

[2] The values of r_2 and r_4 are taken from Table 2.3.3 in [DUJ18].

$p_i > 0, i = 1, \ldots, n, \sum_{i=1}^{n} p_i = n$, or to increase the strength of simultaneity by using $\sum_{i=1}^{n} p_i > n$, such aggregators will be weighted t-norms studied in Chap. 15. Some of the basic algebraic properties of commutative t-norms will be modified, but (for similar reasons we discussed for power means) the applicability of such forms of graded conjunction will not be significantly reduced.

The models of *hyperdisjunction* are assumed to be De Morgan duals of the presented models of hyperconjunction, and t-conorms are defined as $S(x, y) = 1 - T(1 - x, 1 - y)$. Similarly to t-norms, t-conorms support commutativity ($S(x, y) = S(y, x)$), nondecreasing monotonicity ($(a, b) \leq (A, B) \Rightarrow S(a, b) \leq S(A, B)$), associativity ($S(x, S(y, z)) = S(S(x, y), z)$), neutral element ($S(x, 0) = x$) and annihilator ($S(x, 1) = 1$). Generally, if $\overline{\wedge}$ denotes hyperconjunction and $\overline{\vee}$ denotes hyperdisjunction, then we assume that $x_1 \overline{\vee} \ldots \overline{\vee} x_n = 1 - (1 - x_1) \overline{\wedge} \ldots \overline{\wedge} (1 - x_n)$.

References

BEL07 Beliakov, G., A. Pradera, and T. Calvo. 2007. *Aggregation Functions: A Guide for Practitioners*. Springer.

BEL16 Beliakov, G., H. Bustince Sola, and T. Calvo Sanchez. 2016. A Practical Guide to Averaging Functions. *Studies in Fuzziness and Soft Computing* vol. 329, Springer.

DUJ07a Dujmović, J., and H.L Larsen. 2007. Generalized Conjunction/Disjunction. *International Journal of Approximate Reasoning*. 46: 423–446.

DUJ15a Dujmović, J. 2015. Weighted Compensative Logic with Adjustable Threshold Andness and Orness. *IEEE Transactions on Fuzzy Systems* 23(2): 270–290. https://doi.org/10.1109/TFUZZ.2014.2312018.

DUJ18 Dujmović, J. 2018. *Soft Computing Evaluation Logic*. J. Wiley.

DUJ73b Dujmović, J. 1973. A Generalization of Some Functions in Continuous Mathematical Logic—Evaluation Function and its Applications, (In Serbo-Croatian). *Proceedings of the Informatica Conference*, paper d27, Bled, Yugoslavia.

DUJ73c Dujmović, J. 1973. Mixed Averaging by Levels (MAL)—A System and Computer Evaluation Method, (In Serbo-Croatian). *Proceedings of the Informatica Conference*, paper d28, Bled, Yugoslavia.

DUJ74a Dujmović, J. 1974. Weighted Conjunctive and Disjunctive Means and Their Application in System Evaluation. *Journal of the University of Belgrade, EE Department, Series Mathematics and Physics* (483): 147–158.

DUJ74b Dujmović, J. 1974. New Results in the Development of the Mixed Averaging by Levels Method for System Evaluation, (In Serbo-Croatian). *Proceedings of the Informatica Conference 1974*, paper 4.36, Bled, Yugoslavia.

DUJ75a Dujmović, J. 1975. Extended Continuous Logic and the Theory of Complex Criteria. *Journal of the University of Belgrade, EE Department, Series Mathematics and Physics* (537): 197–216.

DUJ87 Dujmović, J. 1987. The LSP Method for Evaluation and Selection of Computer and Communication Equipment. Proceedings of MELECON'87, Mediterranean Electrotechnical Conference and 34th Congress on Electronics (Joint Conference), IEEE/RIENA, Rome. 251–254.

FOD94 Fodor, J., and M. Roubens. 1994. *Fuzzy Preference Modelling and Multicriteria Decision Support*. Dordrecht, The Netherlands: Kluwer Academic Publishers.

GRA09 Grabisch, M., J.-L. Marichal, R. Mesiar, and E. Pap. 2009. *Aggregation Functions*. Cambridge: Cambridge University Press.

MIL56 Miller, G.A. 1956. The Magical Number Seven, Plus or Minus Two: Some Limits on Our Capacity for Processing Information. *The Psychological Review* 63: 81–97.

YAG88 Yager, R. R. 1988. On Ordered Weighted Averaging Aggregation Operators in Multi-Criteria Decision-making. *IEEE Transaction.* SMC 18: 183–190.

Chapter 9
Adjusting the Strength of Conjunction/Disjunction: Andness, Orness, and Duality

In natural human reasoning, as well as in GL, the operations of graded conjunction (simultaneity) and graded disjunction (substitutability) are not separate operations but complementary special cases of GCD. GCD is a fundamental integrated logic operation, which is a model of both simultaneity and substitutability. To adjust the intensity of simultaneity and substitutability, we need quantitative indicators of intensity, called *the conjunction degree* or *andness* (α), and its complement, *the disjunction degree* or *orness* ($\omega = 1 - \alpha$). By adjusting the strength of simultaneity, we also automatically adjust the complementary strength of substitutability. Figure 9.1 illustrates the process of natural human reasoning where a decision-maker selects the desired logic properties of the GCD function by intuitively adjusting the graded percept of balance between conjunctive and disjunctive properties. The same process is used both in selecting a mental (intuitive) GCD function, and in developing analytic logic models. The resulting GCD function can be neutral ($\alpha = \omega$), conjunctive ($\alpha > \omega$), or disjunctive ($\alpha < \omega$).

Andness α and orness ω are the most important indicators in GL, because they are *input parameters* selected by decision-makers to realize desired logic properties of GCD. Figure 9.1 shows a conjunctive GCD function where a decision-maker simultaneously selects the intensity of simultaneity and substitutability by selecting a desired conjunction (or disjunction) degree. To focus strictly on formal logic properties of GCD, both andness and orness are defined assuming that all inputs are equally important. So, our next steps are to introduce analytic forms of andness and orness, and to use them in developing GCD functions in the context of graded propositional calculus.

There are two ways to adjust the level of GCD simultaneity/substitutability: (1) by directly choosing a desired andness/orness level from an appropriate rating scale of andness/orness, and (2) by using numerical procedures for training preferential neurons [DUJ91]. In the case of the first approach, we first select a rating scale of desired cardinality, and then apply a stepwise refinement procedure for selecting the most appropriate degree from the selected rating scale. The second approach is based

J. Dujmović, *Graded Logic*, Intelligent Systems Reference Library 273,
https://doi.org/10.1007/978-3-031-88558-7_9

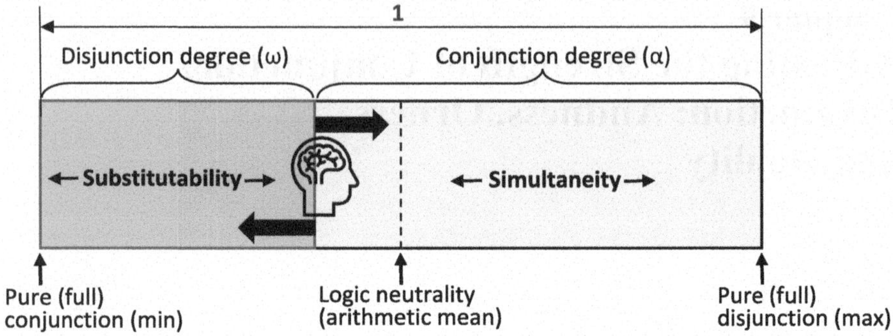

Fig. 9.1 Continuous complementary adjustment of the unidimensional percept of the intensity of simultaneity (andness) and the intensity of substitutability (orness) of an idempotent GCD function

on interpreting GL propositional formulas as neural networks where GCD aggregators have the status of preferential neurons. For training of preferential neurons (GCD or other basic GL functions) the decision-maker must provide a training set, which is a table of desired mapping that in each row contains selected input values and a desired output value. In both cases we are adjusting the parameters of a GCD function $F : I^n \rightarrow I, I = [0, 1]$ that has the status of logic aggregator. Prior to explaining these procedures, we have to show the definition of andness and orness.

A Gentle Introduction Using the Simplest Case of Two Variables

The simplest possible case of logic aggregation is the GCD aggregation of two variables: $z = F(x, y)$. In the GL interpretation, x and y are degrees of truth of individual assertions that each of selected input attributes completely satisfies stakeholder's requirements. For example, x can be the degree of truth of homebuyer's statement "I am completely satisfied with home quality," and y can be the degree of truth of the statement "I am completely satisfied with the suitability of home location." Suppose that we are interested in simultaneous satisfaction of inputs. More precisely, we are not interested in equivalence (the similarity of values of x and y), but in the *simultaneity of high values* of x and y. The resulting aggregated value $z = F(x, y)$ is the degree of truth of the homebuyer's statement "I am completely satisfied with this home." In a general case, z is the degree of truth of statement "I am completely satisfied with the simultaneous high satisfaction of all inputs."

How strong should be the simultaneity? This question assumes the use of an adjustable degree of simultaneity, i.e. the conjunction degree (andness) α, or its complement, the disjunction degree (orness) $\omega = 1 - \alpha$. Such a combination of conjunction and disjunction can be written as follows:

$$\begin{aligned} z = F(x, y) &= \alpha(x \wedge y) + (1 - \alpha)(x \vee y) = (1 - \omega)(x \wedge y) + \omega(x \vee y) \\ &= \alpha(x \wedge y) + \omega(x \vee y),\ x \in I,\ y \in I,\ I = [0, 1]. \end{aligned}$$

So, α decides the impact of simultaneity, and $1-\alpha$ decides the impact of substitutability. From this expression it follows that the andness and orness are not constants but functions of inputs:

$$\alpha(x,y)=\frac{(x\vee y)-F(x,y)}{(x\vee y)-(x\wedge y)},\quad \omega(x,y)=\frac{F(x,y)-(x\wedge y)}{(x\vee y)-(x\wedge y)}.$$

This definition is inconvenient because these forms of andness/orness are not the overall properties of the GCD function F, but pointwise indicators that change from point to point. For example, if $F(x,y)=\sqrt{xy}$, then we have $\alpha(1,0)=1$, $\alpha(0.7,0.2)=0.65$, $\alpha(0.9,0.4)=0.6$, etc. In addition, if $F(x,y)$ is a mean, then $\alpha(x,y)$ and $\omega(x,y)$ are undefined for $x=y$.

Decision-makers need andness as a fixed overall simultaneity indicator of each aggregator. A natural way to define such a form of andness is to use mean values:

$$V=\int_0^1\int_0^1 F(x,y)dxdy=\alpha\int_0^1\int_0^1(x\wedge y)dxdy+(1-\alpha)\int_0^1\int_0^1(x\vee y)dxdy.$$

This approach has geometric interpretation, because the mean value of function F in the unit square I^2 is the same as a volume V under the surface $F(x,y)$ inside the unit cube I^3. Therefore, such constant global forms of andness and orness are:

$$\alpha=\frac{\int_0^1\int_0^1(x\vee y)dxdy-\int_0^1\int_0^1 F(x,y)dxdy}{\int_0^1\int_0^1(x\vee y)dxdy-\int_0^1\int_0^1(x\wedge y)dxdy},$$

$$\omega=\frac{\int_0^1\int_0^1 F(x,y)dxdy-\int_0^1\int_0^1(x\wedge y)dxdy}{\int_0^1\int_0^1(x\vee y)dxdy-\int_0^1\int_0^1(x\wedge y)dxdy}.$$

The mean value of disjunction (the maximum function) is

$$\begin{aligned}\int_0^1\int_0^1(x\vee y)dxdy&=\int_0^1 dx\int_0^x(x\vee y)dy+\int_0^1 dx\int_x^1(x\vee y)dy\\&=\int_0^1 dx\int_0^x xdy+\int_0^1 dx\int_x^1 ydy=\int_0^1 x^2dx+\int_0^1\frac{1-x^2}{2}dx\\&=\frac{1}{3}+\frac{1}{2}-\frac{1}{6}=\frac{2}{3}.\end{aligned}$$

Similarly, the mean value of conjunction (the minimum function) is

$$\int_0^1\int_0^1 (x\wedge y)dxdy = \int_0^1 dx\int_0^x (x\wedge y)dy + \int_0^1 dx\int_x^1 (x\wedge y)dy$$
$$= \int_0^1 dx\int_0^x ydy + \int_0^1 dx\int_x^1 xdy = \int_0^1 \frac{x^2}{2}dx + \int_0^1 (x-x^2)dx$$
$$= \frac{1}{6}+\frac{1}{2}-\frac{1}{3}=\frac{1}{3}.$$

Therefore, in the case of two variables, the andness of the GCD aggregator $F(x, y)$ is

$$\alpha = \frac{2/3 - V}{2/3 - 1/3} = 2 - 3V.$$

The orness ω is a complement of andness: $\omega = 1 - \alpha = 3V - 1$.

Now, we can compute the global andness and orness for several characteristic logic aggregators of two variables, sorted according to increasing andness and decreasing orness[1]:

- The arithmetic mean: $V = \int_0^1\int_0^1 \frac{x+y}{2}dxdy = \frac{1}{2}\int_0^1 (x+\frac{1}{2})dx = \frac{1}{2}$; $\alpha = \frac{1}{2}$, $\omega = \frac{1}{2}$.
- The geometric mean: $V = \int_0^1\int_0^1 \sqrt{xy}dxdy = \frac{2}{3}\times\frac{2}{3} = \frac{4}{9}$; $\alpha = 2 - \frac{4}{3} = \frac{2}{3} = 0.67$, $\omega = \frac{1}{3}$.
- The harmonic mean: $V = \int_0^1\int_0^1 \frac{2xy}{x+y}dxdy = \frac{4-ln16}{3}$; $\alpha = ln16 - 2 = 0.77$, $\omega = 0.23$.
- Conjunction: $V = \int_0^1\int_0^1 (x\wedge y)dxdy = \frac{1}{3}$; $\alpha = 2 - 1 = 1$, $\omega = 0$.
- Product t-norm: $V = \int_0^1\int_0^1 xydxdy = \frac{1}{2}\times\frac{1}{2} = \frac{1}{4}$; $\alpha = 2 - \frac{3}{4} = \frac{5}{4} = 1.25$, $\omega = -0.25$.
- Product of squares: $V = \int_0^1\int_0^1 x^2y^2dxdy = \frac{1}{3}\times\frac{1}{3} = \frac{1}{9}$; $\alpha = \frac{5}{3} = 1.67$, $\omega = -0.67$.
- Drastic conjunction: $V = \int_0^1\int_0^1 xydxdy = 0$; $\alpha = 2$, $\omega = -1$.

Some logic aggregators have adjustable parameters. A typical example is the power mean: $F(x, y; r) = (0.5x^r + 0.5y^r)^{1/r}$. In such cases the volume under the surface of aggregator is a function of the parameter: $V(r) = \int_0^1\int_0^1 F(x, y; r)dxdy$. Consequently, the andness and orness are also functions of the parameter r: $\alpha = 2 - 3V(r)$. Now, we can use numerical approximations to compute the value of parameter r as a function of desired value of andness/orness:

$$r = V^{-1}\left(\frac{2-\alpha}{3}\right) = V^{-1}\left(\frac{1+\omega}{3}\right).$$

[1] This calculation provides values that are used in Tables 9.1 and 10.2.

It is important to emphasize that in GL applications we don't solve *problems of analysis* (the computation of andness/orness for a given GCD aggregator $F(x, y)$), but *problems of synthesis* (the development of logic aggregators that have desired values of andness/orness). Indeed, the andness or orness are inputs, and not outputs in graded logic problems. Consequently, parameterized means are suitable for building graded logic aggregators in the range from the pure conjunction to the pure disjunction.

From the graded logic standpoint, all means can be interpreted as GCD functions. An elementary way to show that for two variables can be based on harmonic, geometric, arithmetic and quadratic means: $h = 2xy/(x+y)$, $g = \sqrt{xy}$, $a = (x+y)/2$, $q = \sqrt{0.5x^2+0.5y^2}$. Then from $0 \le \left(\sqrt{x}-\sqrt{y}\right)^2 = x - 2\sqrt{xy} + y = 2a - 2g \ge 0$ it follows $a \ge g$. Since $h = g^2/a$, we have $g = \sqrt{ha}$ and therefore g is a value between h and a: $h \le g \le a$. From $q^2 = 0.5\left(x^2+y^2\right)$ and $2q^2 = x^2 + y^2 = (x+y)^2 - 2xy = 4a^2 - 2g^2$ it follows $q^2 = a^2 + (a^2 - g^2)$ and since $a \ge g$ we have $q \ge a$. Consequently,

$$x \wedge y = \min(x, y) \le h \le g \le a \le q \le \max(x, y) = x \vee y.$$

This inequality shows that volumes under these surfaces satisfy $V_{\min} \le V_h \le V_g \le V_a \le V_q \le V_{\max}$. From $\omega = 3V - 1$ it follows that $0 = \omega_{\min} \le \omega_h \le \omega_g \le \omega_a \le \omega_q \le \omega_{\max} = 1$. Therefore, basic means create a sequence of logic functions in the range from pure conjunction to pure disjunction along the path of continuous transition from conjunction to disjunction. Indeed, using power mean $M(x, y; r) = (0.5x^r + 0.5y^r)^{1/r}$ we have $x \wedge y = M(x, y; -\infty)$, $h = M(x, y; -1)$, $g = M(x, y; 0)$, $a = M(x, y; 1)$, $q = M(x, y; 2)$, $x \vee y = M(x, y; +\infty)$, and we can use the parameter r to realize a parameterized continuous transition from the pure conjunction to the pure disjunction. All means along that path can be interpreted and used as GCD logic functions.

Of course, means are restricted to modeling only those logic functions that have andness or orness in the basic range [0, 1], and such models are most frequent in applications. Significantly less frequently, human mental logic aggregation uses models of simultaneity that have andness and orness outside the range [0, 1]. In such cases the simultaneity is stronger than conjunction, and the andness is greater than 1, or the substitutability is stronger than disjunction and the orness is greater than 1. Such GCD aggregators are called hyperconjunction and hyperdisjunction.

In the case of hyperconjunction the result of logic aggregation is lower than the minimum input. Examples of hyperconjunction are visible in all logic models with probabilistic interpretation. For example, let x denote a degree of truth defined as the probability that a candidate can solve problems in the area A_x, and let y denote a degree of truth defined as the probability that the candidate can solve problems in a different area A_y. If the candidate is expected to solve problems that simultaneously require expertise in independent areas A_x and A_y, then $z = xy$ denotes the degree of truth of the assertion "the candidate is completely suitable for activities in the intersection of areas A_x and A_y." For example, A_x can be biology, A_y can be computing, and the candidate is expected to work in computational biology.

Another example of hyperconjunction and hyperdisjunction in human reasoning is visible in real estate, where a homebuyer aggregates the suitability of home and the suitability of location to assess the overall suitability of a home offer. If one of the inputs is very low (e.g. 10%) and the other input is good, the homebuyer may consider such a situation unacceptable and assign the overall suitability 0 in order to reject that defective home offer. In a dual situation, if both inputs are unexpectedly rather high (e.g. close to 90%), the homebuyer may be enthusiastic and assign the overall suitability 100%, reflecting the opinion that nothing better can be expected.

The claim that at least one input is very high is equivalent to the claim that it is not possible that both inputs are simultaneously very low. This is a possible verbal interpretation of De Morgan's law $x \nabla y = 1 - (1-x)\Delta(1-y)$. Therefore, if we have a mathematical model of simultaneity, that is sufficient for creating a dual model of substitutability. Similarly, the claim that both inputs are highly satisfied is equivalent to the claim that it is not possible that one or more inputs are low: $x \Delta y = 1 - (1-x)\nabla(1-y) = 1$. All GCD aggregators are dual: disjunctive aggregators can be made as duals of conjunctive aggregators, and vice versa. If $F(x, y)$ is a GCD aggregator and $G(x, y)$ is its dual, then $G(x, y) = 1 - F(1-x, 1-y)$ and $F(x, y) = 1 - G(1-x, 1-y)$.

A General Case of n Variables

Aggregators are defined as GL functions that satisfy basic conditions of nondecreasing monotonicity in each variable and two boundary conditions for truth and falsity, as follows:

$$\begin{aligned}
&\boldsymbol{x} = (x_1, \ldots, x_n), \quad \boldsymbol{y} = (y_1, \ldots, y_n), \quad x_i \in I, y_i \in I, i = 1, \ldots, n \\
&\boldsymbol{x} \le \boldsymbol{y} \Rightarrow F(\boldsymbol{x}) \le F(\boldsymbol{y}) \\
&F(0, \ldots, 0) = 0, \quad F(1, \ldots, 1) = 1.
\end{aligned}$$

If $x_i \le y_i,\ i = 1, ..., n$ implies $F(x_1, ..., x_n) \le F(y_1, ..., y_n)$, then any increase of an input degree of truth cannot cause a decrease of the output degree of truth.[2] That is consistent with human evaluation reasoning (Fig. 6.1) where increasing the truth value of input suitability attributes cannot decrease the overall suitability of an evaluated object/alternative. In simple words, the overall suitability of an object cannot be decreased if we increase the suitability of its components.

In predominantly mathematical literature [FOD94, BEL07, TOR07, GRA09, BEL16] this definition is accepted primarily because it is the least restrictive, and it opens the possibility for studying the widest spectrum of aggregation functions. This also includes functions that may have discontinuities and/or discontinuities of first derivatives. Such properties are extremely rare in human reasoning (see Chap. 2), and to support a human-centric approach, in GL we define *graded logic aggregators* that can (and should) be defined in a more restrictive way.

[2] Of course, this holds only for functions that have the status of aggregator. Functions that use negation, such as implication, abjunction, equivalence, and exclusive disjunction are used in graded propositional calculus, but they do not have the status of aggregator.

One such restriction can be the sensitivity to positive and incomplete truth, used in [DUJ18]: if $x_i > 0, i = 1, ..., n$ then $F(x_1, ..., x_n) > 0$, and if $x_i < 1, i = 1, ..., n$ then $F(x_1, ..., x_n) < 1$. The idea behind these restrictions is that generally, if all inputs are positively satisfied the output should also be to some extent satisfied. Similarly, if no input is completely satisfied, then the output should also be incompletely satisfied. In cases where logic aggregators implement criteria that are inconsistent with restrictive conditions, it is justifiable to use the general definition of aggregators. However, from a human-centric standpoint, all aggregators and aggregation methods that cannot convincingly prove the consistency with observable natural human reasoning, should have the status of strictly theoretical result.

Observations of human reasoning yield a conclusion that human logic functions are continuous, and that microscopic changes of the truth degrees of inputs do not result in a macroscopic change of the truth degree of their aggregated output. In other words, logic aggregation functions should be continuous. That can be used as a restrictive condition in the definition of logic aggregators. In addition, observations do not support the conclusion that first derivatives of mental logic aggregation functions can have discontinuities. However, analytic approximations like min and max operations do have discontinuities of first derivatives. Such functions are undoubtedly useful and necessary in analytic modeling, both in the armamentarium of mathematical formalisms, and in software of decision-support systems. However, it is reasonable to characterize such functions as acceptable abstractions and approximations of mental processes for which we currently don't have better understanding and better models.

Logic aggregators are all functions that are based on graded conjunction, graded disjunction and their combinations. In GL, aggregators are GCD, and all functions based on superpositions of GCD. Such functions are indispensable for modeling mental evaluation processes. On the other hand, the graded propositional calculus can also use functions that are not aggregators (e.g., implication, abjunction, exclusive disjunction, equivalence).

It is useful to note that *nor* and *nand* functions are not aggregators, but they can be used to create all other Boolean logic functions, including logic aggregators:

$$nor(x, y) = not(x \vee y) = 1 - (x \vee y) = \overline{x \vee y}$$

$$nor(x, x) = not(x \vee x) = not(x) = \overline{x}$$

$$nor(nor(x, x), nor(y, y)) = not(not(x) \vee not(y)) = x \wedge y$$

$$nor(nor(x, y), nor(x, y)) = not(nor(x, y)) = x \vee y$$

$$nand(x, y) = not(x \wedge y) = 1 - (x \wedge y) = \overline{x \wedge y}$$

$$nand(x, x) = not(x \wedge x) = not(x)$$

$$nand(nand(x, x), nand(y, y)) = not(not(x) \wedge not(y)) = x \vee y$$

$$nand(nand(x, y), nand(x, y)) = not(nand(x, y)) = x \wedge y.$$

In the above relations we used the involution $not(not(x)) = 1 - (1 - x) = x$ and De Morgan duality $not(x \vee y) = not(x) \wedge not(y)$ and $not(x \wedge y) = not(x) \vee not(y)$. Both the involutive negation and De Morgan's laws hold both in classical Boolean logic and in GL.

The conjunction degree is a real number that characterizes the logic properties of function $F(x_1, ..., x_n)$. The first such indicator was introduced in [DUJ73b] and used in decision models in [DUJ73c]. Unsurprisingly, conjunction degree can be defined in various ways, and nine different indicators were reported in [DUJ06b]. So, it is legitimate to ask whether we really need nine different conjunction degrees. Obviously, we need only one, and all others are contributions to theory.

There are two confronting approaches to andness/orness. For theoreticians, andness and orness are *outputs*: they first define a specific form of the GCD function, and then for that function compute the values of andness and orness. For decision engineers, and for everybody interested in applications, andness and orness are *inputs*: a stakeholder/decision-maker first selects a justifiable value of andness/orness and then designs the GCD function that has such desirable properties. The obtained GCD function is then used in decision-making. In GL we assume that andness or orness are inputs: that is called andness-directedness [DUJ19, DUJ21b, TOR22] and presented in Chap. 15.

The most useful form of andness and orness indicators, introduced in [DUJ74a], is based on the following geometric approach:

- Assuming that all input variables have equal importance (equal weight), the global properties of a GCD function $F(x_1, ..., x_n)$ inside the unit hypercube $I^{n+1} = [0, 1]^{n+1}$, can be geometrically characterized by the volume under the surface of the GCD function inside the unit hypercube: $V(n) = \int_{I^n} F(x_1, \ldots, x_n)dx_1, \ldots, dx_n$. The andness α and the orness ω are defined as linear functions of V. This form of andness and orness are usually called the *global andness* and the *global orness*.
- Orness is a linearly increasing function of the volume under the surface of a GCD aggregator, and andness is a linearly decreasing function of the volume under the surface of the GCD aggregator, as shown in Fig. 9.2.
- The minimum volume $V_{\min} = 0$ corresponds to the drastic conjunction function $\hat{C}(x_1, \ldots, x_n) = \lfloor x_1 \cdots x_n \rfloor$ which is equal 0 in all points except for $x_1 = \ldots = x_n = 1$.
- The maximum volume $V_{\max} = 1$ corresponds to the drastic disjunction function $\hat{D}(x_1, ..., x_n) = 1 - \lfloor (1 - x_1) \cdots (1 - x_n) \rfloor$ which is equal 1 in all points except for $x_1 = \ldots = x_n = 0$.

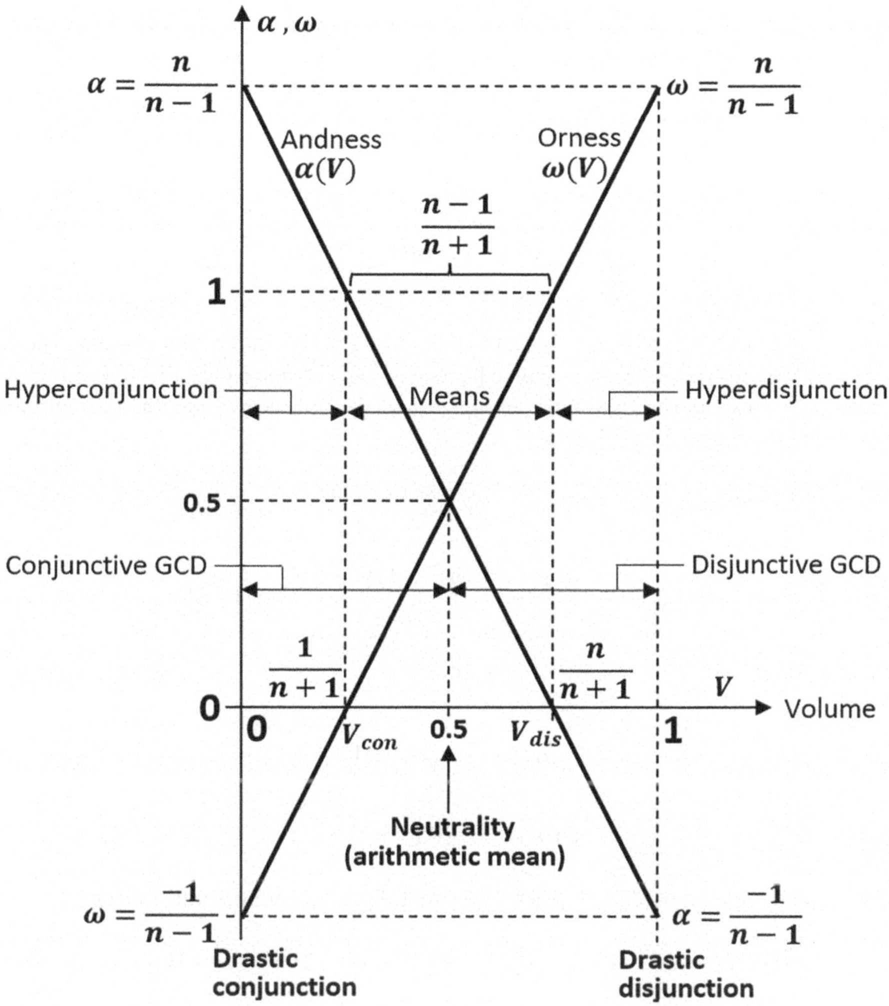

Fig. 9.2 Geometric interpretation of conjunction and disjunction degrees: andness α and orness ω as linear functions of the volume V under the surface of GCD inside the unit hypercube $[0, 1]^n$

- The conjunction degree α for the pure conjunction $\min(x_1, \ldots, x_n)$ is defined to be 1.
- The conjunction degree α for the pure disjunction $\min(x_1, \ldots, x_n)$ is defined to be 0.
- The disjunction degree ω is the complement of the conjunction degree, i.e., $\omega = 1 - \alpha$. Therefore, the disjunction degree ω for the pure conjunction is 0, and for the pure disjunction is 1.
- For hyperconjunction, $\alpha > 1$ and $\omega < 0$. For hyperdisjunction, $\omega > 1$ and $\alpha < 0$.

According to [DUJ73a] the volumes under the pure conjunction and the pure disjunction are the following:

$$V_{con}(n) = \int_{I^n} \min(x_1, \ldots, x_n)dx_1, \ldots, dx_n = \frac{1}{n+1},$$

$$V_{dis}(n) = \int_{I^n} \max(x_1, \ldots, x_n)dx_1, \ldots, dx_n = \frac{n}{n+1}.$$

According to Fig. 9.2, $\alpha(V_{con}) = 1$ and $\omega(V_{dis}) = 1$. Then, since andness and orness are complements ($\alpha + \omega = 1$), we also have $\omega(V_{con}) = 0$ and $\alpha(V_{dis}) = 0$. Using this linear geometric interpretation, the andness α and the orness ω as the linear functions of volume are defined as follows:

$$\alpha(V) = \frac{V_{dis} - V}{V_{dis} - V_{con}} = \frac{n - (n+1)V}{n-1} = \alpha[V; n]$$

$$\omega(V) = \frac{V - V_{con}}{V_{dis} - V_{con}} = \frac{(n+1)V - 1}{n-1} = \omega[V; n]$$

$$n > 1, 0 \le V \le 1$$

$$\alpha(V) + \omega(V) = 1$$

$$\alpha(0.5) = \omega(0.5) = 0.5.$$

For frequent cases $n = 2, 3, 4, 5$ we have the following andness and orness:

$$\alpha[V; 2] = 2 - 3V; \quad \omega[V; 2] = 1 - \alpha[V; 2] = -1 + 3V$$

$$\alpha[V; 3] = 1.5 - 2V; \quad \omega[V; 3] = 1 - \alpha[V; 3] = -0.5 + 2V$$

$$\alpha[V; 4] = (4 - 5V)/3; \quad \omega[V; 4] = 1 - \alpha[V; 4] = (-1 + 5V)/3$$

$$\alpha[V; 5] = 1.25 - 1.5V; \quad \omega[V; 5] = 1 - \alpha[V; 5] = -0.25 + 1.5V.$$

For computation of andness and orness it is necessary to compute multiple integrals. Such computations frequently use the linearity of multiple integrals [POL07]:

$$\int_{I^n} \sum_{i=1}^{n} W_i f_i(x_1, \ldots, x_n)dx_1, \ldots, dx_n = \sum_{i=1}^{n} W_i \int_{I^n} f_i(x_1, \ldots, x_n)dx_1, \ldots, dx_n.$$

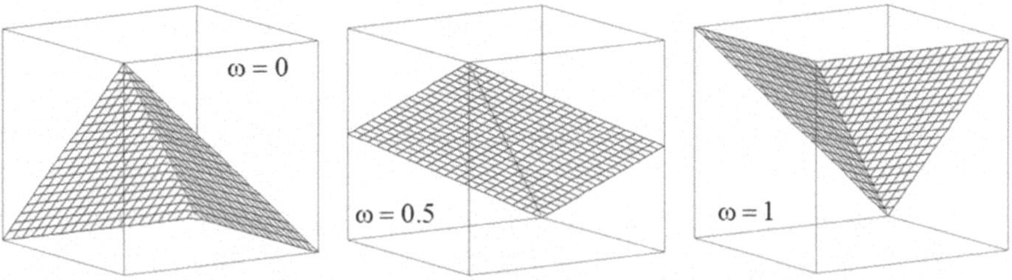

Fig. 9.3 Pure conjunction, neutrality (the arithmetic mean), and the pure disjunction for $n = 2$

For example, if we want to compute the andness of the arithmetic mean $y = \frac{x_1+\ldots+x_n}{n}$, then we have

$$V[Ari; n] = \int_{I^n} \frac{1}{n}\sum_{i=1}^{n} x_i dx_1, \ldots, dx_n = \frac{1}{n}\sum_{i=1}^{n}\int_{I^n} x_i dx_1, \ldots, dx_n = \frac{1}{n}\sum_{i=1}^{n}\frac{1}{2} = \frac{1}{2}.$$

Therefore, $\alpha[Ari; n] = \frac{n-(n+1)V[Ari;n]}{n-1} = \frac{n-(n+1)/2}{n-1} = \frac{1}{2}$ and $\omega[Ari; n] = 1 - \alpha[Ari; n] = \frac{1}{2}$.
The arithmetic mean divides the unit hypercube in two equal parts, as shown in Fig. 9.3. It is the logic neutrality, having 50% of conjunctive properties and 50% of disjunctive properties, because it is the central point in Fig. 9.2. Therefore, in GL we consider that the arithmetic mean (or the logic neutrality) is a logic function that is the centroid of all GCD aggregators.

Figure 9.3 illustrates the concept of De Morgan duality. Let $F_{con}(x_1, \ldots, x_n)$ be any of conjunctive GCD functions, characterized by $F_{con}(x_1, \ldots, x_n) \leq (x_1 + \ldots + x_n)/n$ and $\alpha > 1/2$. Then F_{con} has a dual disjunctive GCD function $F_{dis}(x_1, \ldots, x_n)$ which is defined as follows:

$$F_{dis}(x_1, \ldots, x_n) = 1 - F_{con}(1 - x_1, \ldots, 1 - x_n).$$

The volume under the surfaces of these functions is

$$\begin{aligned} V_d &= \int_0^1 \cdots \int_0^1 F_{dis}(x_1, \ldots, x_n)dx_1 \ldots dx_n \\ &= 1 - \int_0^1 \cdots \int_0^1 F_{con}(1 - x_1, \ldots, 1 - x_n)dx_1 \ldots dx_n \\ &= \begin{vmatrix} 1 - x_1 = y_1 \\ \ldots \\ 1 - x_n = y_n \end{vmatrix} = 1 - \int_0^1 \cdots \int_0^1 F_{con}(y_1, \ldots, y_n)dy_1 \ldots dy_n = 1 - V_c \end{aligned}$$

Since $V_c + V_d = 1$, we have the following relationships between andness and orness of De Morgan duals:

$$\alpha_c = \frac{n-(n+1)V_c}{n-1}; \quad \omega_c = \frac{(n+1)V_c - 1}{n-1}$$

$$\alpha_d = \frac{n-(n+1)V_d}{n-1} = \frac{n-(n+1)(1-V_c)}{n-1} = \frac{(n+1)V_c-1}{n-1} = \omega_c$$

$$\omega_d = \frac{(n+1)V_d-1}{n-1} = \frac{(n+1)(1-V_c)-1}{n-1} = \frac{n-(n+1)V_c}{n-1} = \alpha_c$$

Therefore, the andness of a disjunctive aggregator equals the orness of its conjunctive dual, and vice versa. Similarly, the orness of a disjunctive aggregator equals the andness of its conjunctive dual, and vice versa. Figure 9.2 shows the pure conjunction which has the andness 1, and the orness 0, and its dual,[3] the pure disjunction, which has the andness 0 and the orness 1. Based on duality, the pure conjunction and disjunction can be written as follows:

$$x \wedge y = 1-(1-x) \vee (1-y); \quad x \vee y = 1-(1-x) \wedge (1-y).$$

In the above formulas, V is the volume *under* the GCD surface inside $[0, 1]^{n+1}$, and $1 - V$ is the volume *above* the GCD surface. The volume above the GCD surface determines andness, and the volume below the GCD surface determines orness. From definitions of andness and orness we have

$$\alpha(1-V) = \frac{n-(n+1)(1-V)}{n-1} = \frac{(n+1)V-1}{n-1} = \omega(V)$$

$$\omega(1-V) = \frac{(n+1)(1-V)-1}{n-1} = \frac{n-(n+1)V}{n-1} = \alpha(V).$$

Generally, duality in GL for n variables can be expressed as follows:

$$x_1 \Delta \cdots \Delta x_n = 1-(1-x_1)\nabla \cdots \nabla(1-x_n)$$

$$x_1 \nabla \cdots \nabla x_n = 1-(1-x_1)\Delta \cdots \Delta(1-x_n).$$

These laws have simple verbal interpretation. The first formula can be verbally interpreted as follows: "*if we simultaneously need the high satisfaction (truth values) of n inputs, then it is not acceptable that any one of them is not sufficiently satisfied*."

[3] Bivalent negation, conjunction, and disjunction can be computed using arithmetic operations of addition, subtraction, and multiplication: $\bar{x} = 1-x$, $x \wedge y = xy$, $x \vee y = x+y-xy$; $x \in \{0, 1\}$, $y \in \{0, 1\}$. This notation can be used to prove De Morgan laws: $\overline{\bar{x} \wedge \bar{y}} = 1-(1-x)(1-y) = x+y-xy = x \vee y$, and $\overline{\bar{x} \vee \bar{y}} = 1-[(1-x)+(1-y)-(1-x)(1-y)] = 1-[2-x-y-1+x+y-xy] = xy = x \wedge y$.

Similarly, the second formula can be verbalized as follows: "*if we need at least one sufficiently satisfied input, then it is not acceptable that all of them are simultaneously insufficiently satisfied.*" Obviously, the expressions left of the equality sign are verbally equivalent to the expressions written right of the equality sign. Thus, the duality can be understood as two different ways to express the same logical relationship. In addition, these equivalent interpretations show that De Morgan's duality in naturally present in the commonsense human logic.

As an initial example of computing andness and orness, and using duality, let us use GCD implemented as a geometric mean: $GCD[Geo; n] = x_1 \Delta \cdots \Delta x_n = x_1^{1/n} \cdots x_n^{1/n}$. So, we have the following volume, andness, and orness [DUJ74a]:

$$V[Geo; n] = \int_0^1 x_1^{1/n} dx_1 \cdots \int_0^1 x_n^{1/n} dx_n = \left(\frac{n}{n+1}\right)^n,$$

$$\alpha[Geo; n] = \frac{n - (n+1)V[Geo; n]}{n-1} = \frac{n}{n-1} - \frac{n+1}{n-1}\left(\frac{n}{n+1}\right)^n.$$

$$\omega[Geo; n] = \frac{(n+1)V[Geo; n] - 1}{n-1} = \frac{n+1}{n-1}\left(\frac{n}{n+1}\right)^n - \frac{1}{n-1}.$$

The dual of conjunctive aggregator $GCD[Geo; n]$ is the disjunctive aggregator $GCD\left[\overline{Geo}; n\right] = x_1 \nabla \cdots \nabla x_n = 1 - (1 - x_1)^{1/n} \cdots (1 - x_1)^{1/n}$ (here $\overline{Geo}$ denotes a dual of *Geo*). So, now we have the following andness and orness:

$$\alpha\left[\overline{Geo}; n\right] = \omega[Geo; n] = \frac{n+1}{n-1}\left(\frac{n}{n+1}\right)^n - \frac{1}{n-1},$$

$$\omega\left[\overline{Geo}; n\right] = \alpha[Geo; n] = \frac{n}{n-1} - \frac{n+1}{n-1}\left(\frac{n}{n+1}\right)^n.$$

In a special case for $n = 2$ we have

$$GCD[Geo; 2] = \sqrt{xy}, \; GCD\left[\overline{Geo}; 2\right] = 1 - \sqrt{(1-x)(1-y)},$$

$$\omega\left[\overline{Geo}; 2\right] = \alpha[Geo; 2] = 2/3, \; \alpha\left[\overline{Geo}; 2\right] = \omega[Geo; 2] = 1/3.$$

If the analyzed aggregator is a mean, then $x_1 \wedge \cdots \wedge x_n \le x_1 \Diamond \cdots \Diamond x_n \le x_1 \vee \cdots \vee x_n$ and consequently $\alpha \in [0, 1]$ and $\omega \in [0, 1]$. According to Fig. 9.2, the situation is different when we increase the intensity of simultaneity above the pure conjunction, or the intensity of substitutability above the pure disjunction. To illustrate such forms of GCD, as our second example of computing andness and orness, and using duality, let us use the product t-norm $GCD[Prod; n] = x_1 \Delta \cdots \Delta x_n = x_1 \cdots x_n$ and its dual $GCD\left[\overline{Prod}; n\right] = x_1 \nabla \cdots \nabla x_n = 1 - (1 - x_1) \cdots (1 - x_n)$. Now, we get the following results:

$$V[Prod; n] = \int_0^1 x_1 dx_1 \cdots \int_0^1 x_n dx_n = \left(\frac{1}{2}\right)^n,$$

$$\alpha[Prod; n] = \frac{n - (n+1)V[Prod; n]}{n-1} = \frac{n - (n+1)/2^n}{n-1} = \omega[\overline{Prod}; n],$$

$$\omega[Prod; n] = 1 - \alpha[Prod; n] = \frac{(n+1)/2^n - 1}{n-1} = \alpha[\overline{Prod}; n].$$

If $n = 2$, then $\alpha[Prod; n] = \omega[\overline{Prod}; n] = 5/4$, and $\omega[Prod; n] = \alpha[\overline{Prod}; n] = -1/4$. The GCD aggregators having $\alpha > 1, \omega < 0$ are called the hyperconjunction because they are stronger than the pure conjunction. Similarly, the GCD aggregators having $\alpha < 0, \omega > 1$ are called the hyperdisjunction because they are stronger than the pure disjunction.

As the last example of computing andness and orness and using duality, we show the strongest possible forms of conjunction and disjunction, i.e., their drastic versions. The *drastic conjunction* function (CC) is defined as follows:

$$GCD[CC; n] = \hat{C}(x_1, \ldots, x_n) = \left\{ \begin{array}{l} 1, \textit{if} \sum_{i=1}^{n} x_i = n \\ 0, \textit{if} \sum_{i=1}^{n} x_i < n \end{array} \right\} = \left\lfloor \prod_{i=1}^{n} x_i \right\rfloor.$$

The *drastic disjunction* function (DD) is defined as a dual of the drastic conjunction, as follows:

$$GCD[DD; n] = \hat{D}(x_1, \ldots, x_n) = \left\{ \begin{array}{l} 0, \textit{if} \sum_{i=1}^{n} x_i = 0 \\ 1, \textit{if} \sum_{i=1}^{n} x_i > 0 \end{array} \right\} = 1 - \left\lfloor \prod_{i=1}^{n} (1 - x_i) \right\rfloor.$$

The drastic conjunction has the smallest volume $V_{CC} = 0$, and its andness and orness are

$$\alpha[CC; n] = \frac{n}{n-1} > 1, \quad \omega[CC; n] = \frac{-1}{n-1} < 0.$$

Similarly, the drastic disjunction has the largest volume $V_{DD} = 1$, and its andness and orness are

$$\alpha[DD; n] = \frac{-1}{n-1} < 0, \quad \omega[DD; n] = \frac{n}{n-1} > 1.$$

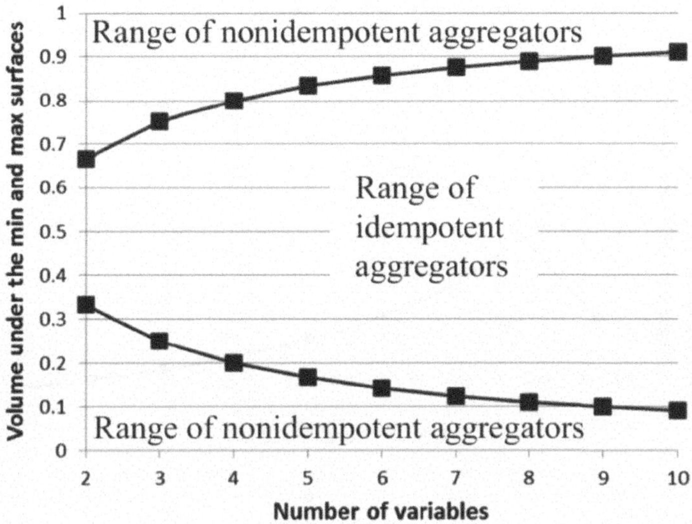

Fig. 9.4 The increasing range of idempotent GCD functions between $V_{con}(n)$ and $V_{dis}(n)$

Therefore, the global andness and orness belong to the interval $I_{\alpha,\omega} = \left[-\frac{1}{n-1}, \frac{n}{n-1}\right]$. In the special case $n = 2$, we have $-1 \le \alpha \le 2$ and $-1 \le \omega \le 2$. For $-1 \le \alpha < 0$, the GCD is a hyperdisjunction. For $0 \le \alpha \le 1$, the GCD is a mean, and for $1 < \alpha \le 2$ the GCD is a hyperconjunction.

If the GCD aggregator is a mean, then $\min(x_1, ..., x_n) \le GCD(x_1, ..., x_n) \le \max(x_1, ..., x_n)$. So, if $x_1 = \ldots = x_n = x$, then $\min(x, \ldots, x) = x \le GCD(x, \ldots, x) \le \max(x, \ldots, x) = x$, and obviously $GCD(x, \ldots, x) = x$, i.e., this form of GCD is an idempotent aggregator. In this case, inside the unit hypercube $I^{n+1} = [0, 1]^{n+1}$, the volume that holds all aggregators that are means (idempotent GCD functions) is the following:

$$V[Means; n] = V_{dis}(n) - V_{con}(n) = \frac{n}{n+1} - \frac{1}{n+1} = \frac{n-1}{n+1} < 1.$$

The strictly increasing function $V_{dis}(n)$ and the strictly decreasing $V_{con}(n)$ are shown in Fig. 9.4.

The volume of unit hypercube that contains nonidempotent aggregators (hyperconjunction and hyperdisjunction) is $V\left[hyper; n\right] = 1 - V[Means; n] = 2/(n+1)$. This is the strictly decreasing function shown in Fig. 9.5. The strictly increasing function shown in Fig. 9.5 is the volume of the fraction of hypercube that contains the idempotent GCD functions (various forms of means). In theory, if $n \gg 1$ then $V[Means; n] \to 1$ and $V\left[hyper; n\right] \to 0$. Both Figs. 9.4 and 9.5 show that with an increasing number of variables the idempotent versions of GCD become a dominant form of aggregation. The nonidempotent aggregators (typically t-norms and t-conorms) become less important because they operate in a shrinking region of

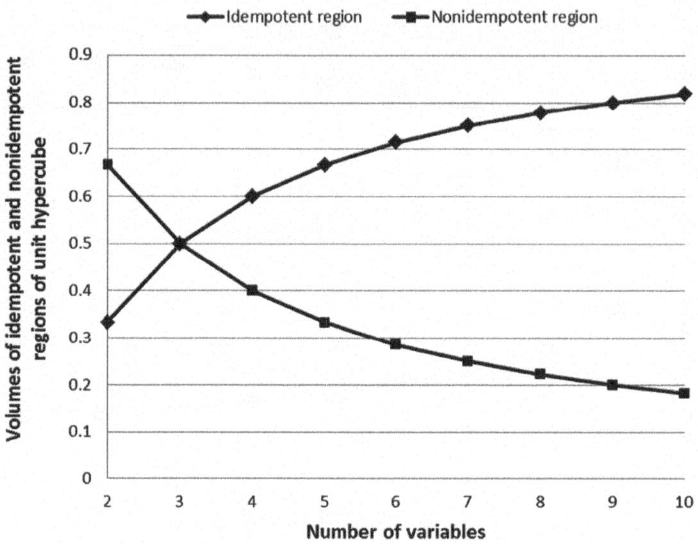

Fig. 9.5 The volumes of regions of unit hypercube that contain idempotent and nonidempotent forms of GCD, as functions of the number of variables

the unit hypercube where differences between individual forms of GCD, caused by different degrees of andness and orness, become rather insignificant.

In practice, the number of input arguments n is regularly small. In natural human reasoning, parameters like andness, orness, and importance of arguments, can be interpreted as unidimensional stimuli (see Fig. 9.1) and for them we should consider Miller's "magical number 7 plus or minus 2" [MIL56] which shows the limits of precision when humans classify continuous percepts into discrete categories. To preserve a reasonable precision, the suggested number of categories is from 5 to 9, and that must be considered in the process of developing and using various rating scales. Experiences with LSP criteria in decision engineering show that it is not advisable to aggregate more than 5 inputs using GCD aggregators.

The number of arguments n affects the volume under the surface of a GCD function; that makes andness and orness the functions of n. The only exception is the arithmetic mean which always divides the hypercube $[0, 1]^{n+1}$ in two equal parts and consequently yields the constant value 0.5 for both andness and orness, for any number of variables $n > 1$. To show the impact of the number of variables, we selected three characteristic numbers of variables ($n = 2, 11$, and $n \gg 1$), and in Table 9.1 we summarize the impact of n on various indicators introduced in this chapter.

Łukasiewicz Conjunctions/Disjunctions in the Context of Many-Valued Logics

In mathematical literature there are many different logics, and all of them are expected to be constructed as *formalized axiomatic deductive systems* [RAS74]. Many-valued logics are based on degrees of truth that belong to the set of m rational numbers

Table 9.1 Volume, andness, and orness of selected GCD functions for three characteristic values of n

Indicator	Values of selected indicator for different number of variables		
	$n = 2$	$n = 11$	$n \to +\infty$
$V_{con}(n)$	1/3	1/12 = 0.0833	0
$V_{dis}(n)$	2/3	11/12 = 0.9167	1
$\alpha[V;n]$	$2 - 3V$	$1.1 - 1.2V$	$1 - V$
$\omega[V;n]$	$3V - 1$	$1.2V - 0.1$	V
$\alpha[CC;n]$	2	1.1	1
$\omega[CC;n]$	−1	−0.1	0
$\alpha[DD;n]$	−1	−0.1	0
$\omega[DD;n]$	2	1.1	1
$I_{\alpha,\omega}(n)$	[−1, 2]	[−0.1, 1.1]	[0, 1]
$V[Means;n]$	1/3 = 0.333	10/12 = 0.833	1
$V[Ari;n]$	1/2	1/2	1/2
$\alpha[Ari;n]$	1/2	1/2	1/2
$\omega[Ari;n]$	1/2	1/2	1/2
$V[Geo;n]$	4/9 = 0.444	0.384	1/e = 0.368
$\alpha[Geo;n]$	2/3 = 0.667	0.639	1 − 1/e = 0.632
$\omega[Geo;n]$	1/3 = 0.333	0.361	1/e = 0.368
$V[Prod;n]$	1/4 = 0.25	0.000488	0
$\alpha[Prod;n]$	5/4 = 1.25	1.099	1
$\omega[Prod;n]$	−1/4 = −0.25	−0.099	0

within the real unit interval $I = [0, 1]$ [GOT22]:

$$T_m = \left\{\frac{k}{m-1}\ ;\ 0 \le k \le m-1\right\},\quad m > 1.$$

Alternatively, the degrees of truth can be all values inside the unit interval:

$$T_\infty = \{x \in \mathbb{R}\ ;\ 0 \le x \le 1\}.$$

Łukasiewicz logic is the only system before the Graded Logic to define two different types of conjunction:

- Strong conjunction: $x\&y = \max(0, x + y - 1)$.
- Weak conjunction: $x \wedge y = \min(x, y)$.

In the case of the strong conjunction the volume under the surface (a pyramid with the base ½ and the height 1) is $V_\& = 1/6$. Therefore, the andness of strong conjunction is $\alpha_\& = 2{-}3V_\& = 1.5$. The andness of weak conjunction (a pyramid with the base 1 and the height 1, and the volume 1/3) is $\alpha_\wedge = 1$. For two variables the andness of drastic conjunction is $n/(n-1) = 2$, and consequently Łukasiewicz proposed conjunctive aggregators at the beginning and in the middle point of the range of hyperconjunction. Based on De Morgan's duality, the corresponding Łukasiewicz strong and weak disjunction are $1-(1-x)\&(1-y)$ and $1-(1-x)\wedge(1-y)$. Łukasiewicz logic also uses negation $\overline{x} = 1-x$ and implication $x \rightarrow y = \min(1, 1-x+y)$.

The proposal of two versions of conjunction/disjunction (i.e. two different value of andness and orness) is the first step towards graded conjunction/disjunction that is necessary in natural human reasoning. Other many-valued logics, like Gödel logics, t-norm logic, 3-valued and 4-valued systems discussed in [GOT22], use for conjunction and disjunction the classic pure conjunction $\boldsymbol{x} \wedge \boldsymbol{y} = \boldsymbol{min}(\boldsymbol{x}, \boldsymbol{y})$ and the pure disjunction $\boldsymbol{x} \vee \boldsymbol{y} = \boldsymbol{max}(\boldsymbol{x}, \boldsymbol{y})$, i.e. andness and orness that are either 0 or 1.

References

BEL16 Beliakov, G., H. Bustince Sola, and T. Calvo Sanchez. 2016. *A Practical Guide to Averaging Functions. Studies in Fuzziness and Soft Computing*, vol. 329, Springer.

BEL07 Beliakov, G., A. Pradera, and T. Calvo. 2007. *Aggregation Functions: A Guide for Practitioners*. Heidelberg: Springer.

DUJ06b Dujmović, J. 2006. Nine Forms of Andness/Orness. *Proceedings of the Second IASTED International Conference on Computational Intelligence*, ed. B. Kovalerchuk, 276–281. ISBN Hardcopy: 0-88986-602-3/CD: 0-88986-603-1.

DUJ18 Dujmović, J. 2018. *Soft Computing Evaluation Logic*. J. Wiley.

DUJ21b Dujmović, J., and V. Torra. 2022. Aggregation Functions in Decision Engineering: Ten Necessary Properties and Parameter-Directedness. Proceedings of the 3rd International Conference on Intelligent and Fuzzy Systems, INFUS 2021 Conference, Izmir, Turkey. In *Intelligent and Fuzzy Techniques for Emerging Conditions and Digital Transformation*, vol. 1 ed Kahraman, C. et al., 173–181. Springer, LNNS 307.

DUJ73a Dujmović, J. 1973. Two Integrals Related to Means. *Journal of the University of Belgrade EE Department, Series Mathematics and Physics* (412–460): 231–232.

DUJ73b Dujmović, J. 1973. A Generalization of Some Functions in Continuous Mathematical Logic—Evaluation Function and its Applications, (In Serbo-Croatian). *Proceedings of the Informatica Conference, paper d27*. Bled, Yugoslavia.

DUJ73c Dujmović, J. 1973. Mixed Averaging by Levels (MAL)—A System and Computer Evaluation Method, (In Serbo-Croatian). *Proceedings of the Informatica Conference, paper d28*. Bled, Yugoslavia.

DUJ74a Dujmović, J. 1974. Weighted Conjunctive and Disjunctive Means and their Application in System Evaluation. *Journal of the University of Belgrade, EE Department, Series Mathematics and Physics* (483): 147–158.

DUJ91 Dujmović, J. 1991. Preferential Neural Networks. Chapter 7. In *Neural Networks—Concepts, Applications, and Implementations*, vol. 2, ed. P. Antognetti and V. Milutinović, 155–206. Prentice-Hall Advanced Reference Series: Prentice-Hall.

DUJ19 Dujmović, J. 2019. Graded Logic for Decision-Support Systems. *International Journal of Intelligent Systems* 34: 2900–2919. https://doi.org/10.1002/int.22177.

1 Fodor, J., and M. Roubens. 1994. *Fuzzy Preference Modelling and Multicriteria Decision Support*. Dordrecht, The Netherlands: Kluwer Academic Publishers.

GOT22 Gottwald, S. Many-Valued Logic. *The Stanford Encyclopedia of Philosophy (Summer 2022 Edition)*, ed. Edward N. Zalta. https://plato.stanford.edu/archives/sum2022/entries/logic-manyvalued/.

GRA09 Grabisch, M., J.-L. Marichal, R. Mesiar, and E. Pap. 2009. *Aggregation Functions*. Cambridge: Cambridge University Press.

MIL56 Miller, G.A. 1956. The Magical Number Seven, Plus or Minus Two: Some Limits on Our Capacity for Processing Information. *The Psychological Review* 63: 81–97.

POL07 Polyanin, A.D., and A.V. Manzhirov. 2007. *Handbook of Mathematics for Engineers and Scientists*. Chapman & Hall/CRC.

RAS74 Rasiowa, H. 1974. *An Algebraic Approach to Non-classical Logics*. PWN and North-Holland.

TOR07 Torra, V., and Y. Narukawa. 2007. *Modeling Decisions*. Springer.

TOR22 Torra, V. 2022. Andness Directedness for t-Norms and t-Conorms. *Mathematics* 10: 1598. https://doi.org/10.3390/math10091598.

Chapter 10
Numerical Computation of Andness and Orness of GCD

Correct computation of andness/orness requires accurate computation of multiple integrals. Analytic computation of multiple integrals is possible for relatively small number of aggregation functions (see [DUJ18]). In most practical cases, the computation of andness and orness must be done numerically.

If the aggregation functions have high values of first derivatives, then accurate numerical procedures for computing multiple integrals can be rather complex and require lot of processor power.[1] Today, each laptop has enough processor power to compute such integrals in a few seconds using the simplistic 5-line Monte Carlo programs shown in Fig. 10.1. That can give three to four accurate decimal digits of the integral and the corresponding andness/orness. That is a sufficient accuracy for applications in GL, because the imprecision of humans is much higher.

Programs shown in Fig. 10.1 compute andness of an aggregator of n variables from the volume under the aggregator surface inside the $(n+1)$-dimensional unit hypercube: $\alpha(V) = \frac{n-(n+1)V}{n-1}$. In the simplest case of $n = 2$ variables the **ANDNESS** program creates $n + 1 = 3$ standard uniform random numbers x, y, and z. Since these values are uniformly distributed inside $[0, 1]$ the triplet (x, y, z) is a uniformly distributed random point inside the cube $[0, 1]^3$. The real random value z can be below or above the surface of the GCD aggregator $F(x, y)$. If we generate the total number of $M \gg 1$ random triplets, let m denote the number of cases where $z < F(x, y)$, i.e., the number of random points that fall under the surface of the analyzed GCD aggregator. If the volume under the aggregator F can have the minimum value 0 and the maximum value 1, then $0 \le m \le M$. So, the quotient m/M denotes a fraction of unit cube that is under the surface of the analyzed aggregator $F(x, y)$. For large values of M, that is exactly the necessary volume V:

[1] The author remembers computing multiple integrals (in up to five dimensions) in 1973/74 using a 16-bit monoprogramming IBM 1130 computer with the main memory of 32 kilobytes and a 1-megabyte removable disk: a sophisticated Fortran program for accurate computation (based on Romberg's method) was running for 36 hours and required the permanent protective presence of the author in the computer center (including sleeping on a very hard desk).

J. Dujmović, *Graded Logic*, Intelligent Systems Reference Library 273,
https://doi.org/10.1007/978-3-031-88558-7_10

```
double urn(void) {return double(rand())/RAND_MAX;}

double ANDNESS(double (*F)(double, double))
{ int Total=10000000, Below, i;   // No integer overflow assuming
  for(Below=i=0; i<Total; i++)    // max int = 2^31-1 = 2.1*10^9
     if(urn() < F(urn(), urn())) Below++;
  return 2.- Below*3.0/Total;
}

double ANDNESS(double (*F)(double, double, double))
{ int Total=10000000, Below, i;
  for(Below=i=0; i<Total; i++)
     if(urn() < F(urn(), urn(), urn())) Below++;
  return 1.5 - Below*2.0/Total;
}

double ANDNESS(double (*F)(double, double, double, double))
{ int Total=10000000, Below, i;
  for(Below=i=0; i<Total; i++)
     if(urn() < F(urn(), urn(), urn(), urn())) Below++;
  return (4.0 - Below*5.0/Total)/3.;
}

double ANDNESS(double (*F)(double,double,double,double,double))
{ int Total=10000000, Below, i;
  for(Below=i=0; i<Total; i++)
     if(urn() < F(urn(), urn(), urn(), urn(), urn())) Below++;
  return 1.25 - Below*1.5/Total;
}
```

Fig. 10.1 C++ functions for computing the andness of GCD functions of 2, 3, 4, and 5 variables using the Monte Carlo method

$$V = \int_0^1 \int_0^1 F(x, y)dxdy = \lim_{M\to\infty} \frac{m}{M}.$$

In a general case of $n \geq 2$ variables, the process of computing the volume is the same as in the case of $n = 2$ variables. We generate n standard uniform random numbers $x_1, ..., x_n$ and m now denotes the number of cases where the standard uniform random value u satisfies $u < F(x_1, ..., x_n)$, i.e., the random point inside $[0, 1]^{n+1}$ is below the surface of aggregator $F(x_1, ..., x_n)$. Thus, we again have

$$V = \int_{I^n} F(x_1, ..., x_n)dx_1, ..., dx_n = \lim_{M\to\infty} \frac{m}{M}.$$

The accuracy of computed volume is modest: if we need k correct significant decimal digits we need approximately $M = 100^k$ random points (in Fig. 10.1 we use $M = 10^7$). With currently available processors it is very easy to get $k = 3$ correct

Table 10.1 Accuracy of andness computed using the Monte Carlo method

Special case of GCD	Indicators	Number of variables			
		2	3	4	5
Product t-norm	Monte Carlo	1.24981	1.25023	1.22937	1.20310
	Exact value	1.25	1.25	1.22917	1.20313
	Absolute error	**−0.00019**	**0.00023**	**0.00020**	**−0.00003**
Pure conjunction	Monte Carlo	1.00029	1.00015	0.999770	0.999840
	Exact value	1	1	1	1
	Absolute error	**0.00029**	**0.00015**	**−0.000230**	**−0.000160**
Geometric mean	Monte Carlo	0.666780	0.656790	0.650701	0.647421
	Exact value	0.666667	0.656250	0.650667	0.647184
	Absolute error	**0.000113**	**0.000540**	**0.000034**	**0.000237**
Arithmetic mean	Monte Carlo	0.500678	0.499872	0.500027	0.499935
	Exact value	0.5	0.5	0.5	0.5
	Absolute error	**0.000678**	**−0.000128**	**0.000127**	**−0.000065**
Pure disjunction	Monte Carlo	0.000159	0.000025	0.000055	0.000259
	Exact value	0	0	0	0
	Absolute error	**0.000159**	**0.000025**	**0.000055**	**0.000259**

Bold: Results of computation

significant decimal digits. A higher accuracy is neither recommended for the Monte Carlo integration method, nor it is necessary in applications where the desired value of andness is approximately (imprecisely) selected by decision engineers, as one of *input parameters* of logic decision models.

The C++ functions in Fig. 10.1 use the variable **`Total`** to denote the value of M, and the variable **`Below`** to denote the value of m. The accuracy of Monte Carlo computation of andness can be verified using various well-known aggregators: the product t-norm, the pure conjunction ($\alpha = 1$), the geometric mean, the arithmetic mean ($\alpha = 0.5$), and the pure disjunction ($\alpha = 0$). The results obtained from programs shown in Fig. 10.1 are compared with analytically computed accurate values and the absolute errors of numerical computations are shown in Table 10.1. All examples have at least three significant decimal digits correct, and some have four correct digits. In some cases, the andness is a function of n and in some special cases it is constant. Using the presented numerical integration, it is possible to compute the andness of variety of functions, and some of them, sorted by decreasing andness, are shown in Table 10.2. We should note that the contraharmonic mean and its De Morgan's dual do not support the nondecreasing monotonicity and therefore they cannot be considered GCD aggregators. All models of hyperconjunction ($\alpha > 1$, e.g., t-norms) and models of hyperdisjunction ($\alpha < 0$, e.g., t-conorms) do not satisfy internality and idempotency, but that does not affect the accuracy of Monte Carlo results.

Table 10.2 The andness of selected conjunctive GCD functions of 2 variables and the orness of their disjunctive duals, sorted in the range from 0.5 to 2

Conjunctive function name	Conjunctive GCD	α	ω	Dual disjunctive GCD
Drastic conjunction	$\lfloor xy \rfloor$	2		$1-\lfloor (1-x)(1-y) \rfloor$
Product-power 5	x^5y^5	1.917		$1-(1-x)^5(1-y)^5$
Product-power 4	x^4y^4	1.880		$1-(1-x)^4(1-y)^4$
Product-power 3	x^3y^3	1.812		$1-(1-x)^3(1-y)^3$
Product-power 2	x^2y^2	1.667		$1-(1-x)^2(1-y)^2$
Product-power 1.5	$x^{1.5}y^{1.5}$	1.520		$1-(1-x)^{1.5}(1-y)^{1.5}$
Łukasiewicz	$\max(x+y-1,0)$	1.5		$\min(x+y,1)$
Product-power 1.2	$x^{1.2}y^{1.2}$	1.380		$1-(1-x)^{1.2}(1-y)^{1.2}$
Einstein sum dual	$\dfrac{xy}{1+(1-x)(1-y)}$	1.318		$\dfrac{x+y}{1+xy}$
Product t-norm	xy	1.25		$x+y-xy$
Hamacher	$\dfrac{xy}{x+y-xy}$	1.13		$\dfrac{x+y-2xy}{1-xy}$
Minimum	$min(x,y) = x \wedge y$	1		$max(x,y) = x \vee y$
Power mean ($r=-4$)	$\left(\dfrac{x^{-4}+y^{-4}}{2}\right)^{-1/4}$	0.885		$1-\left(\dfrac{(1-x)^{-4}+(1-y)^{-4}}{2}\right)^{-1/4}$
Power mean ($r=-2$)	$\dfrac{\sqrt{2}xy}{\sqrt{x^2+y^2}}$	0.8284		$1-\dfrac{\sqrt{2}(1-x)(1-y)}{\sqrt{(1-x)^2+(1-y)^2}}$
Harmonic mean ($r=-1$)	$\dfrac{2xy}{x+y}$	0.7726		$\dfrac{x(1-y)+y(1-x)}{(1-x)+(1-y)}$
Contraharmonic mean dual	$\dfrac{x(1-x)+y(1-y)}{(1-x)+(1-y)}$	0.7726		$\dfrac{x^2+y^2}{x+y}$
Exponential mean ($r=-5.4011$)	$\dfrac{1}{r}ln\dfrac{e^{rx}+e^{ry}}{2}$	0.7504		$\dfrac{-1}{r}ln\dfrac{e^{-rx}+e^{-ry}}{2}$
Power mean ($r=-1/2$)	$\dfrac{4xy}{(\sqrt{x}+\sqrt{y})^2}$	0.7289		$1-\dfrac{4(1-x)(1-y)}{(\sqrt{1-x}+\sqrt{1-y})^2}$
Geometric mean	$\sqrt{xy}$	0.6667		$1-\sqrt{(1-x)(1-y)}$
Quadratic mean dual	$1-\sqrt{0.5(1-x)^2+0.5(1-y)^2}$	0.6232		$\sqrt{0.5x^2+0.5y^2}$
Centroidal mean dual	$1-\dfrac{2}{3}\left(\dfrac{x^2+xy+y^2-3x-3y+3}{2-x-y}\right)$	0.5909		$\dfrac{2}{3}\left(\dfrac{x^2+xy+y^2}{x+y}\right)$
Square root power mean	$(0.5\sqrt{x}+0.5\sqrt{y})^2$	0.5833		$1-(0.5\sqrt{1-x}+0.5\sqrt{1-y})^2$
Heronian mean	$(x+\sqrt{xy}+y)/3$	0.5556		$1-\frac{1}{3}\left[2-x-y+\sqrt{(1-x)(1-y)}\right]$
Generalized Heronian mean	$p\dfrac{x+y}{2}+(1-p)\sqrt{xy}$, $0 \le p \le 1$	$\dfrac{4-p}{6}$		$1-p\dfrac{2-x-y}{2}-(1-p)\sqrt{(1-x)(1-y)}$
OWA/ItOWA	$\alpha(x \wedge y)+\omega(x \vee y)$	α		$\omega(x \wedge y)+\alpha(x \vee y)$
Arithmetic mean	$(x+y)/2$	0.5		$(x+y)/2$

Mean Andness Theorem (MAT)

By definition, both andness and orness are linear functions of the volume V under a GCD aggregator. So, the andness of GCD aggregator is defined as follows:

$$\alpha(V) = a + bV; \quad a = \frac{n}{n-1}, \quad b = -\frac{n+1}{n-1}, \quad n > 1.$$

Suppose now that an aggregator f is a linear combination of $k \geq 2$ aggregators:

$$f(x_1, \ldots, x_n) = \sum_{i=1}^{k} W_i f_i(x_1, \ldots, x_n); \quad \sum_{i=1}^{k} W_i = 1; W_i > 0, i = 1, \ldots, k.$$

Let $V_i = \int_{I^n} f_i(x_1, \ldots, x_n)dx_1, \ldots, dx_n$ be the volume under the component aggregator $f_i(x_1, \ldots, x_n)$. Based on the linearity of integration, the volume under the aggregator f is the following:

$$V = \int_{I^n} f(x_1, \ldots, x_n)dx_1, \ldots, dx_n = \int_{I^n} \sum_{i=1}^{k} W_i f_i(x_1, \ldots, x_n)dx_1, \ldots, dx_n$$
$$= \sum_{i=1}^{k} W_i V_i.$$

The andness α of aggregator f can be computed as follows:

$$\alpha(V) = a + bV = a + b\sum_{i=1}^{n} W_i V_i = \sum_{i=1}^{n} W_i(a + bV_i) = \sum_{i=1}^{n} W_i \alpha_i.$$

This relation proves the following *mean andness theorem* (MAT, introduced in [DUJ05]):

MAT: If $\sum_{i=1}^{k} W_i = 1; W_i > 0, i = 1, \ldots, k$ then the andness of a linear combination of k individual GCD aggregators $f(x_1, \ldots, x_n) = \sum_{i=1}^{k} W_i f_i(x_1, \ldots, x_n)$, is the linear combination of the values of the andness of individual aggregators: $\alpha = \sum_{i=1}^{n} W_i \alpha_i$. Since $\alpha = 1 - \omega$ and $\alpha_i = 1 - \omega_i$ we have $1 - \omega = \sum_{i=1}^{n} W_i(1 - \omega_i)$ and consequently $\omega = \sum_{i=1}^{n} W_i \omega_i$ (the *mean orness theorem*).

Let us show two examples of MAT in the process of creating aggregators with desired values of andness or orness.

Problem 1 Create a GCD aggregator of two variables that has the orness of 60%.

This problem can be solved as a dual of a conjunctive aggregator that has the andness $\alpha = 0.6$. Since the arithmetic mean has the centroid andness $\alpha = 0.5$, and the geometric mean has a higher andness $\alpha = 2/3 = 0.667$, it follows that the desired conjunctive aggregator is located between the arithmetic mean and the geometric mean:

$$f(x, y) = p\frac{x+y}{2} + (1-p)\sqrt{xy}, \quad 0 < \text{p} < 1.$$

According to MAT, the andness of $f(x, y)$ is $\alpha_f = \frac{p}{2} + \frac{2(1-p)}{3} = \frac{4-p}{6}$. This is an example of the interpolation of aggregator $f(x, y)$ between the arithmetic mean and the geometric mean: if $p = 1$ then $\alpha_f = 1/2$, and if $p = 0$ then $\alpha_f = 2/3$. Next, from $\frac{4-p}{6} = 0.6$ we have $p = 4 - 6\alpha = 0.4$ and the result is that aggregator $f(x, y) = 0.2(x + y) + 0.6\sqrt{xy}$ has the andness $\alpha_f = 0.6$. Thus, its dual $h(x, y)$ which has the desired orness $\omega_h = 0.6$ is the following:

$$h(x, y) = 1 - f(1 - x, 1 - y) = 0.6 + 0.2(x + y) - 0.6\sqrt{(1 - x)(1 - y)}.$$

This aggregator satisfies conditions $h(0, 0) = 0, h(1, 1) = 1, h(x, x) = x$. The Monte Carlo program from Fig. 10.1 gives for $h(x, y)$ the approximate orness $\omega = 0.60027$.

It is important to note the following *Dual Volume Theorem (DVT): the volume under the surface of De Morgan dual of any aggregator is a complement of the volume under the surface of the original aggregator*. In the case of two variables, we have:

$$V_f = \int_0^1\int_0^1 f(x, y)dxdy; \quad \alpha_f = 2 - 3V_f, \quad \omega_f = 3V_f - 1,$$

$$V_h = \int_0^1\int_0^1 h(x, y)dxdy = \int_0^1\int_0^1 [1 - f(1 - x, 1 - y)]dxdy = 1 - V_f,$$

$$\alpha_h = 2 - 3V_h = 2 - 3(1 - V_f) = 3V_f - 1 = \omega_f,$$

$$\omega_h = 3V_h - 1 = 3(1 - V_f) - 1 = 2 - 3V_f = \alpha_f.$$

Therefore, in the case of the dual aggregator

$$h(x, y) = 1 - f(1 - x, 1 - y) = 1 - p\frac{2 - x - y}{2} - (1 - p)\sqrt{(1 - x)(1 - y)}, \quad 0 < \text{p} < 1,$$

we have $\omega_h = \alpha_f = \frac{4-p}{6}$ and $p = 4 - 6\omega_h$. Then, the desired aggregator can be written in a form that explicitly shows its orness:

$$h(x, y; \omega_h) = 1 - (2 - 3\omega_h)(2 - x - y) - (6\omega_h - 3)\sqrt{(1 - x)(1 - y)}.$$

Unsurprisingly, there are many other aggregators that can give orness of 60%. According to [REC13], the quadric surface aggregator $q(x, y) = p(x - y)^2 + (x + y)/2$ provides nondecreasing monotonicity for $-0.25 \le p \le 0.25$. The volume under that quadric surface is $V_q = \int_0^1 \int_0^1 q(x, y)dxdy = (p+3)/6$. So, $\omega = 3V_q - 1 = \frac{p+1}{2} = 0.6$, and $p = 2\omega - 1 = 0.2$.

A similar idea is to use the aggregator $A(x, y) = p|x - y| + \frac{x+y}{2}, -\frac{1}{2} \leq \mathrm{p} \leq \frac{1}{2}$. The volume is now $V_a = \int_0^1 \int_0^1 A(x, y)dxdy = \frac{p}{3} + \frac{1}{2}$ and $\omega = 3V_a - 1 = p + \frac{1}{2}$ and $p = \omega - \frac{1}{2} = 0.1$. This aggregator has ultimate simplicity because it is equivalent to the most elementary aggregator $B(x, y) = (1 - \omega)\min(x, y) + \omega \cdot \max(x, y)$.

These aggregators can now be written the following *orness-directed form*:

$$q(x, y; \omega) = (2\omega - 1)(x - y)^2 + (x + y)/2,$$
$$A(x, y; \omega) = \left(\omega - \frac{1}{2}\right)|x - y| + \frac{x+y}{2} = (1 - \omega)\min(x, y) + \omega \cdot \max(x, y),$$
$$h(x, y; \omega) = 6\omega - 3 + (2 - 3\omega)(x + y) - (6\omega - 3)\sqrt{(1 - x)(1 - y)}.$$

According to Tables 2.3.3 and 2.3.4 presented in [DUJ18, pp. 267-268], a similar degree of orness ($\omega = 0.60417$) can be achieved using the following power and exponential means:

$$P(x, y; \omega) = \left(0.5x^r + 0.5y^r\right)^{1/r}, \quad \omega = 0.60417, \quad r = 1.8113,$$
$$E(x, y; \omega) = r^{-1}ln\left(0.5e^{rx} + 0.5e^{ry}\right), \quad \omega = 0.60417, \quad r = 1.747.$$

Because of the same orness 0.6, the presented five aggregators have similar behavior, and it is useful to compute the mean difference D between them, which (for comparison of $P(x, y; \omega)$ and $E(x, y; \omega)$) we can define as follows:

$$D[P; E; \omega] = \int_0^1 \int_0^1 |P(x, y; \omega) - E(x, y; \omega)|dxdy; \quad \omega - 0.6.$$

The results of the analysis of ten mean differences between all pairs of aggregators q, A, h, P, and E are presented in Table 10.3. Because of the same orness, all differences are rather small, mostly around 1%. These differences should be considered in the context of expected precision of arguments, which (based on [MIL56, SAA03, DUJ04a, DUJ15a]) are expected to be in the range from 5% to 15%. Consequently, the precision of aggregators is regularly for an order of magnitude better than the precision of inputs. In such a situation, increasing the precision of aggregation yields no practical benefits. Indirectly, that also yields another observation: from the standpoint of applications, in many cases, the area of aggregation attracts significantly more research attention than it deserves.

Problem 2 Create GCD aggregators of 2, 3, 4, and 5 variables that have andness $\alpha^* = 0.75$.

According to Table 10.2, this GCD function can be interpolated between the geometric and harmonic means. Both the harmonic mean H and the geometric mean

Table 10.3 Mean differences between five concurrent aggregators that have the same orness 0.6

Mean difference	E	P	h	A
q	0.0014	0.0075	0.0105	0.0125
A	0.0125	0.0134	0.0167	
h	0.0110	0.0174		
P	0.0077			

G are special cases of the power mean $P_n(x_1, \ldots, x_n; r)$ as follows:

$$P_n(x_1, \ldots, x_n; r) = \left(\frac{1}{n}\sum_{i=1}^{n} x_i^r\right)^{1/r},$$

$$H_n(x_1, \ldots, x_n) = P_n(x_1, \ldots, x_n; -1),$$

$$G_n(x_1, \ldots, x_n) = \lim_{r\to 0} P_n(x_1, \ldots, x_n; r) = \left(\prod_{i=1}^{n} x_i\right)^{1/n}.$$

The desired GCD function $F_n(x_1, \ldots, x_n)$ is a linear combination of the harmonic and geometric mean and, according to MAT, the corresponding andness is also a linear combination of the andness of the harmonic and geometric means, as follows:

$$F_n(x_1, \ldots, x_n) = p(n)H_n(x_1, \ldots, x_n) + (1 - p(n))G_n(x_1, \ldots, x_n), \quad 0 < p(n) < 1$$

$$\alpha(n) = p(n)\alpha_H(n) + (1 - p(n))\alpha_G(n) = 0.75,$$

$$p(n) = \frac{\alpha(n) - \alpha_G(n)}{\alpha_H(n) - \alpha_G(n)}.$$

The numeric results are presented in Table 10.4. For $n = 2, 3, 4, 5$, the row $\alpha_H(n)$ shows Monte Carlo results of andness of the harmonic mean, and the row $\alpha_G(n)$ shows analytically computed exact values of the andness of geometric mean. Parameters of the interpolative form of function $F_n(x_1, \ldots, x_n)$ are shown in the row $p(n)$. The row $\alpha_F(n)$ contains Monte Carlo results of the andness of the interpolated GCD aggregator $F_n(x_1, \ldots, x_n)$. These results are rather close to the desired value $\alpha^* = 0.75$. The next row contains absolute errors and shows that the andness errors satisfy $|\alpha_F(n) - \alpha^*| < 10^{-3}$.

In the presented example the interpolative GCD aggregator requires the computation of two power means (geometric and harmonic). The computations can be reduced if we use only one power mean, $P_n(x_1, \ldots, x_n; r(n)) = \left(\frac{1}{n}\sum_{i=1}^{n} x_i^{r(n)}\right)^{1/r(n)}$, but in this case the exponent $r(n)$ must be adjusted to achieve the desired andness α^*. The adjusted exponent values taken from [DUJ18] are presented in the row $r(n)$ of Table 10.4. The row $\alpha_P(n)$ contains Monte Carlo results of the andness of the adjusted power mean. Finally, the obtained accuracy is again $|\alpha_P(n) - \alpha^*| < 10^{-3}$. In this example, the maximum relative error of GCD is $100|\alpha_P(n) - \alpha^*|/\alpha^* = 0.06\%$. That

Table 10.4 Numerical comparison of two GCD aggregators that have andness 0.75

n	2	3	4	5
$\alpha_H(n)$	0.772825	0.774012	0.777956	0.781191
$\alpha_G(n)$	0.666667	0.65625	0.650667	0.647184
$p(n)$	0.784992	0.796096	0.780372	0.767244
$\alpha_F(n)$	0.750453	0.750585	0.749647	0.749758
$\alpha_F(n) - \alpha^*$	**0.000453**	**0.000585**	**-0.000353**	**-0.000242**
$r(n)$	-0.7201	-0.7317	-0.7205	-0.7054
$\alpha_P(n)$	0.750264	0.749797	0.750453	0.750008
$\alpha_P(n) - \alpha^*$	**0.000264**	**-0.000203**	**0.000453**	**0.000008**

level of accuracy of the andness of GCD, modeled using power means, is more than sufficient, taking into account the imprecision of arguments provided by stakeholders/decision-makers.

The previous examples show that there are multiple parameterized functions that can give the same andness. Table 10.2 shows that the andness $\alpha = 0.75$ can also be achieved using the parameterized exponential mean $E(x, y; r) = \frac{1}{r} ln \frac{e^{rx}+e^{ry}}{2}$, $r = -5.4011$. Therefore, the GCD aggregator based on the exponential mean has the same andness as the GCD aggregator based on the interpolated function $F_2(x, y; p) = pH_2(x, y) + (1 - p)G_2(x, y)$, $p = 0.784992$. These two functions have the same andness because they have the same volume under their surfaces inside the cube $[0, 1]^3$. However, these are different functions with different properties. The most important difference between them is that for any positive value $0 < t \le 1$ we have $F_2(0, t; p) = F_2(t, 0; p) = 0$ because both $H_2(0, t; p) = H_2(t, 0; p) = 0$ and $G_2(0, t; p) = G_2(t, 0; p) = 0$. The value 0 is an annihilator and such GCD function is called the *hard conjunctive GCD*. On the other hand, $E(0, t; r) > 0$, $E(t, 0; r) > 0$, and the exponential mean does not support the annihilator 0. Such GCD function is called the *soft conjunctive GCD*. From the GL standpoint, that is a fundamental difference. Hard and soft GCD aggregators are studied in the next chapter.

For finite values of r, the exponential mean is always soft. However, the exponential mean has a rather unique property that the dual of function $E(x, y; r)$ is the function $E(x, y; -r) = \frac{-1}{r} ln \frac{e^{-rx}+e^{-ry}}{2}$. So, if the GCD function $E(x, y; r)$ has the andness α, then the function $E(x, y; -r)$ has the andness $1 - \alpha$ (and the orness α). Since $E(x, y; r)$ and $E(x, y; -r)$ are dual, it follows that for exponential means De Morgan duality has the form $1 - E(1 - x, 1 - y; r) = E(x, y; -r)$:

$$\begin{aligned} 1 - E(1-x, 1-y; r) &= 1 - \frac{1}{r} ln \frac{e^{r(1-x)} + e^{r(1-y)}}{2} \\ &= 1 - \frac{1}{r} ln \left(\frac{e^{-rx} + e^{-ry}}{2} e^r \right) \\ &= 1 - \frac{1}{r} \left[ln \left(\frac{e^{-rx} + e^{-ry}}{2} \right) + r \right] = \frac{-1}{r} ln \frac{e^{-rx} + e^{-ry}}{2} \end{aligned}$$

```
Enter a training set containing any number of desired I/O values
(a value outside the [0,1] interval denotes the end of input data)

 x1   x2   x3   x4       y
----------------------------
0.8    1    1    1     0.95
  1  0.5    1    1     0.8
  1    1  0.4    1     0.7
  1    1    1  0.2     0.5
  9    9    9    9     9

 T r a i n i n g ...        Please wait !

THE OPTIMUM VALUE OF PARAMETER :    r  =    -1.0623

 The minimum value of criterion function = 0.133E-01

===================================================
  Training set                   | Trained GCD neuron
---------------------------------+-----------------
  x1   x2   x3   x4       y      |   y        Error
---------------------------------+-----------------
 0.80 1.00 1.00 1.00  0.9500   | 0.9409    -0.91E-02
 1.00 0.50 1.00 1.00  0.8000   | 0.7973    -0.27E-02
 1.00 1.00 0.40 1.00  0.7000   | 0.7228     0.23E-01
 1.00 1.00 1.00 0.20  0.5000   | 0.4904    -0.96E-02
===================================================
```

Fig. 10.2 Results of training of a preferential neuron using the AnSy tool

$$= E(x, y; -r).$$

If the value r gives a conjunctive aggregator, then the value $-r$ gives a symmetric disjunctive aggregator (and vice versa). This property makes the exponential mean a convenient GCD aggregator in applications where there is no need to support hard aggregators.

All previous results in this chapter are based on the concept of andness/orness and adjusting andness/orness by a stakeholder in the process of decision-making. It is important to note that in GL applications andness is an *input*, not an *output*. An alternative numeric procedure to create a GCD aggregator is by considering that GCD is a preferential neuron and propositional formulas are preferential neural networks [DUJ91]. Then we can perform a neuron-by-neuron training by minimizing the error between desired output and achieved output. This process is exemplified in Fig. 10.2 where we use the AnSy tool [SEA10] to perform the training of GCD implemented as the power mean $y = \left(\frac{1}{n}\sum_{i=1}^{n} x_i^r\right)^{1/r}$ for $n = 4$ inputs.

Initially, the user is offered to enter a training set, which is a table of desired input/output values: the inputs x_1, x_2, x_3, x_4 and the desired output y. For example, according to Fig. 10.2, if the inputs are $x_1 = x_2 = x_3 = 1$ and $x_4 = 0.2$, the user would like to generate the output $y = 0.5$. That exposes the individual impact of input x_4 in the case where all other inputs are fully satisfied (completely true). Therefore, it is easy to justify the proposed output value. After four such rows, there is the row with numbers 9, that are greater than 1 and used to signal the end of input data. Then AnSy tries to find the value of exponent r that minimizes the approximation error (the difference between the desired and achieved values). In our example the result is the exponent $r = -1.0623$. This is the following hard conjunctive GCD aggregator: $y = \left(\frac{x_1^{-1.0623}+x_2^{-1.0623}+x_3^{-1.0623}+x_4^{-1.0623}}{4}\right)^{-1/1.0623}$. This GCD aggregator is close to the harmonic mean: we can use the Monte Carlo integration (Fig. 10.1) to compute the andness of this GCD function, and we get $\alpha = -0.7776$.

Andness and orness are defined for commutative versions of the GCD aggregator, and the example presented in Fig. 10.2 shows such a process. However, the training of preferential neurons can be used to simultaneously determine all parameters of GCD, i.e., both the andness and the weights of inputs. That training process is presented in Chap. 13.

Numerical computation of andness and orness, presented in this chapter, is an essential part in the development of logic aggregators that are necessary in graded propositional calculus and applications of GL. The presented aggregator development problems are typical problems in graded propositional calculus, where andness and orness are input parameters specified by decision-makers. Such problems can be used as benchmark problems for evaluation, and comparison of various families of aggregators, such as andness-directed interpolative aggregators, OWA, means, t-norms, and fuzzy integrals. The comparison and selection of families of aggregators, based on logic aggregation benchmark problems, can be found in [DUJ23].

References

DUJ04a Dujmović, J. and W.Y. Fang. 2004. An Empirical Analysis of Assessment Errors for Weights and Andness in LSP Criteria. In *Modeling Decisions for Artificial Intelligence*, ed. V. Torra and Y. Narukawa, 139–150. Springer LNAI 3131.

DUJ05 Dujmović, J. 2005. Seven Flavors of Andness/Orness. In *Proceedings of Eurofuse 2005*, ed. B. De Baets, J. Fodor, and D. Radojević, 81–92. ISBN 86-7172-022-5, Belgrade: Institute Mihajlo Pupin.

DUJ18 Dujmović, J. 2018. *Soft Computing Evaluation Logic*. J. Wiley.

DUJ23 Dujmović, J., and V. Torra. 2023. Logic Aggregators and Their Implementation. In *MDAI 2023, LNAI 13890* ed. V. Torra, and Y. Narukawa, 3–42, Springer.

DUJ91 Dujmović, J. 1991. Preferential Neural Networks. Chapter 7 in Neural Networks—Concepts, Applications, and Implementations. In *Prentice-Hall Advanced Reference Series, Prentice-Hall*, vol II. ed. P. Antognetti and V. Milutinović, 155–206.

DUJ15a Dujmović, J. 2015. Weighted Compensative Logic with Adjustable Threshold Andness and Orness. *IEEE Transactions on Fuzzy Systems* 23 (2): 270–290. https://doi.org/10.1109/TFUZZ.2014.2312018.

MIL56 Miller, G.A. 1956. The Magical Number Seven, Plus or Minus Two: Some Limits on Our Capacity for Processing Information. *The Psychological Review* 63: 81–97.

REC13 Recasens, J. 2013. Aggregation Operators and Quadric Hypersurfaces. In *Aggregation Functions in Theory and Practice*, vol. 228 ed. Bustince, H., J. Fernandez, R. Mesiar, and T. Calvo, 333–343. Springer Advances in Intelligent Systems and Computing.

SAA03 Saaty, T.L., and M.S. Ozdemir. 2003. Why the Magic Number Seven Plus or Minus Two. *Mathematical and Computer Modelling* 38: 233–244.
SEA10 SEAS Co. 2010. *Analysis and Synthesis of Aggregation Operators*. AnSy User Manual V.3.2. SEAS.

Chapter 11
A Second Look at GCD: Selectable Annihilators and the Graded Logic Conjecture

The human-centric approach to GL creates various requirements that GCD aggregators must satisfy. There are two fundamental questions that we must answer:

- What are the main human-centric logic properties that GCD must support?
- What are the necessary and sufficient basic Graded Logic functions?

The purpose of this chapter is to answer these questions.

Five basic types of human-centric logic properties of the GCD function are summarized in Table 11.1. We assume that GCD is a logic aggregator of 2 or more inputs that decision-makers use to express desired logic requirements that reflect simultaneity or substitutability of inputs. We identified four types of logic requirements based on verbal interpretations *"must have all inputs satisfied", "nice to have most inputs satisfied", "nice to have some inputs satisfied"* and *"enough to have any input satisfied"*. In the middle of these four regions is the neutral (centroid) aggregator of arithmetic mean which has balanced conjunctive and disjunctive properties.

The first type specifies mandatory requirements that are verbally classified as the "*must have all inputs satisfied*" requirement. This kind of logic relationship aggregates inputs that are very important for the decision-maker and must *all* be simultaneously satisfied/true as high as possible. Since the decision-maker in this case needs to have high degrees of truth of all inputs, the most influential in the negative sense (i.e., the most critical) are those inputs that have low values. Of course, the highest impact comes from inputs that have the value zero, and in such a case the output is zero regardless of all positive values that other arguments could have. That is a strong requirement, where *all* inputs must be at least partially true. If such a GCD aggregator is $f : [0, 1]^n \rightarrow [0, 1]$, then it supports the annihilator[1] 0:

[1] If a value A of any argument causes the same value A of a function, for all possible values of other arguments, then A is an annihilator (absorbing element) of that function, and the function supports the annihilator. In the case of graded logic functions, annihilators can be the extreme degrees of truth: 0 or 1. For example, the product $p(x, y, z) = xyz$ has the annihilator 0, and its dual $q(x, y, z) = 1 - (1 - x)(1 - y)(1 - z)$ has the annihilator 1.

J. Dujmović, *Graded Logic*, Intelligent Systems Reference Library 273,
https://doi.org/10.1007/978-3-031-88558-7_11

Table 11.1 Four main types of human-centric logic properties supported by GCD

Verbal requirements for satisfied inputs	Type of requirement	Type of aggregator	Supported annihilator	Type of logic connective	Relationship of components
"Must have all"	Mandatory	Hard	0	Conjunctive	Simultaneity
"Nice to have most"	Optional	Soft	None	Conjunctive	Simultaneity
"Nice to have"	Optional	Soft	None	Neutral Con. = Dis.	Balanced Sim. = Sub.
"Nice to have some"	Optional	Soft	None	Disjunctive	Substitutability
"Enough to have any"	Sufficient	Hard	1	Disjunctive	Substitutability

$$\forall i \in \{1, \ldots, n\},\ x_i = 0 \Rightarrow f(x_1, \ldots, x_{i-1}, 0, x_{i+1}, \ldots, x_n) = 0.$$

In other words, a decision-maker uses a mandatory requirement where it is not acceptable that any of inputs is false. Conjunctive aggregators that support annihilators are denoted as *hard* because it is impossible to compensate any zero input, and it is difficult to compensate the impact of low input degrees of truth.

Mandatory requirements are among the most frequent requirements in natural human decision making. Let us see a few examples. In most cases of job selection, a salary above some minimum threshold is a typical mandatory requirement: jobs that offer insufficient compensation will be rejected. A family with working parents and small children will seek a location of residence that must be close to an appropriate school and/or day care, as well as in acceptable vicinity of jobs for parents. Residences that are too far from these points of interest will be rejected. If a family with six members is buying a car, it is unlikely that a car registered for four passengers would be acceptable; six or more seats might be a mandatory requirement. A famous university might have several mandatory requirements that candidates must satisfy to be admitted in a reputable graduate program (e.g., a high undergraduate GPA, and a satisfactory score on a qualifying exam). All these examples are the "*must have*" type, i.e., they reflect hard conjunctive aggregators that must support annihilators: if any of mandatory inputs is zero, the aggregated result must also be zero, rejecting such a candidate. Therefore, a hard conjunctive GCD is one of necessary basic types of Graded Logic functions. Of course, the hard conjunctive GCD has adjustable andness and after a justifiable selection of this type of aggregator, stakeholders adjust its desired strength (andness). It is important to note that andness is an important input parameter in the process of human logical reasoning.

There are many situations where we need conjunctive aggregators that are not hard. These are all cases where we have desired attributes that are not mandatory. For example, a used car buyer would prefer to have a green car, but one that is available (one owner, low mileage, and a very good price) is gray. Accept or reject? It would be nice to have a green car, but other positive properties are sufficient to see that the color is desirable but not mandatory and the gray car would be purchased. Of course, the selection criterion should be designed so that if there are two used cars with similar mandatory properties, one green and another gray, then the green car wins. As another similar example, all individuals and families like to have residences close to specific points of interest. The vicinity to some of them can be defined as mandatory (e.g., job, school, hospital, public transportation) and the vicinity to others (e.g., park, movie theater, public library) can be defined as optional: it is nice to have them, they are desired, but not mandatory (the absence of some of them will not necessarily disqualify a residence location). Thus, soft conjunctive versions of GCD aggregator are also a necessary type of the GCD aggregator.

The most frequently used version of GCD (typically without sufficient justification) is the logic neutrality, i.e., the arithmetic mean. All additive scoring methods belong to this category of logic aggregators, and most frequently the only justification for their use is the simplicity of the arithmetic mean formula (a detailed analysis of various drawbacks of the additive scoring can be found in [DUJ18]). A justified way of using this aggregator is a situation where andness is provably equal to orness, i.e., conjunctive and disjunctive properties must be equally present. The most frequent example of the justified use of logic neutrality is the computation of the average grade (GPA) of students in schools and universities. In this example, we would like simultaneously all high grades (conjunctive requirement), but at the same time we permit the possibility of substitution/compensation of low grades by any other high grades (disjunctive requirement). Both requirements are equally present, making logic neutrality an appropriate aggregator. Another area where neutral aggregators are very frequent is medicine where medical conditions are frequently evaluated using medical rating scales and additive scoring methods [HER96, HUN14, DYC93, DYC05, GRA06a, GRA06b].

Duality is a necessary property in both classical logic and GL. So, disjunctive aggregators are duals of conjunctive aggregators and for all necessary conjunctive aggregators there are symmetric necessary disjunctive aggregators. Examples of soft disjunctive GCD aggregators can be found in all "nice to have" situations where the satisfaction of each input requirement is desired but not sufficient to fully satisfy the stakeholder. Suppose that the vicinity to park, movie theater, and public library are optional requirements that can be either desired but not mandatory (soft graded conjunction) or desired but not sufficient (soft graded disjunction). In the case of soft graded conjunction, the stakeholder is more demanding than in the case of soft graded disjunction, because s/he requests simultaneous satisfaction of all three requirements but will not reject the case where one (or maybe two) of them are missing. On the other hand, in the case of soft graded disjunction, the stakeholder is less demanding because s/he would be satisfied even if only some of inputs are satisfied, but not to the extent where a full satisfaction of one of them could be sufficient to fully satisfy the stakeholder. Therefore, both the soft graded conjunction and the soft graded disjunction are necessary types of GCD aggregators, and between them is the logic neutrality which is selected if it is not clear whether the requirement should be mildly conjunctive or mildly disjunctive. Stakeholders are always assumed to first select (with confidence and justification) the right type of logic aggregator, and after that to adjust more precisely their strength by selecting desired andness or orness.

Symmetric to mandatory requirements are their dual counterpart, sufficient requirements, verbalized as "*enough to have*". They are used in situations where each of several input arguments can fully satisfy a stakeholder. The corresponding GCD model is the hard graded disjunction that supports the annihilator 1. The highest impact comes from inputs that have the truth value one, and in such a case the output is one regardless all low values that other arguments could have. If no input degree of truth is one, then the highest impact comes from those inputs that have high values. If the function $F : [0, 1]^n \to [0, 1]$ is a hard graded disjunction, it supports the annihilator 1 as follows:

$$\forall i \in \{1, \ldots, n\},\ x_i = 1 \Rightarrow F(x_1, \ldots, x_{i-1}, 1, x_{i+1}, \ldots, x_n) = 1.$$

For example, a patient can have motor symptoms (mobility impairments) and/or sensory symptoms (pain or numbness) and the sufficient presence of either motor or sensory symptoms is enough to classify the patient as disabled. Similarly, a disease can cause several observable symptoms, and the presence of each of them is sufficient to identify the presence of the disease. Therefore, the hard disjunctive versions of GCD are a necessary type of the GCD aggregator.

In the above presentation we exemplified four basic types of the GCD aggregator shown in Table 11.1: hard conjunctive, soft conjunctive, soft disjunctive and hard disjunctive, located between the classic functions of the pure conjunction ($\min(x_1, \ldots, x_n)$), logic neutrality $(x_1 + \ldots + x_n)/n$, and the pure disjunction ($\max(x_1, \ldots, x_n)$). In this range, the andness and orness are between 0 and 1, and the GCD aggregator $x_1 \diamond \ldots \diamond x_n$ is a mean:

$$\min(x_1, \ldots, x_n) \le x_1 \diamond \ldots \diamond x_n \le \max(x_1, \ldots, x_n), \quad x_i \in [0, 1], i = 1, \ldots, n.$$

If all inputs are the same, $x_i = x \in [0, 1],\ i = 1, \ldots, n$, then we have

$$x = \min(x, \ldots, x) \le x \diamond \ldots \diamond x \le \max(x, \ldots, x) = x.$$

Therefore, in the range from the pure conjunction to the pure disjunction, where $\alpha \in [0, 1]$, and $\omega \in [0, 1]$, GCD is an idempotent aggregator: $x \diamond \ldots \diamond x = x$. However, GCD in not idempotent for andness and orness inside the range $H = [\frac{-1}{n-1}, \frac{n}{n-1}] \backslash [0, 1]$.

It is important to emphasize that the hard and soft versions of GCD are selectable by the stakeholder. The first two questions in selecting an appropriate version of GCD are: (1) Conjunctive or disjunctive (simultaneity or substitutability)? (2) Hard or soft (with or without annihilator)? For any real stakeholder/decision-maker, these questions are so simple and distinct that there is never problem in selecting answers. After these initial questions there are refinement questions that are used to choose the appropriate andness/orness inside the selected version of GCD. Below, we show the complete process of deciding on the most appropriate GCD aggregator, where the most frequent is the structure of uniform GCD shown in Fig. 11.1.

The range of conjunctive aggregators is obviously the same as the range of dual disjunctive aggregators. However, the range of soft aggregators does not necessarily need to be the same as the range of hard aggregators, and the border between them can be adjustable as proposed in [DUJ15a]. In the absence of convincing proof that the range of hard aggregators should be greater than or less then the range of

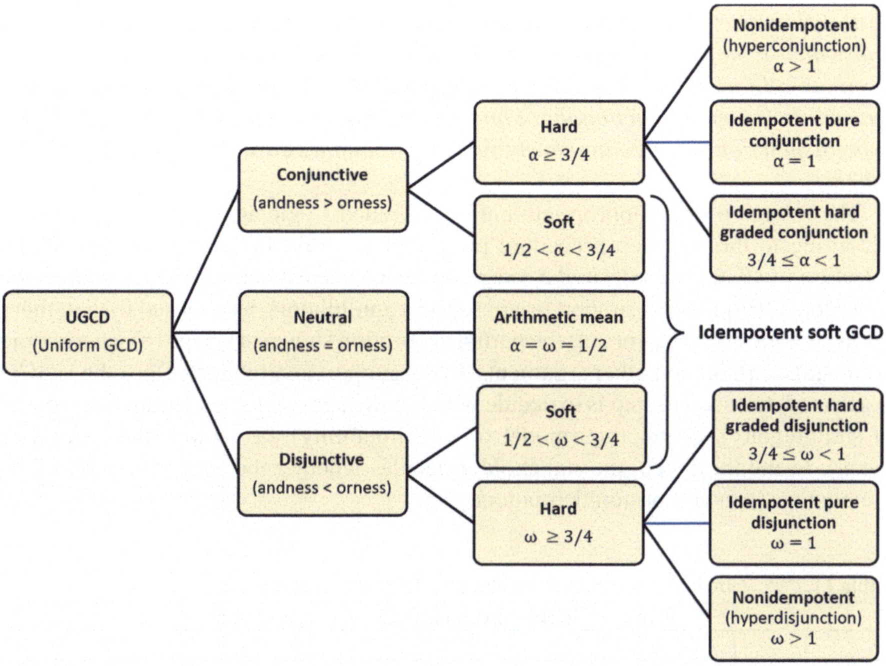

Fig. 11.1 The structure of uniform graded conjunction/disjunction (UGCD)

soft aggregators, it is natural to use the uniform distribution where each of hard conjunctive, soft conjunctive, soft disjunctive, and hard disjunctive versions of GCD has the same size: ¼ of the unit interval [DUJ15b]. We call such aggregator the uniform GCD, abbreviated as UGCD. We also assume that UGCD includes the ranges of andness above the pure conjunction (hyperconjunction, $1 < \alpha \leq \frac{n}{n-1}$) and orness above the pure disjunction (hyperdisjunction, $1 < \omega \leq \frac{n}{n-1}$). That gives the structure of the UGCD aggregator shown in Fig. 11.1.

The fundamental question is whether Fig 11.1 includes all necessary and sufficient versions of the GCD function. According to the human-centric approach to Graded Logic, the presented versions of the GCD function are derived from observations of natural human reasoning. Consequently, there is no strict formal proof that the presented functions are necessary and sufficient. However, a half century of empirical evidence indicates that nothing is missing. In addition, these functions cover all regions of the unit hypercube $[0, 1]^n$, $n > 1$ [DUJ18]. Therefore, we can claim with confidence that the propositional calculus in Graded Logic is based on the following ***Graded Logic Conjecture:***

Necessary and sufficient basic Graded Logic functions include ten characteristic functions: nine logic aggregators and negation. The following list is sorted by increasing orness: (1) hyperconjunction, (2) pure conjunction, (3) hard conjunctive GCD, (4) soft conjunctive GCD, (5) logic neutrality, (6) soft disjunctive GCD, (7) hard disjunctive GCD, (8) pure disjunction, (9) hyperdisjunction, and (10) negation which is not an aggregator. These ten functions are necessary and sufficient to create all formulas of the graded propositional calculus used in natural human reasoning. Hyperconjunction, hard and soft conjunctive and disjunctive GCD, and hyperdisjunction have continuously adjustable andness/orness in their respective ranges.

The selection of an appropriate uniform Graded Logic aggregator is performed according to the five-step procedure presented in Table 11.2. The first step is the selection of UGCD as a desired aggregator. UGCD is a symmetric aggregator in the sense that all arguments support or not support annihilators; as opposed to that, there are asymmetric aggregators (as the partial absorption shown in Chap. 16) where some arguments support and other arguments do not support annihilators. Once the UGCD is selected, the second step is to decide whether the aggregator is conjunctive (model of simultaneity), disjunctive (model of substitutability), or neutral (the arithmetic mean). In the third step, the stakeholder decides whether the aggregator should be idempotent (a mean) or nonidempotent.

Table 11.2 A stepwise procedure for selecting the UGCD aggregator

Five steps in selecting a uniform GCD aggregator					
1	2	3	4	5	
Graded Conjunction/ Disjunction (GCD)	Disjunctive $\omega > \alpha$	Nonidempotent $x\Diamond \ldots \Diamond x > x$	Hyperdisjunction $1 < \omega \le n/(n-1)$	Drastic	
				High	
				Medium (t-conorms)	
				Low	
		Idempotent $x\Diamond \ldots \Diamond x = x$	Hard disjunctive GCD $3/4 \le \omega \le 1$	Pure disjunction	
				Very high	
				High	
				Medium	k
				Low	
				Very low	
			Soft disjunctive GCD $1/2 < \omega < 3/4$	Very high	
				High	
				Medium	k
				Low	
				Very low	
	Neutral: $\omega = \alpha$		Neutrality: $\omega = \alpha = 1/2$		
	Conjunctive $\alpha > \omega$		Soft conjunctive GCD $1/2 < \alpha < 3/4$	Very low	
				Low	
				Medium	k
				High	
				Very high	
			Hard conjunctive GCD $3/4 \le \alpha \le 1$	Very low	
				Low	
				Medium	k
				High	
				Very high	
				Pure conjunction	
		Nonidempotent $x\Diamond \ldots \Diamond x < x$	Hyperconjunction $1 < \alpha \le n/(n-1)$	Low	
				Medium (t-norms)	
				High	
				Drastic	

Idempotent aggregators are the most frequently used, and they cover the range from the pure conjunction to the pure disjunction where both the andness and the orness belong to the interval [0, 1]. If the conjunctive UGCD is selected, then in step 4 the stakeholder decides whether the desired aggregator should be hard or soft. The hard conjunctive UGCD includes the extreme value of the pure conjunction (which is infrequently used), and the hard partial conjunctive UGCD, which is frequently used, same the soft partial conjunctive UGCD.

The final fifth step consists of choosing the intensity of conjunctive aggregation within the selected group. As shown in Table 11.2, each group uses a rating scale that has k values, where k is regularly an odd number that determines the precision, and where the middle value is labeled "*medium.*" The example in Table 11.2 uses $k = 5$. The practical question is what is the most suitable value of k considering that according to Table 11.2 the total number of components in the idempotent UGCD aggregator is $K = 4k + 3$ (such aggregators are symbolically denoted UGCD.K).

If we want to balance sufficient precision and outstanding simplicity, it is convenient to use $k = 3$, yielding the popular UGCD.15 aggregator that has 15 components shown in Fig 11.2. In this case, the stakeholder easily selects between low/medium/high intensity in each group, and the andness/orness difference between adjacent aggregators is 1/14 (approximately 7%). It is easy to note that in the case of disjunctive aggregators the procedure of selecting the most convenient version of UGCD.15 is exactly the same as in the conjunctive case.

Figure 11.2 includes the following symbolic notation of 15 components of UGCD.15, sorted according to decreasing andness, starting from 1 and decreasing with step 1/14: C, HC+, HC, HC-, SC+, SC, SC-, A, SD-, SD, SD+, HD-, HD, HD+, D. Here C denotes conjunction, D denotes disjunction, H denotes hard, and S denotes soft. Suffixes "+" and "-" denote respectively the amplified/stronger version and the attenuated/weaker version of specific aggregator. Without a suffix, the aggregator has the medium position in a selected group.

The design of the UGCD.15 aggregators is presented in Chap. 13. At this point it is useful to note that no natural mean can be used as the UGCD aggregator. As an example of the parameterized mean that includes soft and hard conjunctive aggregators, we can use the power mean $y = \left(\frac{x_1^r+...+x_n^r}{n}\right)^{1/r}$, where in the range $0 < r < 1$ the aggregator is conjunctive and soft, and for $-\infty \le r \le 0$ the aggregator is conjunctive and hard. Unfortunately, the power mean is not uniform: the soft region is smaller than the hard region.

The border between the soft and hard region is $r = 0$ that gives the geometric mean $y = (x_1 \cdots x_n)^{1/n}$. The volume under this surface is $V_g = \int_{[0,1]^n} (x_1 \cdots x_n)^{1/n} dx_1 \cdots dx_n = \left(\frac{n}{n+1}\right)^n$ and the corresponding threshold andness of the power mean is $\alpha_\theta = \frac{n}{n-1} - \frac{n+1}{n-1}\left(\frac{n}{n+1}\right)^n \le \frac{2}{3}$. This is less than the desired threshold andness of UGCD.15, which is $\frac{3}{4}$. So, in the range $\frac{2}{3} \le \alpha < \frac{3}{4}$ the power mean is hard, and cannot model the UGCD.15 function SC+. If we use dual disjunctive operators, then the same problem occurs for SD+, and if we use directly the power mean, then we cannot model the hard partial disjunction, because all disjunctive logic aggregators in the range $1 < r < +\infty$ are strictly soft. Therefore, UGCD.15 and all

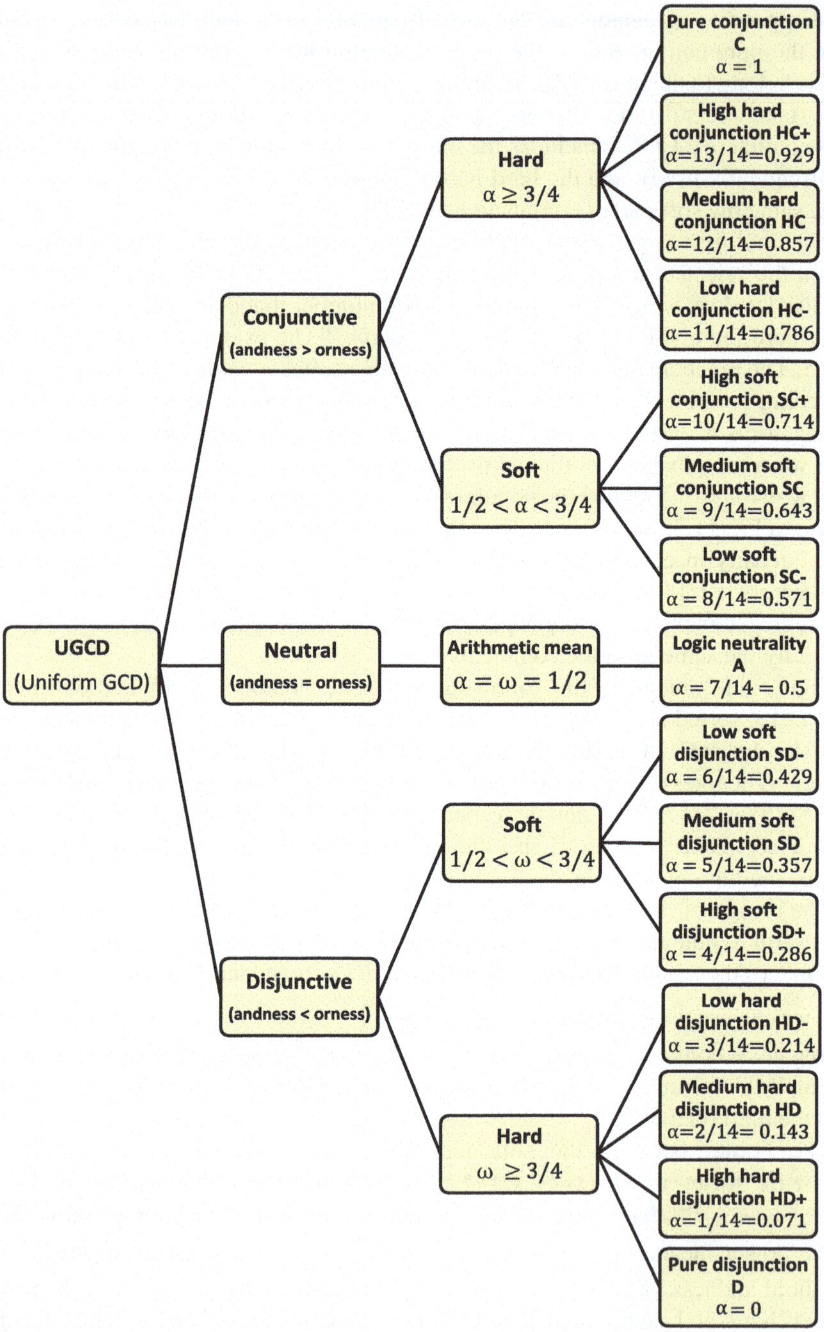

Fig. 11.2 The process of selecting components of the UGCD.15 idempotent aggregator

other idempotent logic aggregators must be designed as multiple connected segments (presented in Chap. 13) that satisfy properties shown in Fig. 11.2. Note that the hard conjunctive aggregators start (as HC-) in the close vicinity of the harmonic mean.

According to Table 11.2, UGCD aggregators are expanded with the symmetric segments of hyperconjunction and hyperdisjunction. The central positions in these regions are reserved for t-norms and their dual t-conorms. The t-norms satisfy commutativity, associativity, nondecreasing monotonicity, continuity and the neutral element 1. The simplest and the most frequently used t-norm is the product $y_{hc} = x_1 \cdots x_n$. Its volume is $V_t = \int_{[0,1]^n} x_1 \cdots x_n dx_1 \cdots dx_n = 2^{-n}$ and the corresponding andness is $\alpha_t(n) = \frac{n-(n+1)/2^n}{n-1}$. So, $\alpha_t(2) = \frac{5}{4} = 1.25$, $\alpha_t(3) = \frac{5}{4} = 1.25$, $\alpha_t(4) = \frac{59}{48} = 1.23$. We can compare these values with the andness of drastic conjunction $\alpha_{dc}(n) = \frac{n}{n-1}$ where $\alpha_{dc}(2) = 2$, $\alpha_{dc}(3) = 1.5$, $\alpha_{dc}(4) = 4/3$.

The middle point of the hyperconjunction region has the andness $\alpha_{mhc}(n) = \frac{1+\alpha_{dc}(n)}{2} = \frac{1}{2} + \frac{n}{2n-2} = \frac{2n-1}{2n-2}$. Therefore, $\alpha_{mhc}(2) = \frac{3}{2} = 1.5$, $\alpha_{mhc}(3) = \frac{5}{4} = 1.25$, $\alpha_{mhc}(4) = \frac{7}{6} = 1.17$, and the product t-norm is close to the middle point of the hyperconjunction area:

$$\frac{\alpha_t(2)}{\alpha_{mhc}(2)} = 0.83, \quad \frac{\alpha_t(3)}{\alpha_{mhc}(3)} = 1, \quad \frac{\alpha_t(4)}{\alpha_{mhc}(4)} = 1.05.$$

The proximity to 1 justifies the status of the product t-norm as a "medium hyperconjunction." Therefore, the region $1 < \alpha < \alpha_t(n)$ can be denoted as "low hyperconjunction" and the region $\alpha_t(n) < \alpha < \alpha_{dc}(n)$ can be denoted as "high hyperconjunction." In these regions, stakeholders can select the most appropriate value of andness.

The andness of a GCD aggregator depends on the volume under its surface: $\alpha(n) = \frac{n-(n+1)V}{n-1}$. The volume under a dual aggregator is $1 - V$ and its andness is

$$1 - \omega_{dual}(n) = \alpha_{dual}(n) = \frac{n-(n+1)(1-V)}{n-1} = \frac{(n+1)V-1}{n-1}$$
$$= \omega(n) = 1 - \alpha(n).$$

Therefore, in a general case, $\omega_{dual}(n) = \alpha(n)$ and $\alpha_{dual}(n) = \omega(n)$.

Using duality, the "medium hyperdisjunction" is $y_{hd} = 1 - (1-x_1) \cdots (1-x_n)$. Since $\omega_d(2) = \frac{5}{4} = 1.25$, $\omega_d(3) = \frac{5}{4} = 1.25$, and $\omega_d(4) = \frac{59}{48} = 1.23$, it follows that the t-conorm can be used as a medium hyperdisjunction. Then, the region $1 < \omega < \omega_d(n)$ can be denoted as "low hyperdisjunction," and the region $\omega_d(n) < \omega < \omega_{dd}(n) = \frac{n}{n-1}$ as "high hyperdisjunction."

According to Table 11.2, in the region from the drastic conjunction to the drastic disjunction we have 4 nonidempotent hyperconjunction functions, 4 nonidempotent hyperdisjunction functions, and $K = 4k + 3$ idempotent versions of the GCD function. Thus, the total number of GCD function types is $N(k) = 4k + 11$. The precision of GCD functions depends on selected value of k: $N(1) = 15$, $N(3) = 23$, $N(5) = 31$, $N(7) = 39$. The value $k = 1$ can be used in beginners demo problems.

The value $k = 3$ can be used in most practical cases. The values $k > 3$ can be used in industrial problems that include highly trained decision support professionals.

References

DUJ15a Dujmović, J. 2015. Weighted Compensative Logic with Adjustable Threshold Andness and Orness. *IEEE Transactions on Fuzzy Systems* 23 (2): 270–290. https://doi.org/10.1109/TFUZZ.2014.2312018.

DUJ15b Dujmović, J. 2015. GCD Aggregators: Scales and Implementations. In *Proceedings of the IEEE International Conference on Fuzzy Systems*, August 2–5, 2015, Istanbul Turkey, Paper No. 15128.

DUJ18 Dujmović, J. 2018. *Soft Computing Evaluation Logic*. Wiley.

DYC93 Dyck, P.J. 1993. Quantitating Severity of Neuropathy. In Peripheral Neuropathy, 3rd ed, ed. P.J. Dyck, P.K. Thomas, J.W. Griffin and J.F. Poduslo, 686–697. Saunders Philadelphia.

DYC05 Dyck, P.J., R.A.C. Hughes, and P.C. O'Brien. 2005. Quantitating Overall Neuropathic Symptoms, Impairments, and Outcomes. In Peripheral Neuropathy, 4th ed, ed. P.J. Dyck and P.K. Thomas, 1031–1052. Philadelphia: Elsevier.

GRA06a Graham, R.C., and R.A.C. Hughes. 2006. A Modified Peripheral Neuropathy Scale: The Overall Neuropathy Limitations Scale. *Journal of Neurology, Neurosurgery & Psychiatry* 77: 973–976.

GRA06b Graham, R.C., and R.A.C. Hughes. 2006. Clinimetric Properties of a Walking Scale in Peripheral Neuropathy. *Journal of Neurology, Neurosurgery & Psychiatry* 77: 977–979.

HER96 Herndon, R.M. 2006. *Hand book of neurologic rating scales*. Second edition.Demos Medical Publishing, 2006.

HUN14 Hunink, M.G.M., M.C. Weinstein, et al. 2014. *Decision-Making in Health and Medicine: Integrating Evidence and Values*, 2nd ed. Cambridge University Press.

Chapter 12
Semantic Aspects: The Fundamental Concept of Adjustable Importance

This is a chapter about importance in a book about logic. The four main pillars of Graded Logic are the human percepts of graded truth, graded simultaneity, graded substitutability, and graded importance. Thus, the purpose of this chapter should be clear without an introductory explanation. However, since this book summarizes a half century of fight for some firm principles of logic and logic decision models, and this chapter explicitly supports the most critical of them, it is useful to start by restating the three main GL ideas. First, logic is an indispensable component of natural human reasoning, and therefore it must be developed in the context of modeling observable, measurable, modelable, and explainable human mental activities, particularly those that are necessary in decision-making, and focused on attaining important stakeholder's goals. Second, the percept of importance is unavoidably and provably present in natural human reasoning and decision-making, and no logic system should ignore the modeling of graded importance of all statements that have a graded degree of truth. Third, all decision methods must be compatible with natural human decision-making, and therefore they should have an explicit relationship with graded logic models, including the concept of importance and its quantification using weights.

For each stakeholder, all goals and interests have particular degrees of importance. The idiom "first things first" clearly reflects the presence of the percept of importance in natural human reasoning. Most human decision-making is motivated by the goal to focus on those activities that are considered the most important. Consequently, the degrees of truth in decision models are always paired with corresponding degrees of importance. One of fascinating questions in classical logic is how is possible that degrees of importance are for many centuries excluded from classical logic models and theories, regardless of their permanent presence and clear visibility in natural human reasoning. The situation is similar in non-classical logic [RAS74, PRI09], modal logic [HUG68], and even in early decision analysis books [KEE76, BUN84], where the book indexes do not include the words "importance," "degree of importance," or "weight," and the impacts of all degrees of truth (or utility) depend only on

J. Dujmović, *Graded Logic*, Intelligent Systems Reference Library 273,
https://doi.org/10.1007/978-3-031-88558-7_12

their magnitude. As opposed to that, among primary goals of Graded Logic is that all logic models must include degrees of importance in a way they are observable in natural human reasoning of specific stakeholder/decision-maker [DUJ15c, BEL02].

A simple way to introduce the concept of importance is to consider a GCD logic aggregator of two variables: $y = L(x, z), x \in [0, 1], z \in [0, 1], y \in [0, 1]$. Suppose that the aggregator gives equal status to both x and z: they are commutative, and the complete values of x and z affect the resulting suitability y. That can be easily changed if we decide to use not the whole values of variables x and z but only their fractions. Let $p \in [0, 1]$ and $q = 1 - p \in [0, 1]$ be multiplicative weights of the degrees of truth x and z. So, px denotes a p-fraction of x, and, qz denotes a q-fraction of z. E.g., if $p = 0.25$ and $q = 0.75$ then only 25% of the value of x and 75% of the value of z participate in the aggregation process $y = L(px, qz)$. Generally, that makes z three times more important than x. Thus, p and q have the status of *importance weights*, i.e., weights that a specific stakeholder/decision-maker uses to adjust a desired relative importance of selected statements in a propositional logic formula of the graded propositional calculus.

Importance weights are different from *logic weights* that are used in aggregators of the OWA family to adjust desired andness [YAG88, YAG97]. The arguments of OWA aggregators are sorted in decreasing order and the weights are selected to emphasize high values (in the case of disjunctive aggregators) or low values (in the case of conjunctive aggregators). Such logic weights adjust logic properties (andness or orness) and not the relative importance of individual arguments. Such weights can also be called *andness weights*. In this chapter we are interested in importance weights that reflect goals and interests of a specific stakeholder.

The Contributor-Aggregator Forms of GCD

The values px and qz in the aggregation process $y = L(px, qz)$ are special cases of the *contributor function* which is used in the following *contributor-aggregator form* of GCD [DUJ07a]:

$$y = A(C(x_1, W_1), \ldots, C(x_n, W_n)), \; A : \mathbb{R}^n \to [0, 1].$$

The *contributor function* $C_i = C(x_i, W_i)$ defines the weighted contribution of the ith input argument using the importance weight W_i. In most cases $0 < W_i < 1$, and sometimes $W_i \in \mathbb{R}$. Theoretically, instead of $W_i \in]0, 1[$, we could use $W_i \in [0, 1]$, but the weights are most frequently sum-normalized ($\sum_{i=1}^{n} W_i = 1$) and the meaning of $W_i = 0$ is that x_i is excluded from the impact on y; similarly, the meaning of $W_i = 1$ is that only x_i affects y and all other inputs are excluded from impact on y. So, $W_i \in \{0, 1\}$ are theoretical extremes and we prefer to use a more realistic condition $0 < W_i < 1, \; i = 1, \ldots, n$.

The aggregator function $A(C_1, \ldots, C_n)$ can have various forms compatible with the contributor function, as in the following examples.

Quasi-arithmetic mean:

$$C_i = C(x_i, W_i) = W_i F(x_i),\ A(C_1, \ldots, C_n) = F^{-1}\left(\sum_{i=1}^{n} C_i\right) = F^{-1}\left(\sum_{i=1}^{n} W_i F(x_i)\right),$$

$$0 < W_i < 1,\ i = 1, \ldots, n,\ \sum_{i=1}^{n} W_i = 1,$$

$F : [0, 1] \to \mathbb{R}$ (F is a continuous and strictly monotonic function).

Weighted power mean:

$$C_i = C(x_i, W_i) = W_i x_i^r,\ A(C_1, \ldots, C_n) = \left(\sum_{i=1}^{n} C_i\right)^{1/r} = \left(\sum_{i=1}^{n} W_i x_i^r\right)^{1/r},$$

$$0 < W_i < 1,\ i = 1, \ldots, n,\ \sum_{i=1}^{n} W_i = 1.$$

Arithmetic mean:

$$C_i = C(x_i, W_i) = W_i x_i,\ A(C_1, \ldots, C_n) = \sum_{i=1}^{n} C_i = \sum_{i=1}^{n} W_i x_i,$$

$$0 < W_i < 1,\ i = 1, \ldots, n,\ \sum_{i=1}^{n} W_i = 1.$$

Geometric mean:

$$C_i = C(x_i, W_i) = x_i^{W_i},\ A(C_1, \ldots, C_n) = \prod_{i-1}^{n} C_i = \prod_{i-1}^{n} x_i^{W_i}$$

$$0 < W_i < 1,\ i = 1, \ldots, n,\ \sum_{i=1}^{n} W_i = 1.$$

Hyperconjunctive product-power:

$$C_i = C(x_i, p_i) = x_i^{p_i},\ A(C_1, \ldots, C_n) = \prod_{i=1}^{n} C_i = \prod_{i=1}^{n} x_i^{p_i}$$

$$0 \leq p_i \leq n,\ i = 1, \ldots, n,\ \sum_{i-1}^{n} p_i = n.$$

Weighted conjunction with implicative weights [YAG87]:

$$C_i = C(x_i, v_i) = v_i \to x_i,\ A(C_1, \ldots, C_n) = \Lambda_{i=1}^{n} C_i = \Lambda_{i=1}^{n} (v_i \to x_i),$$

$$v_i \to x_i = \overline{v}_i \vee x_i,\ 0 \leq v_i \leq 1,\ i = 1, \ldots, n,\ max(v_1, \ldots, v_n) = 1.$$

The presented examples show three fundamental forms of combining importance weights and degrees of truth in contributor functions:

- Multiplicative: $W_i x_i$
- Exponential: $x_i^{p_i}$
- Implicative: $v_i \to x_i$

It is easy to see that all contributor functions are close to the idea of implication: importance plays the role of antecedent and high truth (or high suitability/satisfaction/quality) plays the role of consequent. That can be verbalized in several ways. The most direct implicative form is "it is not acceptable that an argument is important and not sufficiently satisfied," as follows:

$$\begin{aligned} importance \to satisfaction &= \overline{importance \wedge \overline{satisfaction}} \\ &= \overline{importance} \vee satisfaction. \end{aligned}$$

Indeed, if an input is important for a specific stakeholder, then it is desired and expected to be sufficiently satisfied. High importance implies the need for high satisfaction/truth.

In the case of high importance (1) and high satisfaction (1), the presented three contributors behave in the identical way: $1 \to 1 = 1 \times 1 = 1^1 = 1$. The implication and product behave the same way rejecting the combination of high importance (1) and low satisfaction (0): $1 \to 0 = 1 \times 0 = 0$. In addition, the implication and product create similar effects in the case of low importance: $0 \to 1 = 1$, what is a neutral element for conjunction, and $0 \times 1 = 0$, what is a neutral element for additive aggregation.

The presence of multiple forms of contributor-aggregator GCD automatically yields the question of their evaluation and comparison and selection of the best function. Unsurprisingly, that is not a simple problem, and Chap. 13 contains our solution to that problem.

Weight Normalization

The presented examples show three principal forms of weight normalization:

- Sum-one-normalized: $\sum_{i=1}^{n} W_i = 1$
- Count-normalized: $\sum_{i=1}^{n} p_i = n$
- Maximum-one-normalized: $\max(v_1, \ldots, v_n) = 1$.

The sum-one-normalized GCD is related to idempotency of means. Indeed, if we use the quasi-arithmetic means $y = F^{-1}\left(\sum_{i=1}^{n} W_i F(x_i)\right)$ and want to support idempotency $x_1 = \ldots = x_n = y = x \in [0, 1]$, then we have

$$F^{-1}\left(\sum_{i=1}^{n} W_i F(x)\right) = F^{-1}\left(F(x) \sum_{i=1}^{n} W_i\right) = x \quad \Rightarrow \quad \sum_{i=1}^{n} W_i = 1.$$

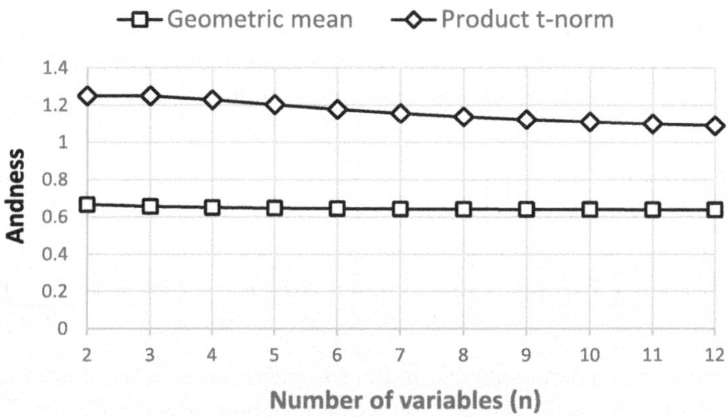

Fig 12.1 Andness of the geometric mean and the product t-norm

This form of sum-one-normalization holds for all descendants of the quasi-arithmetic mean, which (depending on function F) include a variety of means like weighted power means and their special cases (e.g., the arithmetic, geometric, harmonic, and quadratic means), the exponential means, and other similar aggregators.

The exponential weights are used in two important cases: the idempotent sum-one normalized weighted geometric mean $y = \prod_{i=1}^{n} x_i^{W_i}, \sum_{i=1}^{n} W_i = 1$, and the nonidempotent count-normalized product-power GCD $y = \prod_{i=1}^{n} x_i^{p_i}, \sum_{i=1}^{n} p_i = n$. Both the geometric mean and the product-power GCD are hard aggregators that support the annihilator 0. The fundamental difference between them is the sum of weights which determines the logic nature of these aggregators. By increasing the sum of weights, the andness significantly increases. The andness is computed for equal weights. So, we can take the function $y = \prod_{i=1}^{n} x_i^p$ and the volume below this surface inside $[0, 1]^{n+1}$ is $V_p = \prod_{i=1}^{n} \int_0^1 x_i^p dx_i = \left(\frac{1}{p+1}\right)^n$. The corresponding andness is $\alpha = \frac{n-(n+1)V_p}{n-1}$. For the geometric mean $p = 1/n$, $V_g = \left(\frac{n}{n+1}\right)^n$ and the resulting andness is $\alpha_g = \frac{n-(n+1)n^n/(n+1)^n}{n-1}$. For the product-power function (and the product t-norm) $p = 1$, $V_p = \left(\frac{1}{2}\right)^n$ and the andness is $\alpha_p = \frac{n-(n+1)/2^n}{n-1}$. Since $V_p < V_g$, it follows $\alpha_p > \alpha_g$, as shown in Fig. 12.1.

If $p = 1$, then the GCD aggregator $y = \prod_{i=1}^{n} x_i$ is the product t-norm. This aggregator is important because the inputs $x_1, \ldots, x_n$ can be interpreted as probabilities of independent events and then y is the probability of simultaneous occurrence of these independent events. Therefore, the product t-norm is a point of intersection between the probability theory and the Graded Logic. The weighted product GCD aggregator $y = \prod_{i=1}^{n} x_i^{p_i}, \sum_{i=1}^{n} p_i = n$ is a generalization of the t-norm that can be used in cases where we have strong justification for weighted aggregation at this high level of andness (see discussion presented later in this chapter). Of course, the weighted product GCD aggregator is no longer the t-norm but (because of its location in the proximity of the product t-norm) it approximately supports their main properties.

An interesting example of count-normalization of exponential importance weights is the gamma aggregator, introduced in [ZYS79] and analyzed in [ZIM80, DOM82, ZIM87, ZIM96]. This aggregator has the following multiplicative and additive forms:

$$y_{mul} = \left(\prod_{i=1}^{n} x_i^{p_i}\right)^{1-\gamma} \left(1 - \prod_{i=1}^{n} (1 - x_i)^{p_i}\right)^{\gamma},$$

$$y_{add} = (1-\gamma)\left(\prod_{i=1}^{n} x_i^{p_i}\right) + \gamma\left(1 - \prod_{i=1}^{n} (1 - x_i)^{p_i}\right), \quad 0 \le \gamma \le 1, \quad \sum_{i=1}^{n} p_i = n.$$

This aggregator uses a parameter γ (similar to orness) to provide a continuous interpolative transition from the hyperconjunctive weighted product-power $\prod_{i=1}^{n} x_i^{p_i}$ to its hyperdisjunctive dual $1 - \prod_{i=1}^{n} (1 - x_i)^{p_i}$. The importance weights $p_1, \ldots, p_n$ are count-normalized. In a special case $p_1 = \ldots = p_n = 1$ these are commutative aggregators in the range from t-norm to t-conorm. In such cases the andness of gamma aggregator varies in the range $[-0.25, 1.25]$. Unfortunately, special cases of gamma aggregators do not include fundamental operations of pure conjunction, pure disjunction and the centroid operator of logic neutrality based on the arithmetic mean. Consequently, the gamma aggregator cannot be used as a GCD, providing a generalization of the classic Boolean logic. The multiplicative version is strictly hard and the additive version is strictly soft. So, the transition from hard to soft aggregators cannot be performed by adjusting the parameter γ. If $x_i = 0, x_j = 1, i \neq j$ we have $y_{mul} = 0$, and $y_{add} = \gamma$, eliminating the impact of other inputs. This also illustrates that gamma aggregators cannot model a hard partial disjunction, and the multiplicative gamma aggregator cannot model the soft partial conjunction. A very detailed analysis of the restrictive properties of gamma aggregators can be found in [DUJ18].

The maximum-one-normalization is used with implicative weighting. The weights can be initially defined as nonnegative real numbers, and the normalized weights are then obtained by dividing all weights by the weight that has the maximum value. The resulting conjunction of contributors $y = \wedge_{i=1}^{n}(v_i \to x_i), 0 \le v_i \le 1$ can be based on various models of implication. The basic version called weighted conjunction [YAG87] is

$$y_{WC}(\boldsymbol{x}, \boldsymbol{v}) = (\bar{v}_1 \vee x_1) \wedge \cdots \wedge (\bar{v}_n \vee x_n).$$

Using De Morgan's law, the corresponding weighted disjunction is

$$y_{WD}(\boldsymbol{x}, \boldsymbol{v}) = \overline{y_{WC}(\bar{\boldsymbol{x}}, \boldsymbol{v})} = (v_1 \wedge x_1) \vee \cdots \vee (v_n \wedge x_n).$$

The most important weight is 1, and those that are less important are less than 1. For simplicity, suppose that the most important is the first input x_1. Then $v_1 = 1, \bar{v}_1 = 0$, and

$$y_{WC}(\boldsymbol{x}, \boldsymbol{v}) = x_1 \wedge \cdots \wedge (\overline{v}_n \vee x_n) \leq x_1,$$
$$y_{WD}(\boldsymbol{x}, \boldsymbol{v}) = x_1 \vee \cdots \vee (v_n \wedge x_n) \geq x_1.$$

It is easy to see that these aggregators are idempotent:

$$x \wedge \cdots \wedge (\overline{v}_n \vee x) = x,$$
$$x \vee \cdots \vee (v_n \wedge x) = x, 0 \leq x \leq 1.$$

A very low importance yields $\overline{v}_i \cong 1$ and makes the corresponding degree of satisfaction insignificant (i.e., excluded) because for any value of x_i we have $\overline{v}_i \vee x_i \cong 1$. Of course, in Graded Logic the crucial question is what happens in the case of moderate importance, i.e., in all cases that are between the extreme values of importance. Such cases are regularly encountered in the commonsense human reasoning. Looking at the classic implication $v_i \to x_i = \overline{v}_i \vee x_i$ in the case $v_i = 0.5$, the contributor $0.5 \vee x_i$ prevents the impact of the degree of truth x_i as long as $x_i \in [0, 0.5]$. In all applications of implicative weights this property can be questionable, and in the case of modeling evaluation reasoning it is generally unacceptable for two reasons: (1) it is not reasonable to make the aggregated value insensitive to variations of input, and (2) for $\overline{v}_i \geq x_i$ the implicative contributor is the negated importance $v_i \to x_i = \overline{v}_i$ what is a constant value hiding/excluding the impact of a variable input x_i, as long as the input is below the threshold $\overline{v}_i$. As opposed to that, the multiplicative contributor $W_i x_i$ is more convenient because it offers a steady impact of input x_i regardless of its value.

If all weights are the same, then $v_1 = \ldots = v_n = 1$, and the weighted conjunction and the weighted disjunction become the pure conjunction and the pure disjunction. Several aggregation functions (e.g., the weighted power means) have a desirable property of continuous transitions from the weighted soft graded conjunction to the weighted hard graded conjunction and then to pure conjunction as the most conjunctive idempotent aggregator. Such transition is not available for the weighted conjunction/disjunction. On the other hand, there are various forms of implication: in addition to the initial $\max(1 - v, x)$, the most interesting implications are x^v (Yager [YAG04a]) and $1 - v(1 - x)$ (Reichebach [REI34]). A detailed analysis of applicability of various implicative aggregators, including Larsen's AIWA aggregator [LAR03, LAR09], can be found in [DUJ18]. Various forms of GCD in the context of GL is further studied in Chaps. 13 and 15.

The Impact of Small Weights

Multiplicative weights are the simplest and the most frequent form used in quasi-arithmetic and arithmetic means. They can be easily selected from pairwise comparison or arguments. For example, the aggregator $y = 0.1x_1 + 0.2x_2 + 0.3x_3 + 0.4x_4$ shows that x_2 is considered two times more important than x_1 and x_4 is two times more important than x_2. Only 10% of suitability x_1 affects the output y, while 40% of suitability x_4 affects the value of y. This shows that it is necessary to carefully consider inputs that have small importance, because such inputs can be candidates for exclusion from additive logic aggregation criteria. In the case of arithmetic mean and other soft aggregators in its vicinity, the average weight of an input is $1/n$ (0.25 in the above example with four variables). However, if $n = 100$, then the average impact of a single input is only 1%, and if the most important inputs have the impact of a few percent, then some low importance inputs will produce the impact significantly below 1%. That makes the presence of such insignificant inputs highly questionable. In addition, that also shows that the fundamental drawback of all additive scoring and additive utility models is the limited number of inputs. The only way to eliminate such a drawback is by using combinations of soft and hard GCD aggregators to create and use justifiable formulas of graded propositional calculus.

The geometric mean and other hard GCD functions of higher andness show more problems caused by low weights. The hard aggregation and a high andness indicate the stakeholder's request that all inputs must be highly satisfied and consequently such inputs are implicitly classified as very influential and *important*. On the other hand, assigning a very low weight to one or more inputs of a high andness hard GCD shows an opposite claim, that such inputs are *not important*. Therefore, in a general case, a combination of high andness (or high orness) and a very low weight is an obvious contradiction, which inevitably must cause undesirable effects.[1] The primary problem is called hypersensitivity, and we can illustrate it using GCD implemented as the geometric mean.

The geometric mean is the lowest-andness hard special case of the power mean. In the case of two variables, the geometric mean $z = \sqrt{xy}$ has the andness $\alpha = \frac{2}{3}$ (i.e. a relatively low value of 67%). It is a hard aggregator obtained from the power mean $z = (0.5x^r + 0.5y^r)^{1/r}$ in the case where $r \to 0$. If $r < 0$ all resulting aggregators (like the harmonic mean for $r = -1$) are hard and their andness is higher than the andness of the geometric mean. So, the least critical problem of hypersensitivity can be seen in the 5% weight example $z = x^{0.05}y^{0.95}$ where the sensitivity is $\frac{\partial z}{\partial x} = 0.05\left(\frac{y}{x}\right)^{0.95}$. If $y = 1$, then for small values of x we can have a very high sensitivity, not observable in natural human reasoning. For example, for $x = 0.01, 0.02, 0.03, 0.04, 0.05$ we have $z = 0.79, 0.82, 0.84, 0.85, 0.86$; so, for a 1% change of input x in the range $[0, 0.01]$ we have the 79% change of the output z. Such a high sensitivity is not observable in natural human reasoning, and consequently it should not be used in graded logic. For GCD aggregators with higher andness than the geometric mean, the hypersensitivity for low weights is even more critical. Therefore, the combinations of low weight and high andness/orness should be reserved for binary arguments or arguments that have no values is the range of hypersensitivity.

[1] There are exceptions, and one of them is the situation where the input that has a very low weight is a binary (0/1) argument. In such cases, if the input is 0, regardless the low weight, the output is also 0 because the aggregator is hard. On the other hand, if the binary input is 1, then its low weight is leaving space for other arguments to produce more influential effects. So, in this situation the stakeholder/decision-maker is interested only to check that the binary argument is not 0. That is frequently used in evaluation problems where some legal condition must be satisfied (e.g. a delivered equipment must be brand new, or a company must provide 24/7 maintenance service, or a hospital must have some advanced medical equipment, and similar yes/no conditions).

The Percept of Overall Importance and Its Fundamental Components

In decision-making, both intuitive and formalized, stakeholders use GCD functions to aggregate two or more input degrees of truth and generate a resulting degree of truth. Based on stakeholder's goals and interests, each input creates a percept of overall importance. The overall importance reflects the capability of an input to support the attainment of specific stakeholder's goals. More important inputs contribute to stakeholder's goals more than less important inputs. Unsurprisingly, the overall importance is a compound percept, and it can have various components. Typical components that contribute to the percept of overall importance include the indispensable functionality, dominating other inputs, high cost, ability to substitute other inputs, and others. In the graded perceptual calculus problems, we are interested in identifying and controlling the main components of the percept of overall importance. This section is devoted to that crucial percept.

If two degrees of truth are equal, then one that is more important must produce a higher impact on the resulting aggregated degree of truth. Suppose that x, y and z are degrees of truth and z is obtained as a GCD of x and y. If x and y are equally important, then we write $z = x \diamond y$ and assume commutativity $x \diamond y = y \diamond x$. We have equal impact[2] of x and y: for $y = t \in [0, 1]$ the impact $\frac{\partial z}{\partial x}$ is the same as the impact $\frac{\partial z}{\partial y}$ for $x = t$. It is natural to expect that the magnitude of impact directly affects the intuitive percept of importance (and vice versa). In the simplest example, if the andness is $\alpha = 0.5$, and GCD is the arithmetic mean $z = 0.5x + 0.5y$, then $\frac{\partial z}{\partial x} = \frac{\partial z}{\partial y} = 0.5$. For higher andness, if GCD is the pure conjunction $z = x \wedge y$, then the andness is $\alpha = 1$ and for $x < y$ we have $\frac{\partial z}{\partial x} = 1$. In other words, the percept of importance of an argument is higher for high andness than for medium andness. Similarly, if GCD is the pure disjunction $z = x \vee y$, and $x > y$, then $\frac{\partial z}{\partial x} = 1$.

If the andness increases and approaches 1, the inputs become increasingly influential and the percept of importance felt by decision-makers naturally increases. Finally, if $\alpha = 1$, each input has the ultimate power to reduce or annihilate the overall resulting suitability. Likewise, if the orness increases and approaches 1, the inputs again become increasingly influential and the percept of importance that decision-makers feel increases. Finally, if $\omega = 1$, each input has the ultimate compensatory power to increase the overall resulting suitability. Consequently, there is an observable percept of the collective *logic importance* of inputs that is proportional to $|\alpha - \omega|$. For hard conjunctions, the arguments have the "*must have*" status, and that is an obvious reason why they affect the percept of importance in a strong and positive way. Likewise, in the case of hard disjunctions, the arguments have the "*enough to have*" status, and again, that makes them significant and capable of affecting the percept of importance in a strong and positive way, shown in Fig. 12.2. That is also

[2] The general concept of nondecreasing monotonicity permits the use of aggregators that have discontinuities of value, or discontinuities of their derivatives. In practice, however, we predominantly use aggregators that are continuous and differentiable functions; in such cases the nondecreasing monotonicity can be simply expressed using partial derivatives, and we use such notation here and in other places in this book.

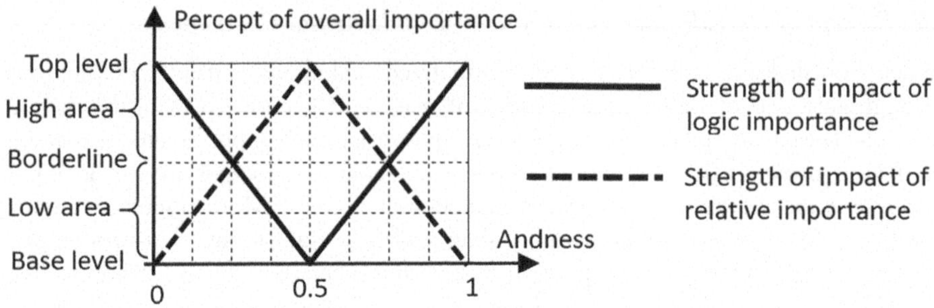

Fig. 12.2 The strength of impact of logic and relative importance

one of the central aspects of the logic duality of GCD functions. The base level corresponds to logic neutrality and the top level to *min* or *max* functions.

The logic importance of arguments is obviously a *collective property* of all GCD inputs. The strength of impact of logic importance is proportional to $|\alpha - \omega|$, as illustrated in Fig. 12.2. In almost all applications of GCD, for $\alpha = 1$ we have the classic unweighted conjunction $\min(x_1, ..., x_n)$, and for $\omega = 1$ the classic unweighted disjunction $\max(x_1, ..., x_n)$; thus, it is possible to conclude that in those cases the logic importance is dominant to the extent that it is the only contributor to the percept of overall importance of GCD inputs. In Fig. 12.2, the base level corresponds to the logic neutrality (the arithmetic mean) and the top level to min or max functions.

In each graded logic propositional calculus problem based on GCD aggregators, decision-makers intuitively create a ranking of inputs of each GCD aggregator according to the percept of their importance. Of course, that is the percept of *overall importance*, which is a compound percept. Indeed, the logic importance is not the only contributor to the overall percept of importance. Decision-makers are also exposed to the effects of relative importance of arguments. Based on their goals and interests, stakeholders naturally compare and rank inputs according to their relative importance. For example, the input x_i can be considered equally important (or two times more/less important) than another input x_j at any level of logic importance.

Since GCD aggregators always have two or more arguments, they share the same andness and orness (i.e. the same logic importance), but their individual impacts do not have to be equal. Indeed, the impact of individual arguments is usually different: a GCD aggregator $z = F(x_1, ..., x_n)$ is generally noncommutative, i.e., structured so that $\frac{\partial z}{\partial x_i} \neq \frac{\partial z}{\partial x_j}, i \neq j$. If $\frac{\partial z}{\partial x_i} > \frac{\partial z}{\partial x_j}$, then the argument x_i is relatively more important than a different argument x_j. In the simplest case of two variables, x and y are generally not equally important, and that can be symbolically written as $z = Wx \diamond \overline{W}y, W \in]0, 1[, \overline{W} = 1 - W, W \neq 0.5$. Here W denotes the degree of relative importance (the *weight*) of the argument x and its complement $\overline{W}$ denotes the degree of relative importance of the argument y (we assume that the meaning of symbolic notation $x \diamond y$ is commutative aggregation, the same as the meaning of symbolic notation $0.5x \diamond 0.5y$). If $W > \overline{W}$, (i.e., $W > 0.5$) then x is relatively more important than y.

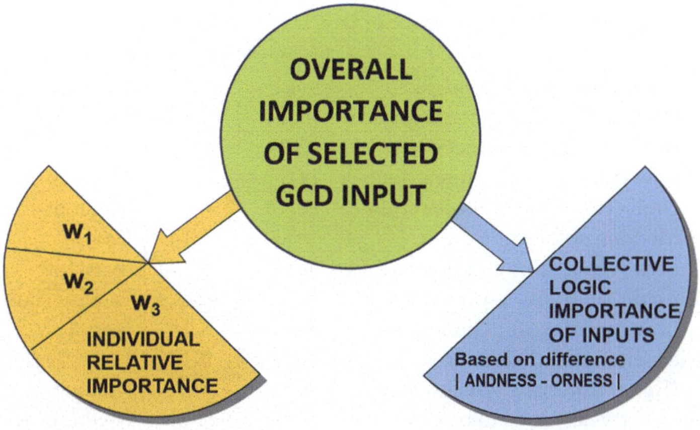

Fig. 12.3 Decomposition of the percept of overall importance of GCD inputs

In the simplest case of arithmetic mean $z = Wx + \overline{W}y$ the weights directly denote the magnitude of impact: $W = \frac{\partial z}{\partial x}$, $\overline{W} = \frac{\partial z}{\partial y}$.

The logic importance is a collective contributor to the compound percept of importance, while the relative importance is an individual property of each GCD input. The logic importance and the relative importance simultaneously contribute to the compound percept of overall importance, as illustrated in Fig. 12.3. It is important to note that in applications trained decision-makers can select weights (individual relative importance) and the degree of andness/orness (collective logic importance) separately, and in any order. That indicates a degree of independence of logic and relative importance, as well as a natural human capability to intuitively decompose and aggregate various percepts of importance.

If $\alpha = \omega = 0.5$, then the GCD is a logic neutrality, and the corresponding percept of logic importance has the lowest base intensity that corresponds to the arithmetic mean, as shown in Fig. 12.2. In such a case the relative importance has the highest impact on the percept of overall importance. If $\alpha \neq \omega$ then the strength of impact of logic importance increases compared to the base level of the arithmetic mean. Since for $|\alpha - \omega| = 1$ the logic importance becomes the only component of the percept of overall importance, it follows that the strength of impact of relative importance is not constant but proportional to $1 - |\alpha - \omega|$, as shown in Fig. 12.2. Thus, the percept of overall importance is a combination of the logic importance, which is a dominant contributor, and the relative importance, which is an auxiliary contributor that has the strength of impact that complements the strength of impact of the dominant contributor.

Logic Importance and Relative Importance in Graded Propositional Calculus Formulas

In a summary of this chapter, we identified three fundamental observable percepts of the importance of inputs of GCD logic aggregators:

Table 12.1 Verbalizing the logic importance of UGCD.15 idempotent aggregators

Symbol	Name	Andness	Orness	Logic importance
C	Pure (full) conjunction	1	0	Highest
HC+	High (strong) hard conjunction	13/14	1/14	Very high
HC	Medium hard conjunction	12/14	2/14	High
HC-	Low (weak) hard conjunction	11/14	3/14	Medium-high
SC+	High (strong) soft conjunction	10/14	4/14	Medium-low
SC	Medium soft conjunction	9/14	5/14	Low
SC-	Low (weak) soft conjunction	8/14	6/14	Very low
A	Logic neutrality	1/2	1/2	Lowest (base level)
SD-	Low (weak) soft disjunction	6/14	8/14	Very low
SD	Medium soft disjunction	5/14	9/14	Low
SD+	High (strong) soft disjunction	4/14	10/14	Medium-low
HD-	Low (weak) hard disjunction	3/14	11/14	Medium-high
HD	Medium hard disjunction	2/14	12/14	High
HD+	High (strong) hard disjunction	1/14	13/14	Very high
D	Pure (full) disjunction	0	1	Highest

(1) *the collective logic importance,*
(2) *the individual relative importance,* and
(3) *the compound overall importance.*

The overall importance is a natural percept created by a decision-maker when thinking about the total impact of a specific input of a GCD aggregator. The overall importance of each input is a compound percept, obtained as a superposition of the collective logic importance of all inputs, and the individual relative importance of the specific input. Modeling importance is not a trivial problem that has a unique solution, and next chapter presents methodology for creating GCD aggregators. Unsurprisingly, the validation of graded logic models must be based on observing natural human reasoning and experiments with human subjects (Chap. 14).

If $\alpha > \omega$ and the andness increases, then the presence of high input truth values becomes necessary for all GCD inputs, increasing their importance status towards the high importance of mandatory inputs of hard aggregators. In a dual way, if $\omega > \alpha$ and the orness increases, then each input value gets the power to compensate for the low values of other inputs. That also yields the status of importance that culminates with the hard disjunctive aggregators where each input becomes very important because it is powerful enough to compensate or substitute all other inputs of a disjunctive aggregator. Consequently, $|\alpha - \omega|$ is an appropriate indicator of the collective logic importance of inputs, above the base level of the arithmetic mean. The logic importance for UGCD.15 aggregators can be verbalized as shown in Table 12.1

The overall importance is not directly visible in mathematical models of GCD. For example, if we use the weighted power mean $y = \left(w_1 x_1^r + \ldots + w_n x_n^r\right)^{1/r}$, then the logic importance is adjusted using the exponent r and the relative importance is adjusted separately and independently using the weights $w_1, \ldots, w_n$. So, the overall importance is a combined effect of superposition of effects caused by the selected

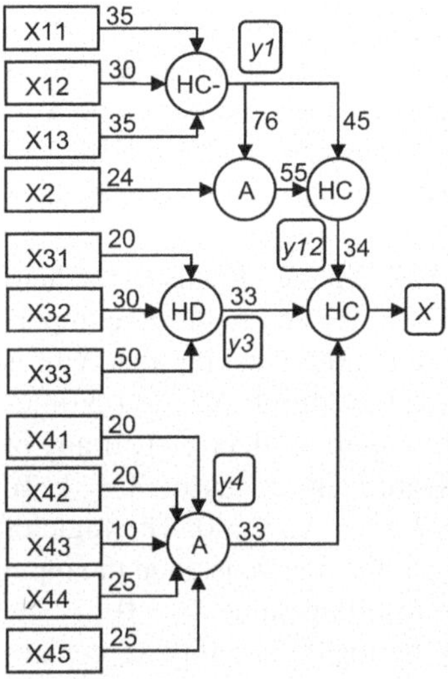

$$
\begin{aligned}
&X(X11, X12, X13, X2, X31, X32, X33, X41, X42, X43, X44, X45)\\
&= \boldsymbol{HC}(0.34y12, 0.33y3, 0.33y4)\\
&= \boldsymbol{HC}(0.34\boldsymbol{HC}(0.45y1, 0.55\boldsymbol{A}(0.76y1, 0.24X2)),\\
&\quad 0.33y3, 0.33y4)\\
&= \boldsymbol{HC}(0.34\boldsymbol{HC}(0.45y1, 0.55\boldsymbol{A}(0.76y1, 0.24X2)),\\
&\quad 0.33\boldsymbol{HD}(0.2X31, 0.3X32, 0.5X33),\\
&\quad 0.33\boldsymbol{A}(0.2X41, 0.2X42, 0.1X43, 0.25X44, 0.25X45))\\
&= \boldsymbol{HC}(0.34\boldsymbol{HC}(0.45\boldsymbol{HC}^{-}(0.35X11, 0.3X12, 0.35X13),\\
&\quad 0.55\boldsymbol{A}(0.76\boldsymbol{HC}^{-}(0.35X11, 0.3X12, 0.35X13),\\
&\quad 0.24X2)),\\
&\quad 0.33\boldsymbol{HD}(0.2X31, 0.3X32, 0.5X33),\\
&\quad 0.33\boldsymbol{A}(0.2X41, 0.2X42, 0.1X43, 0.25X44, 0.25X45))
\end{aligned}
$$

Fig. 12.4 A sample formula of graded propositional calculus (12 inputs and 6 GCD aggregators)

values if exponent r and the weights $w_1, \ldots, w_n$. Since the overall importance is not explicitly visible, we face the problem of decomposing the percept of overall importance in its components and then using the components to create GCD aggregators that are used in propositional calculus formulas. That can be done intuitively (after an appropriate training and experience with andness-directed aggregators), or more formally, using the importance decomposition method; both approaches are presented in Chap. 13.

The result of various techniques for GCD parameter adjustment are GCD aggregators, individually adjusted to satisfy goals and interests of specific stakeholder/decision-maker. The adjusted aggregators are then combined to create a compound propositional calculus formula that reflects stakeholder's needs. Such an example is shown in Fig. 12.4 where we show graphically and analytically a propositional calculus formula that uses six adjusted GCD aggregators. We compute the resulting value X by aggregating 12 inputs denoted $X11, X12, X13, X2, X31, X32, X33, X41, X42, X43, X44$, and $X45$.

For the logic aggregation of degrees of truth, we use GCD aggregators from the UGCD.15 set, introduced in Fig. 11.2 and used in Table 12.1. Each GCD aggregator is denoted using circle containing the symbol of desired andness/orness; the relative importance of inputs is denoted using percent weights that have the sum 100%. For example, the intermediate degree of truth $y1$ is computed using the low hard conjunction aggregator *HC-* which has the medium-high logic importance (slightly above medium and slightly below high).[3] The relative importance of inputs $X11$ and

[3] Note that the UGCD.15 set of aggregators does not include the medium graded conjunction (andness = 0.75) and the medium graded disjunction (andness = 0.25). In the case of UGCD.15, all hard graded conjunctions have andness above 0.75, and all soft graded conjunctions have andness

$X13$ is the same (35%) and relative importance of input $X12$ is little lower (30%). So, the percept of overall importance of inputs $X11, X12, X13$ can be verbalized as slightly below high. The path from inputs $X11, X12, X13$ to the output value X is going through hard conjunctive aggregators (*HC*), and consequently $X11, X12, X13$ are important mandatory inputs: if any of them is 0 the output is $X = 0$ regardless the values of all other inputs.

All other inputs are optional. For example, $X31, X32, X33$ are inputs to the hard disjunction operator (*HD*) that has high compensative power and high logic importance. The relative weights of inputs are 20%, 30%, and 50% showing that $X33$ is considered equally important as $X31$ and $X32$ together. In addition, $X32$ is 1.5 times more important than $X31$, and $X33$ is 2.5 times more important than $X31$. If any of inputs $X31, X32, X33$ is completely true (1), that is sufficient to cause $y3 = 1$. To cause $y3 = 0$, we need all three zero inputs: $X31 = X32 = X33 = 0$. The logic of this formula shows that this is not acceptable, because if that happens, then the input $y3 = 0$ applied to the final HC aggregator causes the resulting value $X = 0$.

The lowest logic importance in this example has the logic neutrality aggregator that computes $y4 = 0.2X41 + 0.2X42 + 0.1X43 + 0.25X44 + 0.25X45$. Therefore, the overall importance of inputs $X41$, $X42$, $X43$, $X44$, and $X45$ is expressed with weights 20%, 20%, 10%, 25% and 25%. Some of these inputs can be 0 without significantly affecting the output X. Only if all of them are simultaneously 0, then $y4 = 0$ and $X = 0$.

Hyperconjunction and Hyperdisjunction: With Or Without Weights?

Most decision models use importance-weighted idempotent aggregators. In such aggregators, the andness α and the orness ω are inside the [0,1] range. Significantly less frequently, decision models use hyperconjunction and hyperdisjunction where $|\alpha - \omega| > 1$. As the andness or orness approaches 1 and increases above 1, the impact of logic importance increases, and the impact of relative importance decreases or disappears. If $|\alpha - \omega| = 1$ the logic aggregator becomes the pure conjunction $\min(x_1, \ldots, x_n)$ or the pure disjunction $\max(x_1, \ldots, x_n)$, which are commutative aggregators where all arguments have the same relative importance. Weighted conjunction and weighted disjunction have properties that can be questionable and infrequently appear in applications. So, the question is whether models of hyperconjunction and hyperdisjunction should support importance weights.

Most frequent reasoning is that if the high andness/orness-based commutativity is achieved already for pure conjunction/disjunction, then it is justifiable to continue in the same way above that threshold. That opens the door for the applicability of commutative t-norms/conorms and commutative interpolative aggregators we study in Chap. 15. The situations where high andness or orness are not sufficient to eliminate the impact of different relative importance of arguments are exceptional but not

below 0.75 (and above 0.5). Likewise, all soft graded disjunctions have andness above 0.25 (and below 0.5) and all hard graded disjunctions have andness below 0.25. That is the reason for using labels medium-high and medium-low.

impossible. Thus, it is useful to have weighted hyperconjunction/hyperdisjunction, such as count-normalized weighted products, discussed earlier in this chapter.

References

BEL02 Belton, V., and T.J. Stewart. 2002. *Multiple Criteria Decision Analysis: An Integrated Approach*. Kluwer Academic Publishers.

BUN84 Bunn, D.W. 1984. *Applied Decision Analysis*. McGraw-Hill Book Company.

DOM82 Dombi, J., and P. Zysno, Comments on the Gamma-Model. In *Cybernetics and Systems Research*, ed. R. Trappl, 711–714. North-Holland.

DUJ07a Dujmović, J., and H.L. Larsen. 2007. Generalized Conjunction/Disjunction. *International Journal of Approximate Reasoning* 46: 423–446.

DUJ15c Dujmović, J. 2015. An Efficient Algorithm for General Weighted Aggregation. In *Proceedings of the 8th International Summer School on Aggregation Operators (AGOP2015)*, ed. M. Baczynski, B. De Baets, and R. Mesiar, 115–120. University of Silesia, ISBN: 978-83-8012-519-3.

DUJ18 Dujmović, J. 2018. *Soft Computing Evaluation Logic*. Wiley.

HUG68 Hughes, G.E., and M.J. Cresswell. 1968. *An Introduction to Modal Logic*. London: Methuen and Co Ltd.

KEE76 Keeney R.L., and H. Raiffa. 1976. *Decisions with Multiple Objectives/Preferences and Value Tradeoffs*. Wiley.

LAR03 Larsen H.L. 2003. Efficient Andness-Directed Importance Weighted Averaging Operators. *International Journal of Uncertainty, Fuzziness and Knowledge-Based Systems* 12 (Suppl.): 67–82.

LAR09 Larsen H.L. 2009. Multiplicative and Implicative Importance Weighted Averaging Aggregation Operators with Accurate Andness Direction. In: *Proceedings of the Joint 2009 International Fuzzy Systems Association World Congress and 2009 European Society of Fuzzy Logic and Technology Conference*, Lisbon, Portugal, July 20–24, 2009, ed. J.P. Carvalho, D. Dubois, U. Kaymak and J.M. da Costa Sousa, 402–407. ISBN: 978-989-95079-6-8.

PRI09 Priest, G. 2009. *An Introduction to Non-Classical Logic*, Second Edition reprinted with corrections. Cambridge University Press.

RAS74 Rasiowa, H. 1974. *An Algebraic Approach to Non-classical Logics*. PWN and North-Holland.

REI34 Reichenbach, H. 1934. Wahrscheinlichtkeitslogic. *Erkenntnis* 5 (1–3): 37–43.

YAG04a YAG04a Yager, R.R. 2004. On Some New Classes of Implication Operators and Their Role in Approximate Reasoning. *Information Science* 167: 193–216.

YAG87 Yager, R. 1987. A Note on Weighted Queries in Information Retrieval Systems. *Journal of the American Society for Information Science* 38 (1): 23–24.

YAG88 Yager, R. R. 1988. On Ordered Weighted Averaging Aggregation Operators in Multi-Criteria Decision-making. *IEEE Transactions on SMC* 18: 183–190.

YAG97 Yager, R.R., and J. Kacprzyk (eds.). 1997. *The Ordered Weighted Averaging Operators—Theory and Applications*. Kluwer Academic Publishers.

ZIM87 Zimmermann, H-J. 1987. *Fuzzy Sets, Decision-Making, and Expert Systems*. Kluwer Academic Publishers.

ZIM96 Zimmermann, H-J. 1996. *Fuzzy Set Theory and Its Applications*. New York: Springer Science+Business Media.

ZIM80 Zimmermann, H.-J., and P. Zysno. 1980. Latent Connectives in Human Decision-making. *Fuzzy Sets and Systems* 4: 37–51.

ZYS79 Zysno, P. 1979. *One Class of Operators for the Aggregation of Fuzzy Sets*. Amsterdam: EURO III Congress.

Chapter 13
A Third Look at GCD: Selection of GCD Aggregators

In this chapter we present techniques for selection of GCD aggregators and their parameters. That is a central topic necessary for all applications of Graded Logic. Indeed, all applications use graded propositional calculus formulas that are based on superposition of carefully adjusted GCD aggregators and (in some cases) negation (e.g., see Fig. 12.4). To select a correct version of a GCD aggregator, it is necessary to adjust its parameters, i.e., the weights and andness. We assume that these parameters are explicitly visible and easily adjustable. Each GCD aggregator is identified by its name which defines a constant value of andness. The weights are associated with input arguments and explicitly visible. The design of such aggregators is presented in Chap. 15.

In the case of n inputs, there are $n + 1$ adjustable parameters of each andness-directed importance-weighted GCD aggregator: the desired andness α, and the set of importance weights $W_1, \ldots, W_n$, shown in Fig. 13.1. This chapter presents methods for selecting GCD parameters in the following five sections:

- Direct specification of GCD parameters.
- The central group of dualized idempotent GCD aggregators.
- Selecting the precision of GCD aggregators.
- Importance decomposition method.
- GCD neuron training method.

Direct Specification of GCD Parameters

According to Fig. 13.1, we must select the desired *andness* and the *importance weights* of the GCD aggregator, and Chaps. 11 and 12 include preliminary material that helps in the selection process. For trained decision analysts, the GCD parameters can be directly selected, based on the specific stakeholder's needs and the knowledge of properties of available aggregators. In this process, the relative importance of input arguments is independent from the desired andness. Consequently, the weights can be selected before or after selecting the desired andness. It is also possible to use

J. Dujmović, *Graded Logic*, Intelligent Systems Reference Library 273,
https://doi.org/10.1007/978-3-031-88558-7_13

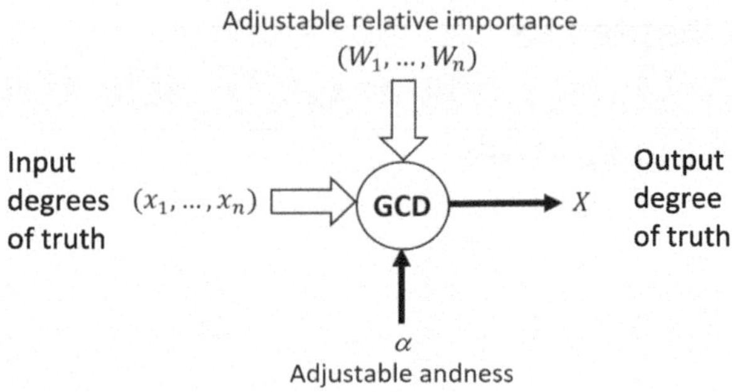

Fig. 13.1 Adjustable parameters of andness-directed importance-weighted GCD aggregators

techniques (presented below) that simultaneously select both the andness and the weights.

The weights are used in the case of general noncommutative GCD aggregators, and they reflect the percepts of individual importance of input arguments $x_1, \ldots, x_n$. Therefore, each GCD aggregator has $n + 1$ parameters that must be provided by the stakeholder/decision-maker. That yields the fundamental requirement for analytic models of GCD aggregators: the parameters must be *explicitly visible and easily adjustable* and not fixed or hidden in some mathematical structure that indirectly determines α and $W_1, \ldots, W_n$. The explicit visibility of all parameters is called the *parameter-directedness* [DUJ21b]. If the andness α is an explicitly visible and easily adjustable parameter of GCD, then such a GCD model is called an *andness-directed aggregator* [LAR03, LAR09, DUJ18, DUJ19, DUJ21b, TOR22].

The direct selection of GCD parameters is based on answering the following questions:

1. What is the type of aggregator? Conjunctive, disjunctive or neutral?
2. Is the idempotency supported or not supported?
3. Are annihilators supported or not supported?
4. What is the desired andness α?
5. What are the desired importance weights $W_1, \ldots, W_n$?

The first three questions are used to restrict the desired range of andness, and the fourth question is used to precisely determine the value of andness inside the selected range, as exemplified in Fig. 13.2. Conjunctive properties are the request for simultaneous high degrees of truth of all inputs, resulting in a high negative impact of any input that has a low degree of truth. The disjunctive properties are the request for high compensability of inputs, where a high truth of any input can be sufficient to significantly increase the resulting output degree of truth.

As discussed in Chap. 9, both conjunctive and disjunctive properties are simultaneously present in GCD aggregators. If $\alpha > 0.5$, then the GCD is conjunctive, i.e., conjunctive properties of GCD are stronger (more present) than disjunctive properties. Similarly, if $\alpha < 0.5$, then the GCD is disjunctive, i.e., disjunctive properties of GCD are stronger than conjunctive properties. In the case of neutrality ($\alpha = 0.5$), the conjunctive and disjunctive properties are perfectly balanced, i.e., equally present.

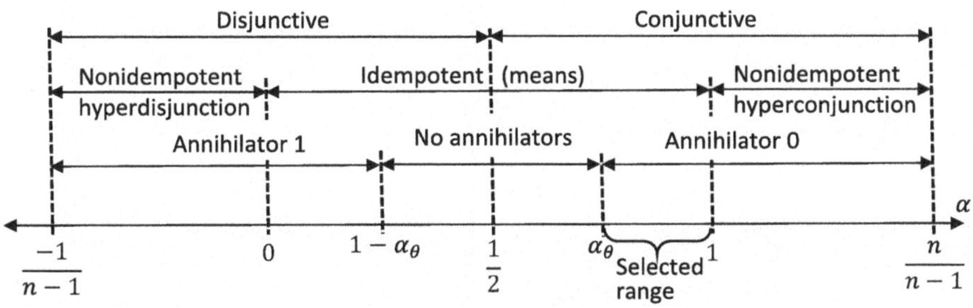

Fig. 13.2 Selecting a range of andness (e.g.: conjunctive, idempotent, with annihilator 0)

Therefore, the first question restricts the range of andness to one half of the total range of andness $\alpha \in \left[-\frac{1}{n-1}, \frac{n}{n-1}\right]$. The next question about idempotency decides whether the desired aggregator belongs to the central group of aggregators that satisfy internality and are modeled using means.

In applications, idempotent aggregators are the most frequently used. In problems where decision-makers use idempotent aggregators, it is necessary to select whether annihilators are supported (hard aggregators) or not supported (soft aggregators). Another important choice is the appropriate precision in selecting the degree of andness/orness. All these decisions are natural, consistent with human reasoning, and can be conveniently specified in a verbalized form illustrated in Table 13.1. The verbalization creates five types of GCD, based on simple concepts "*must have*," "*nice to have*," and "*enough to have.*" Obviously, the group "*must have*" denotes necessary positive satisfaction of all inputs. The meaning of the "*nice to have*" group is that it is very desirable, but not necessary, to have satisfied all inputs; the cases where some inputs are not satisfied are moderately penalized, but still acceptable. Then, in the case of UGCD.15, inside each group, the decision-maker selects one of three (or four) available degrees of strength: *low*, *medium*, *high* (and *highest*). Obviously, the selections among a small number of options (3–5) are simple and easily explainable.

Let us now consider the example shown in Fig. 13.2 where the first two questions specify that the desired GCD aggregator is conjunctive and idempotent. Then, the next step is to decide whether the annihilator 0 is supported (hard GCD) or not supported (soft GCD). Figure 13.2 shows the case of aggregator that is conjunctive, idempotent, and hard. Therefore, the desired andness is a bounded real number: $\alpha \in [\alpha_\theta, 1]$. Inside that relatively narrow range we can select a desired value of andness in one of the following three ways (presented in this chapter): (1) by selecting the most appropriate value from a simple verbalized rating scale, (2) by calculating the value of andness from the importance decomposition method, and (3) by computing the value of andness using a software tool for training preferential neurons.

In the case of UGCD.15, the selection of andness using a rating scale can be based on verbalized options presented in Table 11.2, Fig. 11.2, Tables 12.1 and 13.1, where

Table 13.1 Verbalizing the selection of andness in the case of UGCD.15 aggregators

GCD type	Verbalized logic condition	Level/Strength	Symbol	Andness
Hard graded conjunction	*Must have all* inputs satisfied (mandatory positive truth of all input arguments; a single false input 0 causes a false output 0)	Highest	C	1
		High	HC+	0.9286
		Medium	HC	0.8571
		Low	HC-	0.7857
		UGCD conjunctive threshold		0.75
Soft graded conjunction	*Nice to have most* inputs satisfied (some inputs can be zero)	High	SC+	0.7143
		Medium	SC	0.6429
		Low	SC-	0.5714
Neutrality	*Nice to have* inputs satisfied	Basic	A	0.5
Soft graded disjunction	*Nice to have some* inputs satisfied (other inputs can be zero)	Low	SD-	0.4286
		Medium	SD	0.3571
		High	SD+	0.2857
Hard graded disjunction	*Enough to have any* input fully satisfied (a single input degree of truth 1 is sufficient to cause the output degree of truth 1)	UGCD disjunctive threshold		0.25
		Low	HD-	0.2143
		Medium	HD	0.1429
		High	HD+	0.07143
		Highest	D	0
Meaning of symbols: C = conjunctive, D = disjunctive, H = hard, S = soft, plus = amplified/increased, minus = attenuated/decreased				

Meaning of symbols: C = conjunctive, D = disjunctive, H = hard, S = soft, plus = amplified/increased, minus = attenuated/decreased

analysts can select one of several levels of andness. The verbalization of logic conditions is important because it is accessible to all humans and allows collecting significant inputs for designing graded logic aggregators (and decision criteria) from people who are not familiar with mathematical interpretations. That facilitates communication between stakeholders, domain experts, and decision analysts who cooperate in the development and use of decision-support systems.

The complete process of selection of GCD aggregators is summarized in Fig. 13.3. It contains the whole range of andness $\left[-\frac{1}{n-1}, \frac{n}{n-1}\right]$, including the nonidempotent hyperconjunction and hyperdisjunction, uniform distribution of hard and soft aggregators (UGCD version), and the necessary selection of weights of all inputs.

Table 13.1 shows that conjunctive and disjunctive threshold andness values are not included in the UGCD.15 set of aggregators. Other working sets of GCD aggregators, with more members or less members, and with included/excluded threshold values, are presented in the next section.

The Central Group of Dualized Idempotent GCD Aggregators

Idempotent GCD aggregators are modeled as means. In mathematical literature [MIT69, MIT77, MIT89, BUL03] means are defined as functions $M(x_1, \ldots, x_n)$ that satisfy continuity, internality, monotonicity, idempotency, commutativity, and homogeneity. These properties are useful in decision-making problems, as follows:

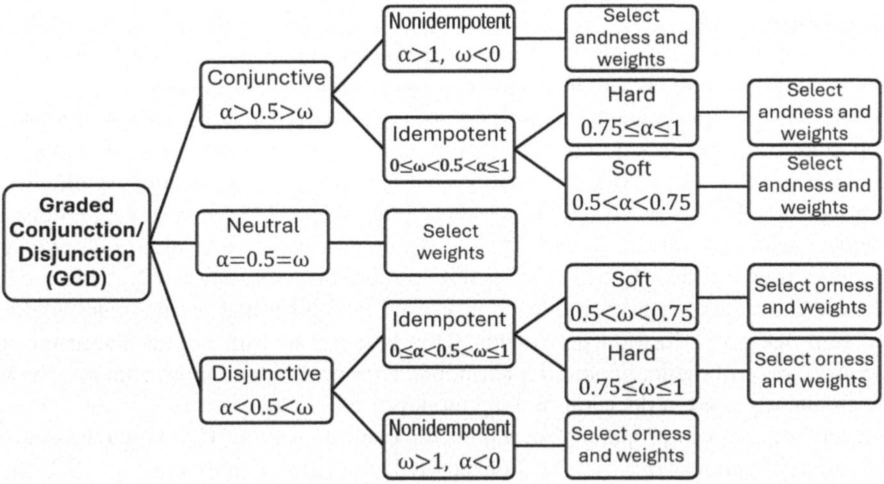

Fig. 13.3 The complete process of selection of the GCD aggregators in the full range of andness

Continuity: $\lim_{\delta x_i \to 0} M(x_1, \ldots, x_i + \delta x_i, \ldots, x_n) = M(x_1, \ldots, x_i, \ldots, x_n)$, $i = 1, \ldots, n$. In decision-making, continuity can be interpreted as an assumption that a microscopic increment of the degree of truth of an argument cannot cause a macroscopic increment of the degree of truth of the corresponding proposition.

Internality: $\min(x_1, \ldots, x_n) \le M(x_1, \ldots, x_n) \le \max(x_1, \ldots, x_n)$. In decision-making, internality means that an object (an evaluated alternative) cannot be better (more suitable) than its best component or worse (less suitable) than its worst component. That can be exemplified by the standard school grading, where the average grade (GPA) cannot be greater than the student's best grade or less than the student's worst grade.

Monotonicity: $(x_1 \le z_1), \ldots, (x_n \le z_n) \Rightarrow M(x_1, \ldots, x_n) \le M(z_1, \ldots, z_n)$. In evaluation of objects/alternatives, which is an indispensable part of decision-making,[1] an increase of the suitability of components of an object cannot cause a decrease of the overall suitability.

Idempotency: $\forall x \in [0, 1], M(x, \ldots, x) = x$. This consequence of internality also yields the idempotent aggregator boundary conditions $M(0, \ldots, 0) = 0$ and $M(1, \ldots, 1) = 1$. In the student grading example, if all individual grades are x, then the GPA should also be x.

Commutativity: $\forall i, j \in \{1, \ldots, n\}$, $M(x_1, \ldots, x_i, \ldots, x_j, \ldots, x_n) = M(x_1, \ldots, x_j, \ldots, x_i, \ldots, x_n)$. In decision-making, this is interpreted as a special case where all arguments are equally important.

[1] See the fundamental stakeholder's question at the end of Chap. 1.

Homogeneity: $M(ax_1, \ldots, ax_n) = aM(x_1, \ldots, x_n), a = const$. In the student grading example, if all individual grades are increased/decreased using a factor *a*, then the GPA should also be increased/decreased by the same factor *a*.

Internality is the fundamental property of GCD aggregators because it has other important consequences. Let us note that in the case of equal arguments the relation $\min(x, \ldots, x) = x \le M(x, \ldots, x) \le x = \max(x, \ldots, x)$ obviously yields the idempotency $M(x, \ldots, x) = x$. If we apply idempotency to the very general class of quasi-arithmetic means $y = f^{-1}\left(\sum_{i=1}^{n} W_i f(x_i)\right)$, where f is a strictly monotone function, then we have $f^{-1}\left(f(x)\sum_{i=1}^{n} W_i\right) = x$ and that is satisfied if $\sum_{i=1}^{n} W_i = 1$. Since the weights are assumed to be positive,[2] it follows that idempotency yields the sum-one-normalization of weights. This is the case with a wide spectrum of means (e.g., arithmetic, geometric, harmonic, exponential, and many others) which are frequently used in decision-making models.

Idempotency is a property that defines the central group of GCD aggregators in the range of andness $0 \le \alpha \le 1$. For the GCD aggregator shown in Fig. 13.1, the stakeholder/decision-maker must answer the following question: if $x_1 = \ldots = x_n = t \in [0, 1]$, what is going to be the value of X? Is it going to be $X = t$, or $X \ne t$? For example, if the arguments $x_1, \ldots, x_n$ denote the grades that a specific student obtained in individual courses, and X denotes student's GPA, then all schools (in the role of stakeholder) select $X = t$, i.e., the GCD must be idempotent. On the other hand, if the arguments $x_1, \ldots, x_n$ denote probabilities that a candidate can solve problems in n independent areas, and X denotes the probability of solving a problem that requires simultaneous skills in all n areas,[3] then $X = t^n \le t$ and consequently, GCD must not be idempotent. If the GCD is idempotent, then $\alpha \in [0, 1]$, and the desired aggregator must be a mean. If $\alpha > 1$, or $\alpha < 0$, then the GCD is not idempotent. If a logic aggregator is idempotent, then the resulting (aggregated) degree of truth cannot be higher than the highest input degree of truth or lower than the lowest input degree of truth.

The next step in selection of GCD aggregator is the decision whether the annihilators are supported (hard aggregators) or not supported (soft aggregators). Of course, in the high range of andness the annihilator 0 is supposed to be supported. In a low range of andness the annihilator 0 should not be supported because low andness and annihilator support are contradictory requirements (the low andness means low logic importance, and annihilator 0 means high logic importance; such a combination is inconsistent with natural human reasoning). The border between these two regions of andness is called the threshold andness and denoted α_θ. The threshold andness

---

[2] Theoretically, weights can be nonnegative. However, if a weight is zero, then the corresponding argument is excluded from the list of inputs. In decision-making we have no reason to use arguments that have no impact. Therefore, in this book we assume that weights are normalized positive values.

[3] For example, a traditional view of computer programming is a combination of three components: (1) the knowledge of algorithms, (2) the knowledge of programming paradigms, and (3) the knowledge of specific programming languages. So, the probability of successful solving of a programming problem can be defined as a product of probabilities that a programmer is familiar with necessary algorithms, paradigms, and languages. Of course, this note assumes a traditional human programmer and not an AI-based machine.

is defined as the lowest value of andness that supports the annihilator 0. Similarly, the threshold orness ω_θ is the lowest value of orness that supports the annihilator 1. If we want to support De Morgan's duality, then we must create aggregators where $\alpha_\theta = \omega_\theta$. Unfortunately, the natural duality of means is rather rare property. As a generalization of Boolean duality $x \vee y = \overline{\overline{x} \wedge \overline{y}}$, the mean $M(\boldsymbol{x}; \boldsymbol{W}, \alpha)$ is naturally dual if $M(\boldsymbol{x}; \boldsymbol{W}, \alpha) = 1 - M(\boldsymbol{1} - \boldsymbol{x}; \boldsymbol{W}, 1 - \alpha)$. An example of naturally dual mean is the exponential mean [DUJ18, Section 2.4.5], and an example of mean that is not naturally dual is the weighted power mean. The simplest way to support De Morgan's duality is to define GCD as a dualized aggregator, based on a parameterized mean $M(\boldsymbol{x}; \boldsymbol{W}, \alpha)$ that is not naturally dual. That can be done in the following two ways:

- Conjunctive-dualized: $GCD(\boldsymbol{x}; \boldsymbol{W}, \alpha) = \begin{cases} M(\boldsymbol{x}; \boldsymbol{W}, \alpha), & 0.5 \leq \alpha \leq 1 \\ 1 - M(\boldsymbol{1} - \boldsymbol{x}; \boldsymbol{W}, 1 - \alpha), & 0 \leq \alpha < 0.5 \end{cases}$
- Disjunctive-dualized: $GCD(\boldsymbol{x}; \boldsymbol{W}, \alpha) = \begin{cases} M(\boldsymbol{x}; \boldsymbol{W}, \alpha), & 0 \leq \alpha < 0.5 \\ 1 - M(\boldsymbol{1} - \boldsymbol{x}; \boldsymbol{W}, 1 - \alpha), & 0.5 \leq \alpha \leq 1 \end{cases}$

The conjunctive-dualized GCD directly uses the conjunctive part of $M(\boldsymbol{x}; \boldsymbol{W}, \alpha)$, and defines the disjunctive part as a dual of $M(\boldsymbol{x}; \boldsymbol{W}, \alpha)$. So, the disjunctive properties of GCD mirror the conjunctive properties, and such an aggregator satisfies duality. In a similar way, the disjunctive-dualized GCD directly uses the disjunctive part of $M(\boldsymbol{x}; \boldsymbol{W}, \alpha)$, and defines the conjunctive part as a dual of $M(\boldsymbol{x}; \boldsymbol{W}, \alpha)$. Again, the conjunctive aggregator properties mirror the disjunctive properties, and such an aggregator satisfies duality.

The reason for dualized definition of GCD is the absence of means that naturally offer hard/soft, and conjunctive/disjunctive properties. For example, in the case of weighted power means (WPM) of two variables $z = (wx^r + (1 - w)y^r)^{1/r}$ if $r > 0$, then the WPM is a soft aggregator. If $r \leq 0$, then the WPM is a hard aggregator. Consequently, the threshold andness aggregator is obtained for $r = 0$, and that is the geometric mean $z = x^w y^{1-w}$, where the threshold andness (assuming $w = 0.5$) is $\alpha_\theta = 2/3$. Unfortunately, the WPM supports the hard disjunctive aggregation only for $r \to +\infty$, i.e., $\omega_\theta = 1$. So, the WPM does not support De Morgan's duality. In addition, WPM has two more limitations: (1) disjunctive aggregation (except the pure disjunction) is strictly soft, and (2) the range of soft conjunctive aggregators is smaller than the range of hard conjunctive aggregators. In the case of two variables, soft conjunctive aggregators are in the range $\frac{1}{2} < \alpha < \frac{2}{3}$ and the range of hard conjunctive aggregators is two times wider: $\frac{2}{3} \leq \alpha \leq 1$. In applications, there are many cases where these limitations are not significant (for example, see [SU87, DUJ03, DUJ06a, DUJ06c, MON16]).

A dualized GCD aggregator can be based on WPM. If we use a conjunctive-dualized version, then the conjunctive part has the threshold andness $\alpha_\theta = 2/3$ and the disjunctive part of GCD is mirroring the conjunctive part: the dual threshold aggregator is $z = 1 - (1 - x)^w (1 - y)^{1-w}$ and the threshold orness is $\omega_\theta = 2/3$. In other words, for $0 < \alpha \leq \frac{1}{3}$ this version of GCD is a hard disjunction, and for $\frac{1}{3} < \alpha < \frac{1}{2}$, the GCD is a soft disjunction. Similarly, if we use a disjunctive-dualized version, then the resulting GCD aggregator is permanently soft (except for the pure conjunction and the pure disjunction).

For each desired value of andness we can compute the corresponding value of WPM exponent. In the case of n variables $\boldsymbol{x} = (x_1, \ldots, x_n)$, weights $\boldsymbol{W} = (W_1, \ldots, W_n)$, and exponent $r_n(\alpha), 0.5 \leq \alpha \leq 1$, we can use the weighted power mean $WPM(\boldsymbol{x}; \boldsymbol{W}, \alpha) = \left(\sum_{i=1}^{n} W_i x_i^{r_n(\alpha)}\right)^{1/r_n(\alpha)}$ as a model of graded conjunction, which supports both the soft and the hard aggregation. To avoid the limitation of strictly soft graded disjunction, and support De Morgan duality, the logic aggregator $GCD(\boldsymbol{x}; \boldsymbol{W}, \alpha)$ can be defined in the following recursive conjunctive-dualized form:

$$GCD(\boldsymbol{x}; \boldsymbol{W}, \alpha) = \begin{cases} WPM(\boldsymbol{x}; \boldsymbol{W}, \alpha), & 0.5 \leq \alpha \leq 1 \\ 1 - GCD(\mathbf{1} - \boldsymbol{x}; \boldsymbol{W}, 1 - \alpha), & 0 \leq \alpha < 0.5 \end{cases}.$$

This form of GCD aggregator can be implemented as a recursive function, based on the following pseudocode:

```
GCD(x; W, α)
{
    if α ≥ 0.5
        then return WPM(x; W, α)
        else return 1 − GCD(1 − x; W, 1 − α)
    endif
}
```

The dualized form of GCD preserves all functionality of theconjunctive form of WPM and mirrors this form to the disjunctive segment of GCD. So, the resulting GCD supports the soft graded conjunction, the soft graded disjunction, the hard graded conjunction, and the hard graded disjunction.

Selecting the Precision of GCD Aggregators

The number of GCD aggregator components, also called the GCD granularity, is selected to adjust the precision of GCD aggregator according to the needs of decision analyst. For the direct selection of GCD aggregator we must divide the andness range [0, 1] into N labeled equidistant values. Then it is possible to select the most suitable andness from these N values. Of course, the granularity N can be too small, or too big, or appropriate. Too small (too coarse) granularity means that the analyst can provide a more precise assessment of the appropriate andness. Too big granularity means that the difference between adjacent values of andness is beyond the analyst's discernibility level. The appropriate value of N is the value that correctly matches the analyst's discernibility level and therefore consistently exploits analyst's expertise. Since various analysts can use GCD aggregators, it is necessary to provide various levels of granularity, so that for each application and each analyst, it is possible to choose the most convenient precision/granularity. Below, we present two andness selection methods, each with three levels of granularity.

Low values of N are suitable for introductory educational experiences with a small number of distinct GCD aggregators. Unsurprisingly, professional-level projects can need a larger number of aggregators. The set of 15 UGCD aggregators, used in previous examples, is selected as a convenient compromise.

Now, let us present two methods for developing working sets of idempotent GCD aggregators: the *uniform insertion method* and the *precision doubling method.*

The *uniform insertion method* is used for creating UGCD aggregators where the range of andness [0, 1] is divided into four equal segments. According to experiments reported in [DUJ15a] the threshold andness should be in the range [0.75, 0.875]. In the absence of an adequate justification for some specific threshold andness value, a default value can be $\alpha_\theta = 0.75$ because this value offers the same number soft and hard GCD aggregators (see Chap. 11). Since UGCD aggregators support duality, the corresponding threshold orness must be $\omega_\theta = 0.75$. Thus, there are two threshold values located at the andness 0.25 and 0.75.

In the uniform insertion method, we organize the working set of idempotent GCD aggregators in a way that uniformly covers the following four segments:

- Hard graded disjunction: $0 \le \alpha \le 0.25$ ("enough to have any positive input").
- Soft graded disjunction: $0.25 < \alpha < 0.5$ ("nice to have some positive inputs").
- Soft graded conjunction: $0.5 < \alpha < 0.75$ ("nice to have most positive inputs").
- Hard graded conjunction: $0.75 \le \alpha \le 1$ ("must have all positive inputs").

The equal sizes of these segments offer equal opportunities for using each of four fundamental types of GCD. According to Table 11.2, in each segment we can insert k aggregators. That gives a working set that includes the total number of $N = 4k + 3$ idempotent GCD aggregators. This includes conjunction, disjunction, and the arithmetic mean as three border aggregators that serve as initial condition in the process of inserting k aggregators in each of four segments. To include the middle point (an "average strength of andness") of each segment we must select k as an odd number: 1, 3, 5. That gives us the granularity 7, 15, 23, labeled as low, medium, and high precision.

The uniform insertion method is shown in Fig. 13.4. The threshold values of andness (1/4 and 3/4) are denoted T, and they are not included in the working set of UGCD aggregators. The hard conjunctive interval is bordered by the pure conjunction and the threshold andness $\alpha_\theta = 3/4$. The soft conjunctive interval is bordered by the threshold andness and the arithmetic mean (A), which is also soft. The disjunctive intervals are symmetric to the conjunctive intervals.

The value $k = 0$ denotes the initial condition (aggregators C, A, and D). The insertion of $4k$ aggregators in segments [C, T], [T, A], [A, T], and [T, D] is based on the following rules:

- The resulting working set of aggregators must include C, A, and D.
- Inserted aggregators must be symmetric with respect to threshold values (T) and the central aggregator A.
- All aggregators in the working set of aggregators must be andness-equidistant.
- Conjunctive and disjunctive aggregators must be dual.

Uniform insertion method (k = group size, N = UGCD granularity)																		
k	Hard conjunctive GCD					Soft conjunctive GCD				Soft disjunctive GCD				Hard disjunctive GCD				N
0	C								A								D	3
1	C		HPC		T		SPC		A		SPD		T		HPD		D	7
ω	0		$\frac{1}{6}$		$\frac{1}{4}$		$\frac{1}{3}$		$\frac{1}{2}$		$\frac{2}{3}$		$\frac{3}{4}$		$\frac{5}{6}$		1	
α	1		$\frac{5}{6}$		$\frac{3}{4}$		$\frac{2}{3}$		$\frac{1}{2}$		$\frac{1}{3}$		$\frac{1}{4}$		$\frac{1}{6}$		0	
	Andness and orness for the low precision case k=1																	

Uniform insertion method (k = group size, N = UGCD granularity)																		
k	Hard conjunctive GCD					Soft conjunctive GCD				Soft disjunctive GCD				Hard disjunctive GCD				N
0	C								A								D	3
3	C	HC+	HC	HC-	T	SC+	SC	SC-	A	SD-	SD	SD+	T	HD-	HD	HD+	D	15
ω	0	$\frac{1}{14}$	$\frac{1}{7}$	$\frac{3}{14}$	$\frac{1}{4}$	$\frac{2}{7}$	$\frac{5}{14}$	$\frac{3}{7}$	$\frac{1}{2}$	$\frac{4}{7}$	$\frac{9}{14}$	$\frac{5}{7}$	$\frac{3}{4}$	$\frac{11}{14}$	$\frac{6}{7}$	$\frac{13}{14}$	1	
α	1	$\frac{13}{14}$	$\frac{6}{7}$	$\frac{11}{14}$	$\frac{3}{4}$	$\frac{5}{7}$	$\frac{9}{14}$	$\frac{4}{7}$	$\frac{1}{2}$	$\frac{3}{7}$	$\frac{5}{14}$	$\frac{2}{7}$	$\frac{1}{4}$	$\frac{3}{14}$	$\frac{1}{7}$	$\frac{1}{14}$	0	
	Andness and orness for the medium precision case k=3																	

Fig. 13.4 Uniform insertion method for selecting the working set of aggregators for $k = 1, 3$

The low precision working set ($k = 1$) has only one representative located in each interval. The inserted aggregators are denoted HPC (hard partial conjunction), SPC (soft partial conjunction), SPD (soft partial disjunction), and HPD (hard partial disjunction). That creates the working set UGCD.7 that includes 7 aggregators: C≺HPC≺SPC≺A≺SPD≺HPD≺D (we use symbol ≺ for ordering nonnumeric objects and aggregator labels according to the increasing orness). The andness-distance between adjacent aggregators is $1/6 = 0.167$. All conjunctive aggregators are symmetric to disjunctive aggregators with respect to A: e.g., the andness of HPC is $5/6$ and the andness of HPD is $1/6$, and $(5/6 + 1/6)/2 = 1/2$, which is the andness of A. The symmetry of HPC and SPC with respect to threshold andness is visible from $(5/6 + 2/3)/2 = 3/4$. It is necessary to note that UGCD aggregators are not located in the middle points of four basic andness segments; for example, the andness of HPC is 5/6, and the middle point of the hard graded conjunction interval is 7/8. In the case of UGCD.7, the distance between aggregators is almost 17% and this low precision is useful only in introductory demo examples; for practical work, we need higher precision.

The medium precision working set ($k = 3$) has three representatives in each of four basic andness segments, as shown in Fig. 13.4. The resulting UGCD.15 aggregators have an inter-argument difference of $1/14 = 0.071$, what is shown to be a rather practical value, both easy to use and usually sufficiently precise. The simplicity of use is based on two easy steps: (1) selection of the basic andness segment, and (2) selection of the *low≺medium≺high* level of andness inside the selected segment. Each disjunctive aggregator is a dual of the corresponding conjunctive aggregator. For example, $\mathbf{HD}(\boldsymbol{x}; \boldsymbol{W}, \alpha) = 1 - \mathbf{HC}(\mathbf{1} - \boldsymbol{x}; \boldsymbol{W}, 1 - \alpha)$.

The duality of UGCD aggregators is visible in Fig. 13.5, where the upper sequence of conjunctive aggregators is dual to the lower sequence of disjunctive aggregators (e.g. HC and HD are dual and located one above the other in the UGCD.15 chart.

Graded conjunction (simultaneity)		
Hard (mandatory)		Soft
C	HPC	SPC
Low precision: **UGCD.7**		A
D	HPD	SPD
Hard (sufficient)		Soft
Graded disjunction (substitutability)		

Graded conjunction (simultaneity)						
Hard (mandatory)				Soft (optional)		
C	HC+	HC	HC-	SC+	SC	SC-
Medium precision: **UGCD.15**						A
D	HD+	HD	HD-	SD+	SD	SD-
Hard (sufficient)				Soft (optional)		
Graded disjunction (substitutability)						

Graded conjunction (simultaneity)										
Hard (mandatory)						Soft (optional)				
C	++HC	+HC	*HC	-HC	--HC	++SC	+SC	*SC	-SC	--SC
High precision: **UGCD.23**										A
D	++HD	+HD	*HD	-HD	--HD	++SD	+SD	*SD	-SD	--SD
Hard (mandatory)						Soft (optional)				
Graded disjunction (substitutability)										

Fig. 13.5 Low, medium, and high precision UGCD aggregators, and their duality

The high precision working set ($k = 5$) has five representatives in each of four basic andness segments, as shown in Fig. 13.5. This is the next level in precision, above the UGCD.15. The resulting set UGCD.23 aggregators have an inter-argument difference of $1/22 = 0.045$. This set of 23 aggregators is acceptable for trained decision analysts because after the selection of the basic andness segment, the second step is the selection of the level of andness from the sequence that is labeled *very_low* $\prec$ *low* $\prec$ *medium* $\prec$ *high* $\prec$ *very_high*. These five levels can be interpreted as a "magical number seven minus two" [MIL56], i.e., a granularity that is acceptable for most analysts. The labels of UGCD.23 aggregators are different from the labels of UGCD.15 and UGCD.7 to make sure that different values of andness have different labels. However, there are obvious similarities between the andness of aggregators placed in equivalent locations. For example, the andness of aggregators located in central positions inside the hard conjunctive segment are rather similar: HPC: 5/6 = 0.833, HC: 12/14 = 0.857, and *HC: 19/22 = 0.864.

The *precision doubling method* for developing working sets of idempotent GCD aggregators is shown in Fig. 13.6. The idea of this procedure is that in each step we insert a new aggregator between each pair of aggregators defined in the previous step. In each step, this process approximately doubles the precision/granularity. According to Fig. 13.6, in the step s, the granularity is $N(s) = 2^s + 1$, or recursively, $N(s) = 2N(s-1) - 1$, with initial condition $N(0) = 2$. These initial two aggregators are the pure conjunction C and the pure disjunction D.

In the first step we insert the aggregator A in the middle point between the initial aggregators A and C. In the second step we insert CA in the middle point between the aggregators C and A, and DA in the middle point between the aggregators D and

s	Precision doubling method (s = step, N = granularity of GCD)																	N
0	C																D	2
1	C								A								D	3
2	C				CA				A				DA				D	5
3	C		C+		CA		C-		A		D-		DA		D+		D	9
4	C	C++	C+	C+-	CA	C-+	C-	C--	A	D--	D-	D-+	DA	D+-	D+	D++	D	17
ω	0	$\frac{1}{16}$	$\frac{1}{8}$	$\frac{3}{16}$	$\frac{1}{4}$	$\frac{5}{16}$	$\frac{3}{8}$	$\frac{7}{16}$	$\frac{1}{2}$	$\frac{9}{16}$	$\frac{5}{8}$	$\frac{11}{16}$	$\frac{3}{4}$	$\frac{13}{16}$	$\frac{7}{8}$	$\frac{15}{16}$	1	
α	1	$\frac{15}{16}$	$\frac{7}{8}$	$\frac{13}{16}$	$\frac{3}{4}$	$\frac{11}{16}$	$\frac{5}{8}$	$\frac{9}{16}$	$\frac{1}{2}$	$\frac{7}{16}$	$\frac{3}{8}$	$\frac{5}{16}$	$\frac{1}{4}$	$\frac{3}{16}$	$\frac{1}{8}$	$\frac{1}{16}$	0	

Graded conjunction (simultaneity)			
Hard		Soft/Hard	Soft
C	C+	CA	C-
Low precision: GGCD.9			A
D	D+	DA	D-
Hard		Soft/Hard	Soft
Graded disjunction (substitutability)			

Graded conjunction (simultaneity)							
Hard (mandatory)				Hard/Soft	Soft (optional)		
C	C++	C+	C+-	CA	C-+	C-	C--
Medium precision: GGCD.17							A
D	D++	D+	D+-	DA	D-+	D-	D--
Hard (sufficient)				Hard/Soft	Soft (optional)		
Graded disjunction (substitutability)							

Graded conjunction (simultaneity)											
Hard (mandatory)						Hard/Soft	Soft (optional)				
C	CS++	CS+	CS	CS-	CS--	CA	CW++	CW+	CW	CW-	CW--
High precision: GGCD.25											A
D	DS++	DS+	DS	DS-	DS--	DA	DW++	DW+	DW	DW-	DW--
Hard (sufficient)						Hard/Soft	Soft (optional)				
Graded disjunction (substitutability)											

Fig. 13.6 Precision doubling method for selecting the working set of aggregators

A. The sequence C≺CA≺A≺DA≺D has little applicability because of insufficient granularity (5) and the uncertainty whether the inserted aggregators CA and DA should be soft or hard.

In the third step we insert the aggregators C+, C-, D-, and D+. Based on their location, we assume that C- and D- are soft aggregators and C+ and D+ are hard aggregators. However, the aggregators CA and DA are exactly in the middle of the conjunctive and disjunctive range and they can be declared as either soft or hard. More precisely, the precision doubling method has the advantage that in each step all aggregators from the previous step remain included, but the procedure does not suggest where the border is between soft and hard aggregators, what is clear in the case of UGCD aggregators. Therefore, the precision doubling procedure is general, in the sense that the analyst can assign the border between soft and hard aggregators in any desired position. Consequently, this family of GCD aggregators is considered general and denoted GGCD.

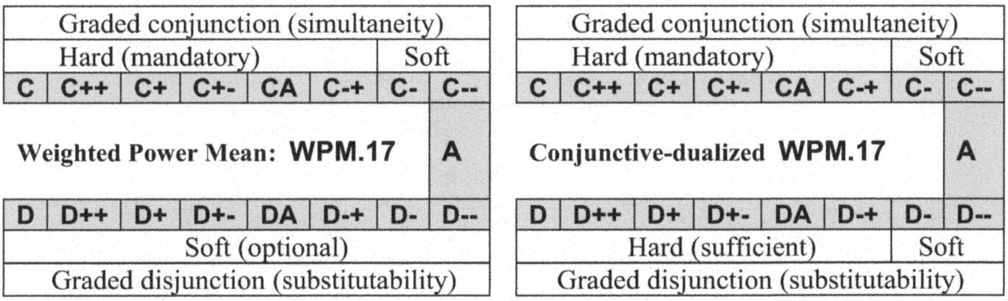

Fig. 13.7 GGCD.17 model based on native WPM and conjunctive-dualized WPM

In steps 3 and 4 we generate the sets GGCD.9 and GGCD.17. In both cases the aggregators CA and DA can be declared to be hard or soft, and considering the experiments from [DUJ15a], it is likely to use them as the threshold hard aggregators. The set GGCD.9 has similar applicability as UGCD.7, i.e., the granularity is too low for professional projects. However, GGCD.17 has a history of many successful applications, particularly in the context of using the weighted power means.

The precision doubling method can be continued with step five, and in that case the number of resulting aggregators would be $N(5) = 2^5 + 1 = 33$. This number is significantly greater than the granularity 23 we used in UGCD.23. The increased granularity contributes to precision, but it also contributes to complexity of requirements that an analyst must satisfy. In GGCD.33 we would have 7 levels of andness in each fundamental segment, plus 5 border aggregators (C, CA, A, DA, D). To reduce the complexity, in Fig. 13.6 we presented GGCD.25, where we inserted 5 levels of andness (very low, low, medium, high, very high) in each of four fundamental segments of andness located between the five border aggregators. In this way we approximately equalized the effort of analysts using UGCD.23 and GGCD.25. Note that GGCD.25 uses notation based on strong (S) and weak (W) aggregators, where CS and DS are equivalent to C+ and D+, and CW and DW are equivalent to C- and D-.

The GCD aggregator can be modeled using directly the native andness-directed weighted power mean $GCD(\boldsymbol{x}; \boldsymbol{W}, \alpha) = \left(\sum_{i=1}^{n} W_i x_i^{r_n(\alpha)}\right)^{1/r_n(\alpha)}$, where the function $r_n(\alpha)$ provides the value of exponent that implements the desired andness α (see [DUJ18] for tables of this function). The resulting aggregator is shown in Fig. 13.7. This form of GCD has only two soft conjunctive GCD aggregators: C-- and C- . Depending on n, the aggregator C-+ gives small negative values for $r_n(11/16)$. So, C-+ is similar to the geometric mean, and CA is an aggregator close to the harmonic mean. However, all disjunctive WPM aggregators have positive exponents and therefore, they are soft. Consequently, WPM.17 does not support duality. If we want to have WPM0.17 that supports duality then we can create conjunctive-dualized version of WPM.17, which is also shown in Fig. 13.6. Then, each disjunctive aggregator is a dual of the corresponding conjunctive aggregator. For example, $\mathbf{DA}(\boldsymbol{x}; \boldsymbol{W}, \alpha) = 1 - \mathbf{CA}(\mathbf{1} - \boldsymbol{x}; \boldsymbol{W}, 1 - \alpha)$.

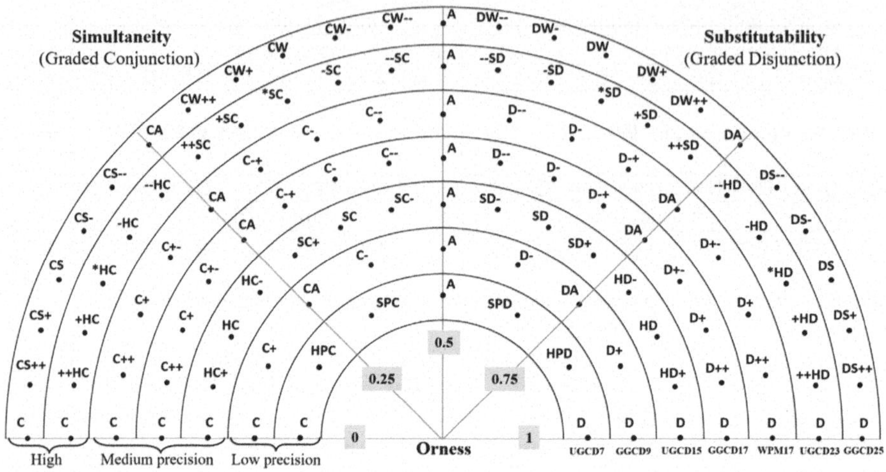

Fig. 13.8 A summary of seven families of idempotent GCD aggregators (low, medium, and high precision) [SHE21b]

For all versions of idempotent GCD aggregators, the suggested value of threshold andness is $\alpha_\theta = 3/4$, giving equal opportunities for using hard and soft GCD. Generally, in the range of andness $0.75 = \alpha_\theta \leq \alpha \leq n/(n-1)$ the GCD is conjunctive and hard, i.e., it supports the conjunctive annihilator 0. In a dual way, for $-1/(n-1) \leq \alpha \leq 1 - \alpha_\theta = 0.25$, GCD is disjunctive and hard, i.e., it supports the disjunctive annihilator 1. It is necessary to note that the threshold andness α_θ can be adjustable [DUJ15a]. However, if $\alpha_\theta \neq 3/4$, then it is necessary to show a convincing logic justification for why the number of hard aggregators should be different from the number of soft aggregators. We need both soft and hard aggregators and in the extreme cases of GCD aggregators that are only hard or only soft, it is not possible to justify their general use in Graded Logic. Of course, there are special cases where they can be used.

In this section we presented seven families of aggregators: UGCD.7, UGCD.15, UGCD.23, GGCD.9, GGCD.17, GGCD.25, and WPM.17. All of them are summarized in Fig. 13.8, and fully supported in professional decision-support systems [SEA17]. This level of detail is not excessive because the idempotent aggregators are most frequent both in human reasoning and in professional applications. In addition, the direct selection of aggregators from a justifiably selected working set is most frequently used in the process of development of propositional calculus formulas. In the next sections we present other approaches to the development of GCD aggregators.

Importance Decomposition Method

The desired andness and all weights of a GCD aggregator can be selected jointly using a method based on the decomposition of overall importance, shown in Fig. 12.3.

Table 13.2 Selecting the level of overall importance of inputs for a conjunctive GCD

Symbol	Type	Overall importance	Level L		Input 1	Input 2	Input 3	Input 4	Input 5
			C	D					
C	Hard Con.	Highest	14	0					
HC+		Very high	13	1	✓			✓	
HC		High	12	2					✓
HC-		Medium-high	11	3		✓			
SC+	Soft Con.	Medium-low	10	4			✓		
SC		Low	9	5					
SC-		Very low	8	6					
A		Base level	7	7					
SD-	Soft Dis.	Very low	6	8					
SD		Low	5	9					
SD+		Medium-low	4	10					
HD-	Hard Dis.	Medium-high	3	11					
HD		High	2	12					
HD+		Very high	1	13					
D		Highest	0	14					

If a stakeholder/decision- maker is comfortable with providing percepts of overall importance, then it can be suitable to apply the procedure presented below.

According to Fig. 12.3, the percept of overall importance of GCD inputs is a combination of the collective logic importance and the individual relative importance of inputs. In the special case where all inputs have equal relative importance, and GCD is commutative, the overall importance reduces to the logic importance. Consequently, in a general case, it is justifiable to interpret andness (or orness) as a mean value of the overall importance of inputs, as proposed in [DUJ12]. The result of increasing the overall importance is an increase in global andness (or orness).

In the case of idempotent aggregators, we can use the technique exemplified for UGCD.15 aggregators in Table 13.2. The importance of GCD inputs is verbalized as in Table 12.1. In this case we use 15 levels of overall importance, from $L = 0$ to $L = L_{\max} = 14$. Table 13.2 can be used for conjunctive GCD aggregators (column C) and disjunctive GCD aggregators (column D).

Decision-makers select the verbalized overall importance to select the appropriate numerical importance level L for each input. In the example shown in Table 13.2 the selected five levels are 13, 11, 10, 13, 12. We assume that GCD uses up to $n = 5$ inputs[4] and the andness is determined as the following relative mean overall importance:

$$\alpha = \frac{L_1 + \ldots + L_n}{n} / L_{\max} = \frac{13 + 11 + 10 + 13 + 12}{5 \times 14} = \frac{59}{70} = 0.843.$$

[4] The number of inputs is restricted by human limitations in assessment of importance using weights [MIL56]. Experience with evaluation models yields advice to avoid more than 5 inputs in GCD aggregators [DUJ18].

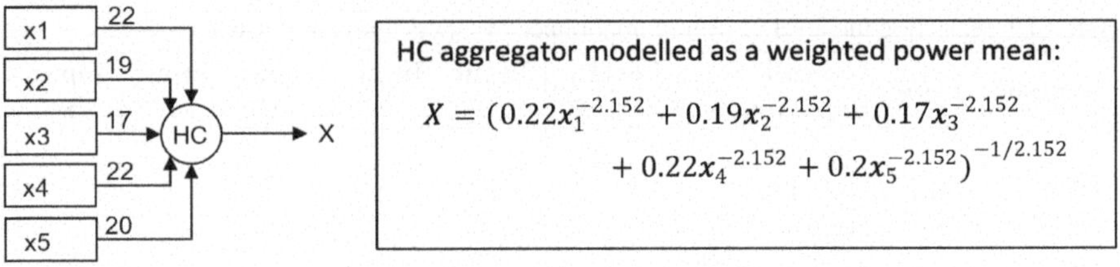

Fig. 13.9 The HC aggregator that corresponds to the example in Table 13.2

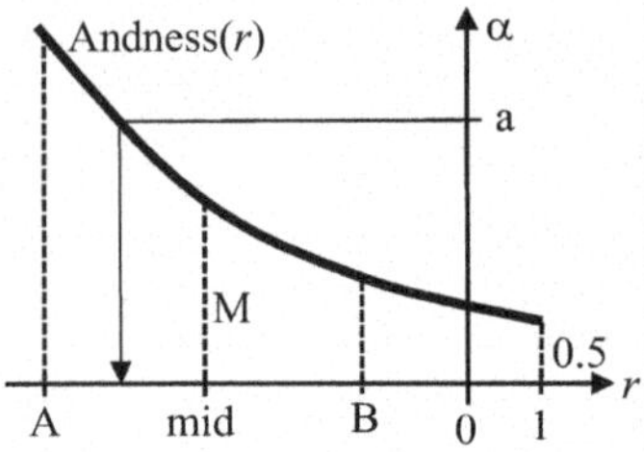


```
// Bisection method for solving equation Andness(r) = a
// a = desired andness;   r = exponent of power mean
// Conditions: A < B; Andness(A) > a, Andness(B) < a
// MaxError is the maximum error of exponent r
// (typical value of MaxError is 0.001)

double R(double A, double B, double a, double MaxError)
{ if(A > B) return 1.e30; // Error message
  double mid = (A+B)/2., M = Andness(mid);
  if(M == a || B-A < MaxError) return mid;
  if(M > a) A = mid; else B = mid;
  return R(A, B, a, MaxError);
}
```

Fig. 13.10 The bisection method for computing the WPM exponent r from the desired value of andness

Since $(L_1 + \ldots + L_n)/L_{\max} = 59/5 = 11.8 \cong 12$, the appropriate UGCD.15 aggregator is HC. The sum-normalized weights can be computed using the overall importance as follows:

$$W_i = \frac{L_i}{L_1 + \ldots + L_n} = \frac{L_i}{\alpha n L_{\max}};$$

$$W_1 = 0.22,\ W_2 = 0.19,\ W_3 = 0.17,\ W_4 = 0.22,\ W_5 = 0.20.$$

This calculation yields $W_{\min} = 0.17, W_{\max} = 0.22$ and $\frac{W_{\max}}{W_{\min}} = \frac{L_{\max}}{L_{\min}} = \frac{13}{10} = 1.3$. So, all weights are located in a 30% range. If that modest range is acceptable for the decision-maker, then the resulting aggregator, shown in Fig. 13.9, can be modeled using the weighted power mean $X = \left(0.22x_1^r + 0.19x_2^r + 0.17x_3^r + 0.22x_4^r + 0.2x_5^r\right)^{1/r}$, where the exponent r corresponds to the operator HC ($\alpha = 0.8571$ and $r = -2.152$).

A more precise aggregator can be obtained using the desired andness $\alpha = 0.843$ and the bisection algorithm shown in Fig. 13.10. The result of numerically solving equation $Andness(r) = 0.843$ is $r = -1.869$. So, the GCD aggregator is $X = \left(0.22x_1^{-1.869} + 0.19x_2^{-1.869} + 0.17x_3^{-1.869} + 0.22x_4^{-1.869} + 0.2x_5^{-1.869}+\right)^{-1/1.869}$. In the bisection algorithm, the necessary andness of aggregators can be computed using the Monte Carlo method shown in Fig. 10.1.

If the same pattern of importance levels in Table 13.2 is set three levels lower, then the results for a conjunctive GCD would be the following:

$$\alpha = \frac{L_1 + \ldots + L_n}{nL_{\max}} = \frac{10 + 8 + 7 + 10 + 9}{5 \times 14} = \frac{44}{70} = 0.629.$$

$$(L_1 + \ldots + L_n)/L_{\max} = 44/5 = 8.8 \cong 9 \text{ (aggregator SC)}$$

$$W_i = \frac{L_i}{L_1 + \ldots + L_n} = \frac{L_i}{44};$$

$$W_1 = 0.23,\ W_2 = 0.18,\ W_3 = 0.16,\ W_4 = 0.23,\ W_5 = 0.20;$$

$$\frac{W_{\max}}{W_{\min}} = \frac{L_{\max}}{L_{\min}} = \frac{10}{7} = 1.43.$$

The range of weights is now expanded from 30 to 43%. Decision-makers sometimes have requirements for specific relationships of weights. In such cases, the obtained weights can be additionally modified. Methods for modifying weights, for the same level of andness, can be found in [DUJ15a]. A simple modification can be based on linear transformation $w_i = aW_i + b$ where constants a and b are determined from $\Sigma_{i=1}^{n} w_i = 1$ and an additional relation between the new weights w_i. For example, we can keep the same andness by using unchanged r and replace the weights W_i with new weights w_i, so that the highest weight is two times more important than the lowest weight. This requirement is $w_4 = 2w_3$ and the corresponding calculation is the following:

$$w_i = aW_i + b,\ i = 1, 2, 3, 4, 5$$

$$\sum_{i=1}^{5} w_i = a\sum_{i=1}^{5} W_i + 5b = a + 5b = 1$$

$$w_4 = aW_4 + b = 2w_3 = 2aW_3 + 2b \Rightarrow a(W_4 - 2W_3) = b = -0.09a.$$

$$a = 1 - 5b = 1 + 0.45a \Rightarrow a = \frac{1}{0.55} = 1.8182;\ b = -0.09a = -0.16364.$$

Therefore, the new weights are the following:

$$w_1 = aW_1 + b = 0.23a + b = 0.2545$$

$$w_2 = aW_2 + b = 0.18a + b = 0.1636$$

$$w_3 = aW_3 + b = 0.16a + b = 0.1273$$

Table 13.3 The structure of training set for a GCD neuron

Training Set				
Input values				Desired output
x_1	x_2	$\cdots$	x_n	y
x_{11}	x_{21}	$\cdots$	x_{n1}	y_1
x_{12}	x_{22}	$\cdots$	x_{n2}	y_2
$\cdots$	$\cdots$	$\cdots$	$\cdots$	$\cdots$
x_{1k}	x_{2k}	$\cdots$	x_{nk}	y_k

$$w_4 = aW_4 + b = 0.23a + b = 0.2545$$

$$w_5 = aW_5 + b = 0.20a + b = 0.2.$$

The resulting GCD aggregator with desired andness and adjusted weights is the following:

$$X = \begin{pmatrix} 0.2545x_1^{-1.869} + 0.1636x_2^{-1.869} + 0.1273x_3^{-1.869} \\ + 0.2545x_4^{-1.869} + 0.2x_5^{-1.869} \end{pmatrix}^{-1/1.869}.$$

GCD Neuron Training Method

Another method for joint selection of andness and weights of idempotent GCD is based on the concept of training neural networks. GCD aggregators used in a propositional calculus formula can be interpreted as neurons in a neural network [DUJ91]. In most decision models, preferential neural networks have a tree structure. Each node in the tree has a clearly defined role, where both inputs and the output have a known semantic identity. That can be used to organize the training process of the whole network sequentially, neuron by neuron. If the GCD aggregator from Fig. 13.1 is interpreted as a preferential neuron, then it is possible to specify a training set of desired input/output values shown in Table 13.3 and then to develop a numerical procedure for selecting the parameters $\boldsymbol{W}$ and α so that the aggregator $GCD(\boldsymbol{x};\ \boldsymbol{W}, \alpha)$ realizes the mapping specified in Table 13.3. In the AnSy software tool [SEA10] that we use in this section, the GCD aggregator is implemented as a weighted power mean $WPM\,(\boldsymbol{x};\ \boldsymbol{W}, r)$ and the training process consists of selecting the optimum values of the weights $\boldsymbol{W}$ and the exponent r.

The selection of the optimum set of parameters for *WPM* ($\boldsymbol{x}$; $\mathbf{W}$, r) consists of the simultaneous satisfaction of the following three conditions:

(1) Minimization of the mean square error function
$$e(\mathbf{W}, r) = \frac{1}{k}\sum_{i=1}^{k}\left[y_i - \left(\sum_{j=1}^{n} W_j x_{ji}^{r}\right)\right]^2.$$

(2) Satisfaction of weight normalization condition $\sum_{j=1}^{n} W_j = 1$.

(3) Satisfaction of weight range condition: $0 < W_j < 1, \;\; j = 1, \ldots, n$.

To satisfy these conditions, we introduce the following auxiliary error functions:

$$p(\mathbf{W}) = \left|\sum_{i=1}^{n} W_i - 1\right|$$

$$t(W_i) = \begin{cases} 1 & W_i \leq 0 \\ 0, & 0 < W_i < 1 \\ 1, & W_i \geq 1 \end{cases}$$

$$q(\mathbf{W}) = \sum_{i=1}^{n} t(W_i).$$

The compound error function that must be minimized in the process of supervised learning is the following:

$$E(\mathbf{W}, r) = e(\mathbf{W}, r) + \gamma p(\mathbf{W}) + q(\mathbf{W}), \quad \gamma = 0.1, 1, 2, 4, 8, 16, \ldots$$

The positive constant γ is used to adjust the influence of the weight normalization function $p(\mathbf{W})$. AnSy minimizes $E(\mathbf{W}, r)$ using the simplex algorithm [NEL64] and $\gamma = 0.1$. That returns the optimum parameters $\mathbf{W}^*$ and r^* that yield the minimum error $E(\mathbf{W}^*, r^*)$. If $E(\mathbf{W}^*, r^*) < 10^{-4}$, then $\mathbf{W}^*$ and r^* are accepted as the final results. Otherwise, AnSy uses $\mathbf{W}^*$ and r^* as initial conditions, increases the γ parameter to the next value $\gamma = 1$, and performs the simplex minimization that returns the improved values $\mathbf{W}^*$ and r^*. If $p(\mathbf{W}^*) \neq 0$ then the minimization is repeated with the next value $\gamma = 2$. Eventually, AnSy returns the optimum values $\mathbf{W}^*$ and r^* that give the minimum value $E(\mathbf{W}^*, r^*)$ and satisfy $p(\mathbf{W}^*) = 0$. Three examples of the GCD training using AnSy are presented in Figs. 13.11, 13.12, and 13.13.

The first example, shown in Fig. 13.11, determines parameters of the GCD aggregator with 4 inputs, implemented as a weighted power mean. The training set, provided by the AnSy user, can contain any number of desired input–output values. The simplest strategy for defining the input–output values is to keep all inputs constant (e.g., 1), except one that has a reduced value (e.g., 0.5). In this way we isolate the impact of an individual input and facilitate the explanation and justification of desired output values. Of course, the training set can contain arbitrary (justifiable) values that can be partially inconsistent and AnSy is supposed to return

```
THE SYNTHESIS OF GRADED CONJUNCTION/DISJUNCTION
                         W1
 x1 ------>---------.
                         W2    |        GCD Neuron
 x2 ------>-------.    |
 .                      | |
 .                     .---.
 .     ------>----->| r |------>  y
 .                     '---'
 .                       Wn   |
 xn ------>--------'

Enter the number of input variables:    n =  4
This function has the following parameters:

Identifier:     1    2    3    4    5
-------------------------------------
Parameter :    W1   W2   W3   W4   4

How many parameters do  you  want  to  determine ? 5

Enter a training set containing any number of desired I/O values
(a value outside the [0,1] interval denotes the end of input data)

x1  x2  x3  x4      y
----------------------
.5  1   1   1      .8
1  .5   1   1      .85
1   1  .5   1      .9
1   1   1  .5      .7
9   9   9   9      9

T r a i n i n g ...      Please wait !

THE RESULTING OPTIMUM VALUES OF PARAMETERS

W1 =     0.259443
W2 =     0.183951
W3 =     0.116299
W4 =     0.440307
r  =    -0.861743

The minimum value of criterion function = 0.584E-07
The number of function computations      =  335
W1+...+Wn = 0.1000E+01,  Max error of result = 0.941E-06

==================================================
 Training set                    | Trained GCD neuron
---------------------------------+------------------
 x1    x2    x3    x4       y    |  y         Error
---------------------------------+------------------
0.50 1.00 1.00 1.00  0.8000  | 0.8000    -0.21E-07
1.00 0.50 1.00 1.00  0.8500  | 0.8500     0.13E-07
1.00 1.00 0.50 1.00  0.9000  | 0.9000    -0.95E-07
1.00 1.00 1.00 0.50  0.7000  | 0.7000    -0.71E-08
==================================================
```

Fig. 13.11 Finding all parameters of the weighted power mean used as GCD aggregator with 4 inputs

the most suitable model. In Fig. 13.11 the resulting aggregator is approximately $y = \left(0.26x_1^{-0.862} + 0.18x_2^{-0.862} + 0.12x_3^{-0.862} + 0.44x_4^{-0.862}\right)^{-1/0.862}$. At the end of the report, AnSy shows the table containing the desired outputs, the achieved outputs, and the resulting errors. In addition to optimum values of weights and exponent, the

```
THE SYNTHESIS OF GRADED CONJUNCTION/DISJUNCTION
                     W1
  x1 ------>---------.
                     W2    |        GCD Neuron
  x2 ------>-------.       |
  .                |       |
  .                .---.
  .    ------>----->| r |------>  y
  .                 ---'
  .                 Wn  |
  xn ------>--------'

 Enter the number of input variables:   n =  3
 This function has the following parameters:

 Identifier:     1    2    3    4
 --------------------------------
 Parameter :    W1   W2   W3   r

 How many parameters do  you  want  to  determine ? 3
 Enter the identifier(s) of selected parameter(s) : 2 3 4

 Enter fixed values of remaining parameters:
 W1 = 0.5

 Enter a training set containing any number of desired I/O values
 (a value outside the [0,1] interval denotes the end of input data)

 x1  x2  x3     y
------------------
 0   1   1      0
 1  0.5  1      0.75
 1   1  0.5     0.9
 9   9   9      9

 T r a i n i n g ...     Please wait !

 THE RESULTING OPTIMUM VALUES OF PARAMETERS

 W2 =     0.370764
 W3 =     0.129236
 r  =   -0.533207

 The minimum value of criterion function = 0.127E-07
 The number of function computations     =  181
 W1+...+Wn = 0.1000E+01,  Max error of result = 0.179E-06

=============================================
  Training set              | Trained GCD neuron
----------------------------+------------------
  x1    x2    x3        y   |  y        Error
----------------------------+------------------
 0.00 1.00 1.00  0.0000  | 0.0000    0.00E+00
 1.00 0.50 1.00  0.7500  | 0.7500    0.72E-08
 1.00 1.00 0.50  0.9000  | 0.9000    0.75E-08
=============================================
```

Fig. 13.12 Using AnSy to find two weights and exponent of a weighted power mean

report also shows the sum of weights (that must be 1), and the resulting criterion function errors.

AnSy offers the possibility that selected parameters can be fixed by the user. In Fig. 13.12 we show an example where the user has reasons to assign 50% as the relative importance of the first input of a conjunctive aggregator that has three inputs.

```
THE SYNTHESIS OF GRADED CONJUNCTION/DISJUNCTION
                    W1
  x1 ------>---------.
                    W2    |        GCD Neuron
  x2 ------>-------.  |
   .                |  |
   .               .---.
   .   ------>----->| r |------>  y
   .               '---'
   .                Wn  |
  xn ------>--------'

 Enter the number of input variables:    n =  3
 This function has the following parameters:

 Identifier:     1    2    3    4
 --------------------------------
 Parameter :     W1   W2   W3   r

 How many parameters do  you  want  to  determine ? 3
 Enter the identifier(s) of selected parameter(s)  : 1 2 3

 Enter fixed values of remaining parameters:
 r  = -1.0

 Enter a training set containing any number of desired I/O values
 (a value outside the [0,1] interval denotes the end of input data)

 x1  x2  x3     y
------------------
 0   1   1      0
 1  0.5  1     0.75
 1   1  0.5    0.9
 9   9   9      9

 T r a i n i n g ...      Please wait !

 THE RESULTING OPTIMUM VALUES OF PARAMETERS

 W1 =    0.555556
 W2 =    0.333333
 W3 =    0.111111

 The minimum value of criterion function = 0.168E-06
 The number of function computations     =  145
 W1+...+Wn = 0.1000E+01,  Max error of result = 0.617E-06

=============================================
  Training set              | Trained GCD neuron
----------------------------+------------------
  x1    x2    x3        y   |   y         Error
----------------------------+------------------
 0.00 1.00 1.00  0.0000  | 0.0000     0.00E+00
 1.00 0.50 1.00  0.7500  | 0.7500     0.17E-06
 1.00 1.00 0.50  0.9000  | 0.9000     0.23E-06
=============================================
```

Fig. 13.13 An example of training a harmonic mean

So, it is necessary to determine the remaining two weights and the WPM exponent r.

The training set in Fig. 13.12 illustrates another frequent strategy for the specification of the training set. The user knows that this aggregator must be a model of hard graded conjunction, and therefore it supports the annihilator 0. So, to make sure that this condition is satisfied, the first row of the training set table contains one input that has the value of 0 and all others that have the maximum value 1. The requested output is 0, which is the annihilator of hard conjunction and the reason for the resulting negative exponent r. Other rows of the training set table are based on the standard strategy of isolating the impact of a single input. The resulting hard conjunctive aggregator is $y = \left(0.5x_1^{-0.533} + 0.37x_2^{-0.533} + 0.13x_3^{-0.533}\right)^{-1/0.533}$.

The example shown in Fig. 13.13 illustrates the situation where the training is applied to the harmonic mean of three inputs. So, the user decided to use $r = -1$, and the problem is to find the weights. The training set specification strategy is the same as in the example shown in Fig. 13.12: the first condition reinforces the support for the conjunctive annihilator 0, and the remaining conditions specify isolated impact of individual inputs. Similarly to the previous examples, the resulting aggregator is $y = \left(0.56x_1^{-1} + 0.33x_2^{-1} + 0.11x_3^{-1}\right)^{-1}$.

All examples in Figs. 13.11, 13.12, and 13.13 create results that have the criterion error 0. It is easy to note that if we have k conditions and K parameters that must be determined, then the case where $k = K$ is similar to solving k equations with K unknown variables. So, if the solution exists, then the resulting error will be 0. If $k < K$, then we can have more than one zero-error solution, and one of them will be found and reported by AnSy. If $k > K$, then in most cases the conditions will not be fully consistent, and the resulting error of the minimization criterion will be positive.

References

BUL03 Bullen, P.S. 2003, 2010. *Handbook of Means and Their Inequalities*. Kluwer.

DUJ91 Dujmović, J. 1991. *Preferential Neural Networks. Chapter 7 in Neural Networks—Concepts, Applications, and Implementations*, vol. II, ed. P. Antognetti and V. Milutinović, Prentice-Hall Advanced Reference Series, Prentice-Hall, 155–206.

DUJ03 Dujmović, J. 2003. Optimizing Computer System Configurations. *Journal of Automatic Control* 13 (2): 22–33.

DUJ12 Dujmović, J. 2012. Andness and Orness as a Mean of Overall Importance. In *Proceedings of the IEEE World Congress on Computational Intelligence, Brisbane, Australia, June 10–15, 2012*, 83–88.

DUJ15a Dujmović, J. 2015. Weighted Compensative Logic with Adjustable Threshold Andness and Orness. *IEEE Transactions on Fuzzy Systems* 23(2): 270–290. https://doi.org/10.1109/TFUZZ.2014.2312018.

DUJ18 Dujmović, J. 2018. *Soft Computing Evaluation Logic*. Wiley.

DUJ19 Dujmović, J. 2019. Graded Logic for Decision-Support Systems. *International Journal of Intelligent Systems* 34: 2900-2919. https://doi.org/10.1002/int.22177.

DUJ06c Dujmović, J., and H. Bai. 2006. Evaluation and Comparison of Search Engines Using the LSP Method. *ComSIS* 3(2): 31–56.

DUJ06a Dujmović, J., and H. Nagashima. 2006. LSP Method and its Use for Evaluation of Java IDE's. *International Journal of Approximate Reasoning* 41 (1): 3–22.

DUJ21b Dujmović, J., and V. Torra. 2022. Aggregation Functions in Decision Engineering: Ten Necessary Properties and Parameter-Directedness. Proceedings of the 3rd International Conference on Intelligent and Fuzzy Systems, INFUS 2021 conference, Izmir, Turkey. In *Intelligent and Fuzzy Techniques for Emerging Conditions and Digital Transformation*, vol. 1, ed. C. Kahraman et al. Springer, LNNS, vol. 307, 173–181.

LAR03 Larsen, H.L. 2003. Efficient Andness-Directed Importance Weighted Averaging Operators. *International Journal of Uncertainty, Fuzziness and Knowledge-Based Systems* 12 (Suppl.): 67–82.

LAR09 Larsen, H.L. 2009. Multiplicative and Implicative Importance Weighted Averaging Aggregation Operators with Accurate Andness Direction. In *Proceedings of the Joint 2009 International Fuzzy Systems Association World Congress and 2009 European Society of Fuzzy Logic and Technology Conference, Lisbon, Portugal, July 20–24, 2009*, ed. J.P. Carvalho, D. Dubois, U. Kaymak, and J.M. da Costa Sousa, 402–407. ISBN: 978-989-95079-6-8.

MIL56 Miller, G.A. 1956. The Magical Number Seven, Plus or Minus Two: Some Limits on Our Capacity for Processing Information. *The Psychological Review* 63: 81–97.

MIT69 Mitrinović, D.S., and P.M. Vasić. 1969. *Means (In Serbo-Croatian), Mathematical Library*, vol. 40. Belgrade.

MIT89 Mitrinović, D.S., and J.E. Pečarić. 1989. *Mean Values in Mathematics*. Belgrade: Scientific Book Publishers (in Serbo-Croatian: Srednje vrednosti u matematici, Naučna knjiga, Beograd, 1989).

MIT77 Mitrinović, D.S., P.S. Bullen, and P.M. Vasić. 1977. *Means and Related Inequalities (In Serbo-Croatian)*. Belgrade: Publications of the University of Belgrade EE Dept., No. 600.

MON16 Montgomery, B., S. Dragićević, J. Dujmović, and M. Schmidt. 2016. A GIS-Based Logic Scoring of Preference Method for Evaluation of Land Capability and Suitability for Agriculture. *Computers and Electronics in Agriculture* 124: 340–353.

NEL64 Nelder, J.A., and R. Mead. 1964. A Simplex Method for Function Minimization. *Computer Journal* 7(4): 308–313.

SEA10 SEAS Co. 2010. Analysis and Synthesis of Aggregation Operators. *AnSy User Manual V.3.2*. SEAS.

SEA17 SEAS Co. 2017. LSP.NT—LSP Method for Evaluation over the Internet. *LSP.NT User Manual*. SEAS.

SHE21b Shen, S., S. Dragićević, and J. Dujmović. 2021. GIS.LSP: A New Soft Computing Method and Tool for Geospatial Suitability Analysis. *Transactions in GIS* 25(3): 1228–1248.

SU87 Su, S.Y.W., J. Dujmović, D.S. Batory, S.B. Navathe, and R. Elnicki. 1987. A Cost-Benefit Decision Model: Analysis, Comparison, and Selection of Data Management Systems. *ACM Transactions on Database Systems* 12 (3): 472–520.

TOR22 Torra, V. 2022. Andness Directedness for t-Norms and t-Conorms. *Mathematics* 10: 1598. https://doi.org/10.3390/math10091598.

Chapter 14
Experimental Measurement and Modeling of Subjective Conjunction

In physics, chemistry, and engineering, it is normal to accept mathematical models of physical phenomena if and only if the models can be experimentally verified. Nobody would trust Maxwell's equations unless their validity can be convincingly proved by experiments in classical electromagnetism. That is a general scientific approach that is accepted in all areas of mathematical modelling of natural phenomena. Human mental processes, including the commonsense logical reasoning and decision making, are also natural phenomena. They are observable, measurable, and modelable.

Since all forms of logic and all decision methods are mathematical models of natural phenomena, they need experimental validation. If a convincing experimental verification is missing, then both logic and decision models have the status of abstract mathematical theory, unrelated to humans and their mental processes. Such theories are not developed as precise mathematical models of observable natural phenomena. Of course, abstract mathematical theories are fully legitimate in mathematics, but it must be perfectly clear that the applicability of such models is not a goal of their development. Therefore, the uncritical use of such models in decision engineering, frequently proposed by fuzzy set theoreticians, can be very questionable (in [HIS88], experimentally unverified fuzzy logic models are brutally criticized as solutions of nonexistent problems).

The need for experimental validation of human subjective logic operations is analyzed by many authors including [GOG69, ODE77, THO79, ZYS79, ZIM80, HIS88, KOV92, DUJ18, DUJ22]. A particularly visible problem in this area is the modeling of intersection of fuzzy sets. This is the problem of modeling subjective conjunction[1] that many authors in the fuzzy logic area try to reduce to using selected t-norms. In most cases, the proposed aggregators are only the minimum and product,

[1] In GL, De Morgan duality $x\Delta y = 1-(1-x)\nabla(1-y)$ and $x\nabla y = 1-(1-x)\Delta(1-y)$ are two equivalent ways to express the same logic relationship (Δ = graded conjunction, ∇ = graded disjunction). So, the properties of disjunction depend on the properties of conjunction, and vice versa. Thus, modeling of conjunction is sufficient for modeling disjunction, and the only necessary basic model of logic aggregators is conjunction.

J. Dujmović, *Graded Logic*, Intelligent Systems Reference Library 273,
https://doi.org/10.1007/978-3-031-88558-7_14

and sometimes that is the range between the Łukasiewicz weak and strong conjunction like in the Frank and similar t-norms [ALS83, MAY91, KLI95, KLE99, RAD00]. The corresponding andness range is $1 \leq \alpha \leq 1.5$, as follows:

- Strong conjunction: $x\&y = max(0, x + y - 1)$, $\alpha_{\&} = 1.5$.
- Product t-norm: $P(x, y) = xy$, $\alpha_P = 1.25$.
- Weak conjunction: $x \wedge y = min(x, y)$, $\alpha_{\wedge} = 1$.

In the case of two variables, the range of andness of conjunctive aggregators implemented as GCD is $0.5 < \alpha \leq 2$. So, the size of this range is 1.5, while the size of the range of andness of popular t-norms is 0.5 or 0.25. Obviously, it is highly unlikely that there are some mysterious reasons that restrict humans to use for subjective conjunction only 1/3 or 1/6 of the available range of andness. Indeed, experiments convincingly show that humans use subjective conjunctions in the whole conjunctive range of GCD andness.

Regardless of obvious concerns, theoreticians regularly ignore the need for the experimental validation of logic and decision models. For example, in [WIK25] the Fuzzy Logic conjunction and disjunction are modeled using $\min(x, y)$ and $\max(x, y)$ without any proof that humans really use min and max in their commonsense logical reasoning, and that nothing else is needed. In the case of t-norm fuzzy logics [WIK22], t-norms are used as models of conjunction without any experimental proof that humans really use them. On the other hand, there are experimental results [THO79, DUJ22] showing that the minimum and product t-norms are not suitable for modeling some forms of conjunction. Experiments reported in [THO79] show that the minimum function is more suitable than product, but both are unsuitable. An opposite conclusion (that the product t-norm is more suitable than the minimum) is suggested in [ODE77]. Experiments presented in [DUJ22] and in this chapter show that t-norms cannot be used for modeling the simultaneity of observable properties of evaluated objects. Therefore, the models of simultaneity in human commonsense reasoning need experimental verification.

The goal of Graded Logic is to be applicable in decision engineering. GL uses a human-centric approach: we assume that all logic and decision-making models must be experimentally verified as models of natural phenomena of human commonsense reasoning. Thus, we assume that all authors of applied logic and decision models have the responsibility to experimentally demonstrate that their models are valid as models of natural phenomena of human reasoning. That is not a trivial problem. However, experimentally verified models are the only way to provide the credibility of corresponding decision engineering results.

The subjective conjunction has a variety of characteristic properties and for each property it is necessary to organize a specific measurement. So, that makes conjunction measurements a delicate problem. Another sensitive problem is the selection of respondents and their preparation. Generally, respondents can be very different people with different ages, education, interests, and experiences. These properties can affect the results of experiments. Obviously, subjective conjunction and commonsense reasoning are general human properties. So, one of problems in the subjective conjunction measurement is to find a way to eliminate or minimize the impact of respondent's specific personality, or social/economic status, on various results of commonsense logic measurements. According to [DUJ22], to achieve that goal, it is useful to separate two types of measurements of subjective conjunction:

(1) Simultaneity of independent properties (or attributes) of analyzed objects.
(2) Simultaneity of realization of desired goals of a specific stakeholder/decision-maker.

In the first case, respondents are objective neutral observers, and in the second case, respondents are requested to play a clearly defined role of a specific stakeholder. In both cases the impact of specific respondent's personality is minimized, maximizing the usability of results of measurement.

Characteristic Properties of Subjective Conjunction

In human reasoning, the most frequent logic connective is simultaneity (subjective conjunction): we are frequently interested in objects that simultaneously have some desired properties, or in actions that simultaneously satisfy some desired goals. In GL, the mathematical model of simultaneity is graded conjunction (denoted Δ) as an andness-directed logic aggregator in the range between the arithmetic mean and the drastic conjunction. By observing simultaneity in human communication and intuitive logic reasoning it is possible to identify a variety of characteristic special cases and necessary properties that include the following:

- *Conjunctive aggregator*: logic aggregator where $\alpha > \omega$ (global andness is greater than global orness). By definition of logic aggregators, $0\Delta 0 = 0$ and $1\Delta 1 = 1$.
- *Idempotent conjunction*: all cases where $x\Delta x = x$. In such cases the subjective conjunction is modeled using means: $\min(x, y) \leq x\Delta y < (x+y)/2$, $x > 0,\quad y > 0,\quad x \neq y$, For example, $x\Delta y = 2xy/(x+y)$.
- *Full (pure) conjunction*: $x\Delta y = \min(x, y) = x\wedge y$; this is the strongest idempotent form of simultaneity.
- *Partial conjunction*: $\min(x, y) < x\Delta y < (x+y)/2,\quad x > 0,\quad y > 0,\quad x \neq y$.
- *Hyperconjunction*: all nonidempotent conjunctive aggregators stronger than the full conjunction. So, $x\Delta y < \min(x, y)$, except for $x = y = 0 = 0\Delta 0$ and $x = y = 1 = 1\Delta 1$. A typical example is $x\Delta y = xy$.
- *Commutative conjunction*: the special case where x and y are equally important: $x\Delta y = y\Delta x$.
- *Noncommutative conjunction*: all cases where x and y are not equally important. E.g., $x\Delta y = x^w y^{1-w}$ and $w \in \,]0, 1/2[\ \cup\ ]1/2, 1[$. In applications, this is the most frequent special case (weighted simultaneity where $x\Delta y \neq y\Delta x$).
- *Hard conjunction*: full support for annihilator 0. Thus, $\forall x > 0, \forall y > 0 \Rightarrow 0\Delta y = 0, x\Delta 0 = 0$. The hyperconjunction is always hard, and the partial conjunction can be either hard or soft. E.g., $x\Delta y = \sqrt{xy}$ is hard.
- *Soft idempotent conjunction*: no support for annihilator 0. Thus, $\forall x > 0,\ \forall y > 0 \Rightarrow 0\Delta y > 0,\ x\Delta 0 > 0$. E.g., $x\Delta y = (0.5\sqrt{x} + 0.5\sqrt{y})^2$ or $x\Delta y = 0.25x + 0.25y + 0.5\sqrt{xy}$.

- *Threshold conjunction*: rejection of cases where arguments are less than a specific minimum threshold value. So, $\forall x < x_\theta \Rightarrow x\Delta y = 0$ and $\forall y < y_\theta \Rightarrow x\Delta y = 0$.
- *Asymmetric conjunction*: this is a noncommutative idempotent case of simultaneity implemented as the conjunctive partial absorption function [DUJ79, DUJ18], where both x and y are simultaneously desirable, but have different support for annihilator 0: $\forall y \geq 0 \Rightarrow 0\Delta y = 0$ and $\forall x > 0 \Rightarrow x\Delta 0 > 0$ (x is a hard input and y is soft).
- *Drastic conjunction*: the ultimate simultaneity where $x\Delta y = 1$ if and only if $x = y = 1$ and in all other cases $x\Delta y = 0$. In other words, $x\Delta y = \lfloor xy \rfloor$.

In human reasoning it is easy to identify all the above basic types of subjective conjunction, as well as similar aggregators of more than two variables, and their combinations. If we want to create a mathematical model that supports all the above properties, then the result is GCD, or a combination of two GCD aggregators in the asymmetric conjunction case (Chap. 16).

Simultaneity of Independent Properties of an Object

The simplest form of simultaneity is the simultaneity of properties of analyzed objects. We assume that such properties are graded, i.e., an analyzed property can be fully present or partially present, or not present. If the property is that an object is *round* (i.e., circular or spherical) then coins of most currencies (like \$0.10 or €0.10) are (almost) completely round, while rectangular paper bills (like \$10 or €10) are not round (or very little round). Similarly, an object can be hard (for human hands) and in such a case, the \$0.10 coin is (almost) completely hard, while the \$10 bill is completely not hard. Now, we can ask to what extent an object (e.g. a tennis ball) is simultaneously hard and round. In such cases, we have simultaneous presence of two (or more) independent properties, where all properties are objectively present and fully observable. So, all human observers can be neutral and behave in a similar way. Thus, the results of such experiments can be accepted with confidence.

Examples of problems in the "*simultaneity of properties*" group, applied to an arbitrary object T, are the following [THO79, KOV92, DUJ22]:

$$\text{Metallic container(T)} = \text{metallic(T)}\ \Delta\ \text{container(T)}.$$
$$\text{Heavy ball(T)} = \text{heavy(T)}\ \Delta\ \text{ball(T)}.$$
$$\text{Soft and round}(T) = \text{soft(T)}\ \Delta\ \text{round(T)}.$$

In these examples, two properties, like "metallic" and "container," or "heavy" and "ball," or "soft" and "round" are independent and uncorrelated properties of object T. So, any respondent can be considered a *neutral or passive observer* of object T, who can select proper degrees of membership of object T in the "metallic set," "container set," and the "metallic container" set without visible goal or personal involvement and interest. It is only necessary to observe and assess the simultaneous presence of two independent properties. It is realistic to expect that these properties should be commutative, and the resulting weights of aggregators should be close to 0.5.

Now, let us present the results of modeling the simultaneity of properties using GCD. The first of well-known experiments, performed by Thole et al. [THO79], analyzed whether the minimum function (andness $\alpha = 1$) and product operator (andness $\alpha = 1.25$) can be suitable models for the metallic container suitability. In GL, the andness range of simultaneity is $^1/_2 < \alpha \leq 2$. Unsurprisingly, the chances that the continuously adjustable andness of an arbitrary simultaneity operator could be very close to one of these two fixed values (either 1 or 1.25) are rather low. So, the conclusion of [THO79] is that neither the product nor the minimum can fit measurements very well, but the minimum is a better fit. This can be examined by adjusting the andness of the GCD function to provide the optimum fit of measurements reported in [THO79]. For necessary calculations we use the preferential neuron training tool ANSY [DUJ18]. That gives the following GCD version of the weighted metallic container aggregator $WMC(M, C)$ and its global andness (computed using the volume under $WMC(M, C)$):

$$WMC(M, C) = \left(0.502009M^{-2.31474} + 0.497991C^{-2.31474}\right)^{-0.432014}$$
$$\alpha_w = 0.840964, E_{mean} = 0.03668, r = 0.986822.$$

The accuracy of this model is evaluated using the mean absolute error E_{mean} between the aggregated values obtained from the experiment and the aggregated values generated by the above GCD model. The accuracy of the model is also assessed using the coefficient of linear correlation r between the measured and GCD values. These indicators (low error and high correlation) show that idempotent GCD provides slightly better accuracy than the gamma operator proposed in [ZYS79] and subsequently used in [ZIM80, ZIM87, ZIM96]. Note that the optimum weights are close to 0.5, showing the independence of M and C.

The global andness is strictly defined for equal weights, and optimum fitting of the GCD exponent gives the following results:

$$MC(M, C) = \left(0.5M^{-2.3118} + 0.5C^{-2.3118}\right)^{-0.432563}$$
$$\alpha = 0.84127, E_{mean} = 0.03665, r = 0.986799.$$

As expected, the differences between these results and weighted results are negligible. Since the real andness of this aggregator is 0.84, that explains why the minimum function (andness 1) is not a satisfactory model, but better than the hyperconjunctive product, where andness is 1.25.

The second well known experiment introduced (together with the gamma aggregator) in [ZYS79] and subsequently analyzed in [ZIM80] is the model of industrial tile quality as a conjunctive aggregate of its solidity (S) and dovetailing shape (D). Using the same procedure as in the metallic container case, the *GCD/WSD* function is the following:

$$SWD(S, D) = \left(0.448229S^{0.376647} + 0.551771D^{0.376647}\right)^{2.65501}$$

$$\alpha_w = 0.604065,\ E_{mean} = 0.0396367,\ r = 0.978477.$$

The commutative version of this aggregator is

$$SD(S, D) = \left(0.5S^{0.3577} + 0.5D^{0.3577}\right)^{2.79564}$$
$$\alpha = 0.608607,\ E_{mean} = 0.0458292,\ r = 0.972153.$$

The third experiment, performed by Kovalerchuk and Taliansky [KOV92], is based on objects that can be heavy (H), and shaped as a ball (B). Following is the weighted heavy ball GCD model:

$$WHB(H, B) = \left(0.613339H^{-1.181986} + 0.386661B^{-1.181986}\right)^{-0.846034}$$
$$\alpha_w = 0.778532,\ E_{mean} = 0.057505,\ r = 0.953719.$$

This GCD aggregator is hard and very close to the harmonic mean. Its commutative version has slightly higher andness:

$$HB(H, B) = \left(0.5H^{-1.7555} + 0.5B^{-1.7555}\right)^{-0.569638}$$
$$\alpha = 0.817886,\ E_{mean} = 0.0595935,\ r = 0.948253.$$

The presented experiments show simultaneity models that belong to the partial conjunction. The metallic container and heavy ball are hard aggregators with very consistent results of andness that is slightly above 0.8. The tiles example yields the soft partial conjunction with andness approximately 0.6.

To expand these results, we designed experiments for testing whether objects are hard and round ($H \Delta R$) or soft and round ($S \Delta R$) and applied to different groups of respondents. The unpaid subjects were students (a group of 59 for HR, and a group of 42 for SR) who submitted responses anonymously and after a very short preparation. A sample request is shown in Fig. 14.1.

After collecting initial data, we found that 62.7% respondents provided data that show conjunctive aggregation, 13.6% of respondents decided to systematically apply the arithmetic mean, and 23.7% gave responses that resulted in disjunctive aggregators. For the whole population of 59 subjects, this experiment generated the following GCD aggregator:

$$WHR(H, R) = \left(0.450204H^{0.358819} + 0.549796R^{0.358819}\right)^{2.78692}$$
$$\alpha_w = 0.607458,\ E_{mean} = 0.015675,\ r = 0.997255.$$

$$HR(H, R) = \left(0.5H^{0.3745} + 0.5R^{0.3745}\right)^{2.67023}$$
$$\alpha = 0.605334,\ E_{mean} = 0.021135,\ r = 0.993896.$$

HARD AND ROUND

In this experiment we would like to see how humans reason about simultaneity by combining the concepts of "hard object," "round object," and "hard and round object". Following is the meaning of these concepts:

HARD: an object that is (*to some extent*) hard for human hands, e.g., almond, baseball ball, glass, potato

ROUND: an object that is (*to some extent*) round in two or three dimensions, e.g., ball, pizza, car tire, tea cup.

HARD and ROUND: an object that is (*to some extent*) simultaneously hard and round.

The meaning of ***"to some extent"*** is from 0 (not at all) to 100% (completely).

Please fill the table with numbers 0 to 100 where each number denotes the extent to which the object is hard, round, and simultaneously hard and round. To help you start, we included an example of watermelon.

Object	**HARD** (0 – 100%)	**ROUND** (0 – 100%)	**HARD and ROUND** (0 – 100%)
Watermelon [example]	50	95	70
Tennis ball			
Walnut			
Feather pillow			
Brick			
Raw potato			
Cherry			
Needle			
Rose (flower only)			
Avocado			
Gold wedding ring			
Tomato			
Mattress			
1 Dollar bill			
Peeled hard-boiled egg			
Apple			
Sponge			
Mashed potato			
Pancake			
Car tire			
Strawberry			

Fig. 14.1 Description of the hard and round experiment

It is possible to disqualify questionable data of subjects who misinterpreted the experiment and provided data that do not correspond to conjunctive aggregators. If we include only conjunctive aggregators, then the resulting andness is $\alpha = 0.686213$ which is still in the range of soft partial conjunction.

The experiment with the same objects can be performed with different subjects (the group of 42 participants) for soft and round attributes. The resulting GCD aggregators are:

$$WSR(S, R) = \left(0.473821S^{0.514929} + 0.526179R^{0.514929}\right)^{1.94202}$$
$$\alpha_w = 0.580547,\ E_{mean} = 0.02256,\ r = 0.991169.$$

$$SR(S, R) = \left(0.5S^{0.4915} + 0.5R^{0.4915}\right)^{2.03459}$$
$$\alpha = 0.585169,\ E_{mean} = 0.023595,\ r = 0.989711.$$

These results are very similar to the case of the hard and round experiment, validating the aggregation functions. As in the previous case, if we include only conjunctive aggregators, then the resulting andness is $\alpha = 0.727756$. For relations between the *hard* Δ *round* and *soft* Δ *round* object experiments, see results presented in Chap. 5 (Tables 5.2 and 5.3).

The presented five experiments support the conclusion that the subjective conjunction used as a model of simultaneity of object properties yields GCD aggregators that have the expected andness approximately in the range $\alpha \in [0.58, 0.85]$. The expected values of andness are in the vicinity of the standard threshold andness $\alpha_\theta = 0.75$ which is used for uniform interpolative GCD aggregators, offering equal ranges for soft and hard aggregators (Chap. 13). All experiments show that GCD can fit the experimental data with good accuracy (expected mean error of a few percent and expected coefficient of correlation above 0.95). All models are in the range of soft and hard partial conjunctions. Both hyperconjunctive aggregators (e.g., t-norms) and the pure conjunction *cannot be used* for modeling simultaneity of properties. This is a convincingly verified definitive conclusion that could not be reached without making experiments. So, this conclusion should be taken seriously into account by researchers and practitioners in fuzzy logic.

Local Andness Analysis in the Case of Two Variables

The analysis of human commonsense logic aggregators should start with the simplest case of two variables. So, let us investigate a GCD aggregator $z = A(x, y) = A(m, M)$ where $m = \min(x, y) = x \wedge y$ and $M = \max(x, y) = z \vee y$. Generally, $0 \le m \le M \le 1$ and $0 \le z \le 1$. If $0 \le z < m$ then $A(m, M)$ is a hyperconjunction, and if $M < z \le 1$ then $A(m, M)$ is a hyperdisjunction. In logical aggregation experiments, respondents are given a pair (x, y). Then, according to specific problem description, respondents provide the most appropriate commonsense aggregated value z. Consequently, the properties of commonsense logic aggregators can be studied from the sequence of triplets (x, y, z). By the definition of aggregator, $A(0, 0) = 0$ and $A(1, 1) = 1$; consequently, the pairs $(0, 0)$ and $(1, 1)$ must be excluded from (x, y) values used in experiments.

In the case of n variables the global andness α is in the range $-1/(n-1) \le \alpha \le n/(n-1)$ and if $n = 2$ then $-1 \le \alpha \le 2$. To analyze logic properties of individual results collected from respondents we use the local (subjective) andness a which we define as a function of x, y, z shown in Fig. 14.2. The analytic form of the proposed local (subjective) andness of two variables has the following four options:

- Equal values ($a = 0.5$):

$$a(x, y, z) = \frac{1}{2}, \quad 0 < m = z = M < 1, \quad m = M$$

- Means ($0 \le a \le 1$):

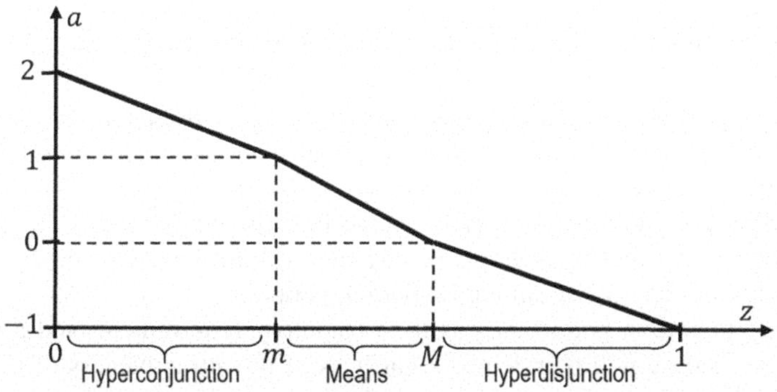

Fig. 14.2 Local (subjective) andness $a(x, y, z)$ as a function of respondent's input z

$$a(x, y, z) = \frac{M - z}{M - m}, \quad 0 \le m \le z \le M \le 1, \quad m < M$$

- Hyperconjunction ($a > 1$):

$$a(x, y, z) = 2 - \frac{z}{m}, \quad 0 \le z < m, \quad m > 0$$

- Hyperdisjunction ($a < 0.$):

$$a(x, y, z) = -\frac{z - M}{1 - M}, \quad M < z \le 1, \quad M < 1.$$

Simultaneity of Realization of Stakeholder's Goals

The analysis of the simultaneity of object properties puts respondents in the position of neutral or passive observers, leading to rather uniform responses. More frequently, however, humans use subjective conjunction to aggregate the truthfulness of attributes which express stakeholder's goals and interests. Typical examples of problems in the "simultaneity of goals" group are the following:

$$\text{Suitability(job)} = \text{good\_salary(job)}\ \Delta\ \text{attractiveness(job)}$$

$$\text{Suitability(product)} = \text{high\_quality(product)}\ \Delta\ \text{good\_price(product)}$$

$$\begin{aligned}\text{Convenience(home\_location)} = {} & \text{proximity\_to\_job(parents)} \\ & \Delta\ \text{proximity\_to\_school(children)}\end{aligned}$$

$$\text{Suitability(home)} = \text{quality(home)}\ \Delta\ \text{convenience(home\_location)}$$

$$\text{Satisfaction(home)} = \text{suitability(home)} \,\Delta\, \text{affordability(home)}.$$

In these examples we can see obvious goals of a decision-maker: the goal to find employment, the goal to buy a product, and the goal to have a suitable home. The goals can be achieved if we *simultaneously* satisfy sub-goals: e.g., to have a convenient home location for a family, the location must be sufficiently close to both the job locations of parents and the schools or day care of children. Obviously, all these goals can significantly differ among individual people.

For many people, a good salary is the mandatory requirement, and a job offer that has a low salary will be rejected regardless of the attractiveness of job, which is desirable, but frequently only optional. Thus, this type of simultaneity is not a strict conjunction, but a compound logic function called conjunctive partial absorption (Chap. 16) that aggregates mandatory and optional inputs (the mandatory input supports the annihilator 0 and the optional input does not support the annihilator).

In the case of buying products the buyer can apply the threshold conjunction, rejecting products below the threshold of quality and below the threshold of affordability (or above the threshold of price). For acceptable quality-affordability combinations, the buyer can apply one of noncommutative partial conjunction operators.

In all presented examples of attribute simultaneity, we cannot assume commutativity, arguments can be correlated (not independent) and the selected importance weights differ from person to person. For example, for some respondents, the suitability of a rented apartment location will be primarily affected by the satisfaction of needs of children and much less by the distance from jobs of parents, yielding significant differences in weights. Therefore, the results of experiments can be distributed with substantial deviation, respondents should be carefully prepared for experiments, and some responses will have to be disqualified. To reduce the effects of individual interests of respondents and the variability of responses, we can ask respondents to *play the role of a specific stakeholder*. In such cases we expect that the impact of individual interests of respondents will be reduced, and the results will primarily depend on the subjective conjunction which is the goal of our experiment.

As a typical subjective conjunction case we can observe a specific homebuyer who evaluates homes by combining two necessary and equally important inputs:

- Q = *overall quality of home* (total area, number of bedrooms, bathrooms, etc.)
- L = *overall quality of home location* (proximity to all points of interest, safety, etc.).

The homebuyer simultaneously needs both Q and L to find a suitable home. Therefore, the overall suitability of home S is a conjunction $S = Q \Delta L$. Our goal is to organize an experiment that shows the mental simultaneity function that is in this case used by decision-makers.

A natural way to make the experiment is to give to respondents strategically selected values of H and L, and to ask them to intuitively assess the most appropriate, justifiable value of S. Thus, we must first decide how to specify H, L, and S.

The simplest approach to specifying Q, L, and S is to use numeric values. These values are degrees of satisfaction of homebuyer requirements. So, they are in the range from no satisfaction (0%) to complete satisfaction (100%). Let us now analyze the problem of specification of (Q, L) data points.

Decisions are made by humans (e.g., young children) before they learn to use numbers. Similarly, decisions are also made by animals, obviously without using numeric coding of the magnitude of inputs. Therefore, logic operations can be intuitively performed by the brain without the use of numbers. Numbers are not naturally necessary, but they are education-based convenient way to specify degrees of satisfaction. So, it seems reasonable to conclude that human decision-makers must first be capable of converting given numeric values of Q and L to natural intuitive percepts (intensity of the feeling of satisfaction). Then, decision-makers intuitively logically combine such percepts to create the resulting percept of the overall satisfaction S. Finally, the resulting intuitive percept S is converted to its numeric value that is returned as the result of the homebuyer experiment.

We now face a natural question to explain the process of converting numbers to intuitive percepts and the intuitive percepts back to numbers. It is easy to note the following: (1) respondents must be familiar (educated and skilled) to work with numbers, (2) all percept-number conversions take time, and (3) intuitive conversions are performed with limited accuracy (they cannot be performed without errors). Therefore, the most convenient respondents are people who in their life or profession often use numeric values (so, they cannot be young children, adults without education, or people from professions that don't use numeric quantifications). While intuitive percepts are assumed to be selected values in the unit interval ([0, 1] or [0, 100%]), their conversion to numbers (and vice versa) always creates conversion errors (differences between the degree of feeling and the numeric expression of its value). In addition, intuitive conversions of round values like 20% are easier and need less effort than conversions of precise decimal values like 21.5421%. In all cases, Miller's "magical number 7 plus or minus 2" [MIL56] should be considered when estimating the resulting accuracy. Consequently, imprecisions and errors in the range from 5 to 15% should be expected as normal, and the use of precise numeric values with more than 2 significant decimal digits is not sufficiently justified. Thus, it is reasonable to use round values with step of 5% or 10% (e.g., 0, 5, 10, 15, ..., 95, 100) to specify Q and L.

The process of number-percept conversion can be avoided by using computer interfaces based on sliders or similar tools that automatically and precisely convert a numeric value (a real number from a unit interval) to geometrically visualized quantity (e.g. a length, or an area). The mechanism of conversion of visualized quantity to intuitive percept of satisfaction, without the use of numbers, might be more natural and more straightforward than the bidirectional conversions of a selected number to the corresponding intuitive percept of the degree of satisfaction. In addition, the resulting quantity selected by a respondent can be converted to numeric value automatically and precisely. Experiments reported in [DUJ04a, DUJ04b] indicate that assessment of importance simulated as lengths of linear segments can be higher than the magical number 7 reported in [MIL56] and [SAA03]. However, at this time, the

Imagine you are buying a home and there are two **equally** important factors in finding your perfect home:

- **Quality of Home** (square footage, organization of rooms, material, age of the home)
- **Location of Home** (the proximity of home to all your points of interest).

Each **equally** important factor is expressed by a numeric value ranging from 0 (completely unsatisfied) to 100% (completely satisfied).

We will provide 20 Quality/Location pairs, and assuming they are **equally** important, ask you to enter your overall satisfaction **S** with the combination on the potential home purchase.

We assume that you want to have both a good home **and** a good location. Your overall satisfaction must range from 0 to 100%.

Please note that in this experiment we need your **intuitive feeling** (percept) of the overall satisfaction, obtained without using any formula, calculations, and calculators.

Fig. 14.3 Request for measurement of subjective conjunction for home satisfaction experiment

comparison of numeric and visual approaches to the experimental analysis of human mental logic operations remains to be a topic for future work.

To analyze logic aggregation in the case of simultaneity of goals, we performed two evaluation experiments: (1) suitability of residence (selecting a home or apartment), and (2) suitability of an arbitrary purchased product. In both cases respondents try to simultaneously attain two obvious goals. In the case of the experiment with residence evaluation, the goals of respondents are to have a quality home in a convenient location. These two requests must be simultaneously satisfied and the overall satisfaction with the value of home is not affected by financial components. In the case of buying a desired product, the financial aspect is present, and the goal of respondents is to get a "good deal," i.e. a conveniently priced good product. Obviously, such goals can resonate in a different way in the mind of individual respondents. The request for home suitability experiment is shown in Fig. 14.3. In the product buying experiment we use the similar request, shown in Fig. 14.4. We want to compare the results of these similar experiments.

There are six main steps in the design and organization of experiments with the simultaneity of goals: (1) selection of respondents, (2) preparation of respondents, (3) selection of questions and input data, (4) collection of results, (5) filtering of results, and (6) processing and interpretation of results.

Unpaid respondents that usually participate in such experiments are students who cannot devote a long time for careful and precise preparation of responses. Their preparation is regularly insufficient, because (to facilitate the collection of results) experiments are organized on the Internet and for some participants it is reduced to quick reading a short description exemplified in Figs. 14.3 and 14.4. The results are collected using an Internet application and the interface exemplified in Fig. 14.5.

In this experiment we are asking you to *play the role of an average buyer of an arbitrary product for which you have an average need.* In this role you want **at the same time** a good quality product and a good (affordable) price. These two values are **equally important** to you.

Assuming that you play the role of an average buyer, we will give you the degree of your satisfaction with the **product quality (Q)** and the degree of your satisfaction with the **product price (P)** for 20 products. Both Q and P can be any value between 0% (completely unsatisfactory) and 100% (completely satisfactory).

For each combination of Q and P values, we will ask you to enter your intuitive feeling of the **overall satisfaction (S)** with the given Q, P pair. The value S should be in the range from 0% (unacceptable) to 100% (an excellent deal).

We assume that you want to have both a good quality **and** a good price of each product. Please note that in this experiment we need your **intuitive feeling of the overall satisfaction,** obtained without using any formula, calculations, and calculators.

Fig. 14.4 Request for measurement of subjective conjunction for product satisfaction experiment

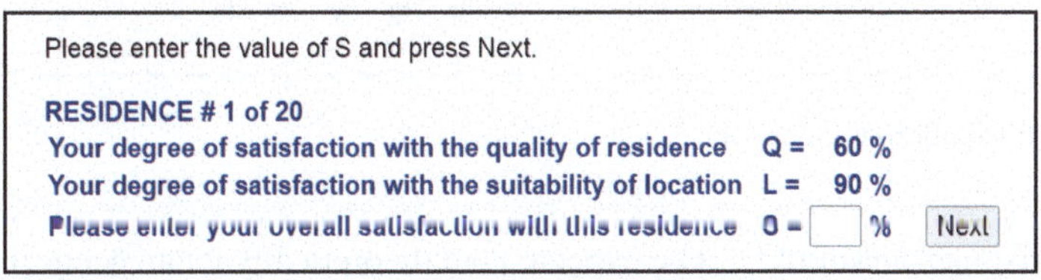

Fig. 14.5 User interface for collecting the residence satisfaction responses

The preparation of input data is a central point in the design of these experiments. We selected 20 values (x, y pairs), expressed as percentages, and shown in Fig. 14.6. The selected pairs of (x,y) values are given to respondents in a random order. Figure 14.5 illustrates a random sequence that starts with the pair #7. The (x,y) values are selected to satisfy the following requirements:

- To keep respondents focused, and the experiment time low, the number of (x,y) pairs given to each respondent must be moderate. We selected 20, which is consistent with experiments previously reported in the literature [THO79, KOV92].
- Data should be symmetric: in 9 cases $x < y$, and in 9 cases $x > y$. So, the cases $x \leq y$ and $x \geq y$ are balanced.
- Idempotency is tested in 2 measurements (#1 and #2).
- Hyperconjunction is tested with small input values (primarily #1), but theoretically it can be visible in 18 cases of nonzero inputs.

#	1	2	3	4	5	6	7	8	9	10
x [%]	**10**	**80**	**90**	**65**	**50**	**85**	**60**	**55**	**100**	**100**
y [%]	**10**	**80**	**80**	**80**	**30**	**60**	**90**	**90**	**60**	**55**
$\|x - y\|$	0	0	10	15	20	25	30	35	40	45

#	11	12	13	14	15	16	17	18	19	20
x [%]	**90**	**40**	**40**	**20**	**75**	**30**	**90**	**15**	**85**	**0**
y [%]	**40**	**90**	**95**	**80**	**10**	**100**	**15**	**95**	**0**	**90**
$\|x - y\|$	50	50	55	60	65	70	75	80	85	90

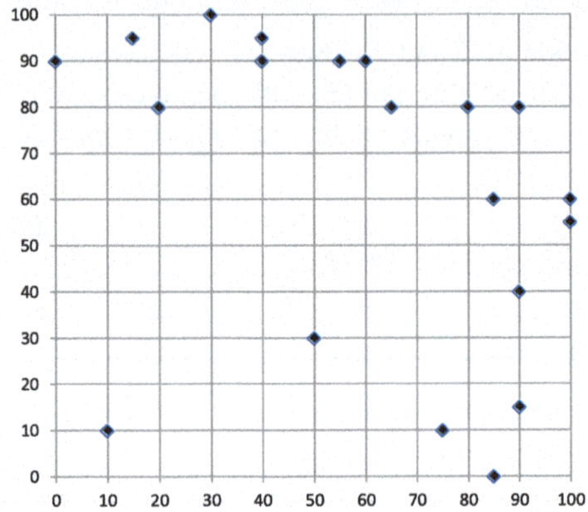

Fig. 14.6 Distribution of input data

- Hyperdisjunction should not be expected but theoretically it can occur in 17 cases where inputs do not include 100, primarily in situations where inputs have high values.
- Commutativity is expected and tested in #11 and #12.
- Hard aggregation (annihilator 0) is expected and explicitly checked in two measurements (#19 and #20).
- For easier intuitive aggregation, all input values are rounded so that (x mod 5) = (y mod 5) = 0.
- Conjunctive aggregation is expected in 18 cases (#3–20).
- Conjunctive aggregation is expected to be present regardless of the difference between the inputs x and y. Consequently, the inputs are selected to create a regularly increasing sequence of positive differences $|x - y| = 10, 15, 20, 25, 30, 35, 40, 45, 50, 55, 60, 65, 70, 75, 80, 85, 90$.
- To emphasize the effects of simultaneity it is desirable to distribute points so that they are not close to the idempotency line $x = y$ (see Fig. 14.6).

Unexpected and unexplainable opinions are a visible phenomenon in experiments with human reasoning. For example, regardless of explicit requests, typically more than 10% of respondents cannot resists temptation to use the arithmetic mean formula strictly and permanently; each their answer is $(x + y)/2$ and the resulting andness is

Table 14.1 Possible, expected, and measured cases in subjective conjunction experiments

Property	Condition $m = x \wedge y, M = x \vee y$	Possible cases	Expected cases	Measured cases
Idempotent	$x = y = z$	2	0	I
Conjunctive	$m \leq z < (0.5x + 0.5y) < M$	18	18	C
Neutral	$m < (z = 0.5x + 0.5y) < M$	18	0	N
Disjunctive	$m < (0.5x + 0.5y) < z \leq M$	18	0	D
Hyperconjunctive	$z < m \leq M$	18	1	CC
Hyperdisjunctive	$m \leq M < z$	17	2	DD
Annihilator 0	$z = m = 0 =$ annihilator	2	2	CH
Annihilator 1	$z = M = 1 =$ annihilator	3	0	DH

exactly 0.5. It is fully justifiable to detect and disqualify such inputs, because they do not reflect the intuitive feeling of andness, i.e., they only reflect the knowledge of a trivial formula.

Sometimes, respondents give the same answer to each question. Regardless of the motivation for such results collected from anonymous volunteer respondents, there is an obvious need to have the possibility to detect and reject measurements that reflect significant inconsistency with the description of experiment provided to all respondents. This is the *problem of filtering*: before processing collected data, it is necessary to remove those inputs that explicitly violate the spirit and the main requirements of the experiment.

If x and y are inputs of logic aggregator $z = A(x, y)$, then we can classify types of aggregation as shown in Table 14.1. For each property we have the possible number of cases and the expected number of cases that reflect the correct understanding of the experiment. For example, in the case of idempotency we selected very small values (10%) and rather large values (80%); thus, from some respondents we could also expect the hyperconjunction and hyperdisjunction (see Fig. 14.11).

For each response we can measure the frequencies *I, C, N, D, CC, DD, CH*, and *DH*. These indicators are not independent and can be used for computing the following experiment understanding error:

$$E = I + |C - 18| + 2N + 3D + |CC - 1| + |DD - 2| + |CH - 2| + DH.$$

The errors are weighted to differentiate their impact. In the ideal case of correct understanding of experiment, we have $E = 0$. High error scores can be used for disqualifying and rejecting inappropriate responses. E.g. if the respondent strictly uses the arithmetic mean, then $I = 2$, $C = 0$, $N = 18$, $D = 0$, $CC = 0$, $DD = 0$, $CH = 0$, $DH = 0$ and consequently $E = 2 + 18 + 36 + 1 + 2 + 2 = 61$. If the rejection threshold is set at 20 (permitting up to10 arithmetic mean answers), this measurement would be convincingly rejected.

In the context of these measurements, it is advisable to compute the coefficient of linear correlation r between the sequence of 20 respondent's values S and the corresponding sequence of 20 arithmetic means $(Q + L)/2$ or $(Q + P)/2$. The high values of r (the values close to 1) detect the unacceptable extensive use of the arithmetic mean. The unacceptable use of constant values is easily detected if the standard deviation of S values is equal to or close to 0. Such responses must be eliminated. Of course, the filtering process must be used with caution, avoiding any form of bias.

Results of the Goal Realization Experiment

In the case of the home satisfaction experiment, based on request shown in Fig. 14.3, the total number of respondents was 92. They provided 1840 aggregators. Our second experiment (product satisfaction) is performed with 166 respondents (different people from the first experiment and different time of experiment). They provided 3320 aggregators. The request for this experiment is shown in Fig. 14.4. Input data were the same as in the first experiment. This provides the opportunity to compare the two experiments as shown in Table 14.2 [DUJ22].

The Applicability of GCD Aggregators

Measurements collected in the home satisfaction and the product satisfaction experiments can be used to evaluate the applicability of GCD aggregators. Let $s(Q, L)$ be the mean value of the 92 home satisfaction values collected for a given Q, L pair. This function can be modeled using the following GCD aggregator:

$$GCD(Q, L) = (0.533Q^{0.414} + 0.467L^{0.414})^{2.41}, \alpha = 0.599.$$

According to Table 14.3, this model fits measured values with a mean absolute error of 2.39%. The global andness of this model is $\alpha = 0.599$ which is consistent with the overall mean local andness for this experiment shown in Table 14.2.

In the case of product satisfaction, the mean overall suitability (s) is a function of product quality (Q) and product affordability/price (P). It can be modelled as the following GCD aggregator:

$$GCD(Q, P) = (0.567Q^{0.386} + 0.433P^{0.386})^{2.59}, \alpha = 0.603.$$

This model fits measured values with a mean absolute error of 3.47%.

It is easy to see that the results of both experiments are very similar. The weights of inputs are rather close, showing that the specifications of experiments sufficiently preserve commutativity. The mean local andness for both experiments is consistent with the global andness of resulting GCD analytic models. Both experiments were performed *with all collected data, i.e. without filtering.* That included questionable disjunctive and hyperdisjunctive aggregators, and they reduced the overall andness close to 0.6 and generated soft analytic models but showed that the GCD analytic models can fit experimental data with high accuracy. If we reject the strict use of the arithmetic mean (approximately 15% of aggregators) and approximately 25% disjunctive aggregators, then the andness enters in the range above 0.75. As expected, that is the region of hard aggregators.

The experiments with simultaneity of goal realization show the presence of hyperconjunctive aggregators, and in that region, the mean measured values of andness were close to 1.5 (Łukasiewicz t-norm); that is stronger than the product t-norm aggregator which has the global andness of 1.25. GCD versions of logic aggregators with the global andness of 1.25 and 1.5 are shown in Fig. 14.7.

An Empirical Analysis of the Diagonal Section of Hyperconjunctive Aggregators

The diagonal section function of an aggregator $A(x_1, \ldots, x_n)$ is defined as the function that shows the value of the aggregator along the diagonal $x_1 = \ldots = x_n = x$ of the hypercube $[0, 1]^n$. The diagonal section $\delta_A(x) = A(x, \ldots, x)$ shows the fundamental character of the aggregator since for idempotent aggregators $\delta_A(x) = x$. Hyperconjunctive aggregators are noncommutative and for them $\delta_A(x) < x$ for $0 < x < 1$.

Table 14.2 The results of subjective conjunction experiments

Conditions: same inputs, different respondents	Subjective conjunction experiments	
Measured values based on local andness of 2 variables	Home satisfaction	Product satisfaction
Total number of respondents	92	166
Total number of analyzed logic aggregators	1840 (100%)	3320 (100%)
Overall mean local andness for this experiment	0.6195	0.6603
Total number of averaging aggregators (means)	1550 (84.24%)	2667 (80.33%)
Mean local andness of averaging aggregators (means)	0.57	0.5892
Total number of partial conjunction aggregators	657 (35.71%)	1102 (33.19%)
Mean local andness of partial conjunction aggregators	0.711	0.7216
Total number of conjunctive averaging aggregators	801 (43.53%)	1415 (42.62%)
Mean local andness of conjunctive averaging aggregators	0.763	0.7832
Total number of hyperconjunctive aggregators	206 (11.2%)	483 (14.55%)
Mean local andness of hyperconjunctive aggregators	1.4917	1.5065
Total number of (all) conjunctive aggregators	1007 (54.73%)	1898 (57.17%)
Mean local andness of all conjunctive aggregators	0.912	0.9673
Total number of pure conjunction aggregators	144 (7.83%)	313 (9.43%)
Total number of logic neutrality aggregators	283 (15.38%)	476 (14.34%)
Total number of pure disjunction aggregators	58 (3.15%)	107 (3.22%)
Total number of partial disjunction aggregators	408 (22.17%)	669 (20.15%)
Mean local andness of partial disjunction aggregators	0.3209	0.3366
Total number of disjunctive averaging aggregators	466 (25.33%)	776 (23.37%)
Mean local andness of disjunctive averaging aggregators	0.281	0.2902
Total number of hyperdisjunctive aggregators	84 (4.57%)	170 (5.12%)
Mean local andness of hyperdisjunctive aggregators	-0.6073	-0.6292
Total number of (all) disjunctive aggregators	550 (29.89%)	946 (28.49%)
Mean local andness of all disjunctive aggregators	0.1453	0.125

An empirical analysis of the diagonal section of hyperconjunctive aggregators used in human intuitive reasoning can be based on request that respondents estimate the overall suitability of a hyperconjunctive aggregator. Our experiment is based on two inputs: the quality of home Q and the suitability of home location L. The respondents play the role of a homebuyer where Q and L are assumed to be equally important.

Obviously, the overall suitability of home S is a conjunctive function of Q and L: every homebuyer simultaneously wants a good home in a suitable location. The corresponding aggregator $S = A(Q, L)$ is obviously hard: nobody would buy a home with zero quality or a home in completely unacceptable location; thus, $A(0, L) = A(Q, 0) = 0$. In addition, if Q is extremely low, or L is extremely low, many homebuyers would reject such a home, assigning $Q = 0$ even in cases where $Q > 0$ and $L > 0$. If $S < \min(Q, L)$ the aggregator is hyperconjunctive and this example is convenient for observing the use of hyperconjunction in human reasoning. In the first step, we would like to investigate the diagonal section function, i.e., the special cases of equal values $Q = L$ (or $Q \approx L$). Therefore, the respondents are given the Table 14.4 of equal values for Q and L.

To prepare the respondents, we supply the following guidelines:

Table 14.3 The accuracy of GCD model for the home satisfaction experiment

#	Q	L	s [%]	GCD [%]	Error [%]
1	80	80	83.7	80	−3.66
2	40	90	59.8	60.4	0.631
3	85	0	19.1	18.6	−0.486
4	90	80	86.9	85.2	−1.7
5	60	90	73.9	73.1	−0.797
6	100	55	72	77	5.06
7	40	95	59.6	62.3	2.65
8	55	90	66.3	70.1	3.82
9	20	80	43.1	42.2	−0.9
10	100	60	76.6	79.8	3.28
11	75	10	36.7	35.7	−0.914
12	90	15	43.9	45.7	1.82
13	10	10	6.21	10	3.79
14	85	60	71.4	72.7	1.3
15	15	95	40.5	42.3	1.8
16	30	100	51.7	56.7	5.01
17	50	30	34.7	39.9	5.23
18	90	40	62.3	63.7	1.42
19	65	80	70	71.8	1.77
20	0	90	16.1	14.3	−1.77

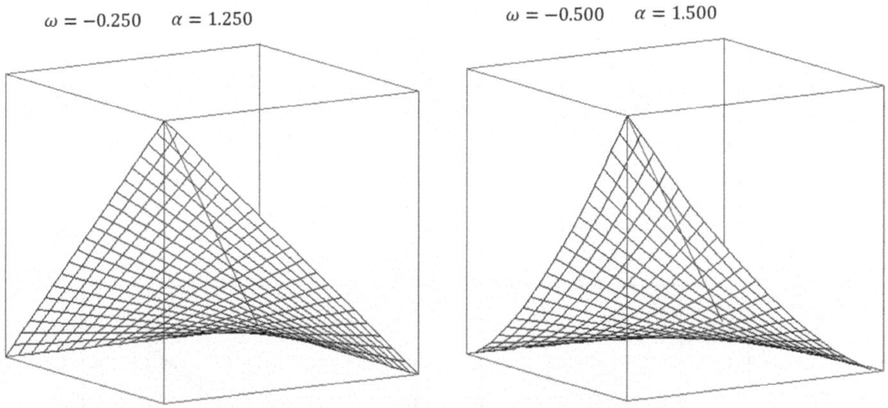

Fig. 14.7 Two hyperconjunctive GCD aggregators with andness 1.25 and 1.5

Table 14.4 Request for evaluation

Q [%]	2	4	6	8	10	12	14	16	18	20	25	30	35	40	45	50	60	70	80	90
L [%]	2	4	6	8	10	12	14	16	18	20	25	30	35	40	45	50	60	70	80	90
S [%]																				

Please, insert in row S your overall satisfaction with the (Q, L) values. The values of S should reflect your opinion in the assumed role of homebuyer who simultaneously wants a sufficiently good home in a sufficiently suitable location. Of course, you can answer any values that you feel are appropriate, but please explain and justify the idea of your reasoning. The reason for this experiment is the following: people sometimes want to put a high penalty in the cases of low satisfaction. This can be done in the following two ways:

(a) For some very small positive Q, L values you might say "this Q, L pair is so low that it is completely unacceptable for me. Therefore, I will reject it with confidence and assign $S = 0$."
(b) For some low Q, L values you might feel that you are satisfied less than Q, L and assign a smaller positive value $0 < S < Q = L$, because you want to penalize, but not to reject, that combination of values.

Since we need simultaneously high Q and high L, we are supposed to be satisfied with high values; in such cases, you may want to assign $S = Q = L$. Assigning $S = 100\%$ in cases $Q = L < 100\%$ is unacceptable because you simultaneously want high quality Q and the high suitability L, and it is not reasonable to equalize the cases $Q = L < 100\%$ and $Q = L = 100\%$. In this situation it is not justifiable to consider that the cases $Q = L < 100\%$ and $Q = L = 100\%$ are equivalent.

Based on this preparation we collected valid responses from 28 respondents. The mean values of their responses (S) are shown in Table 14.5 and in Fig. 14.8. For all

Table 14.5 Average values of 28 responses

Q [%]	2	4	6	8	10	12	14	16	18	20
L [%]	2	4	6	8	10	12	14	16	18	20
S [%]	0	0.143	0.25	0.857	2.21	3.75	4.82	6	7.54	10.8

Q [%]	25	30	35	40	45	50	60	70	80	90
L [%]	25	30	35	40	45	50	60	70	80	90
S [%]	15.3	20.2	25.4	31.6	37.9	44.6	54.6	67.1	79	89.3

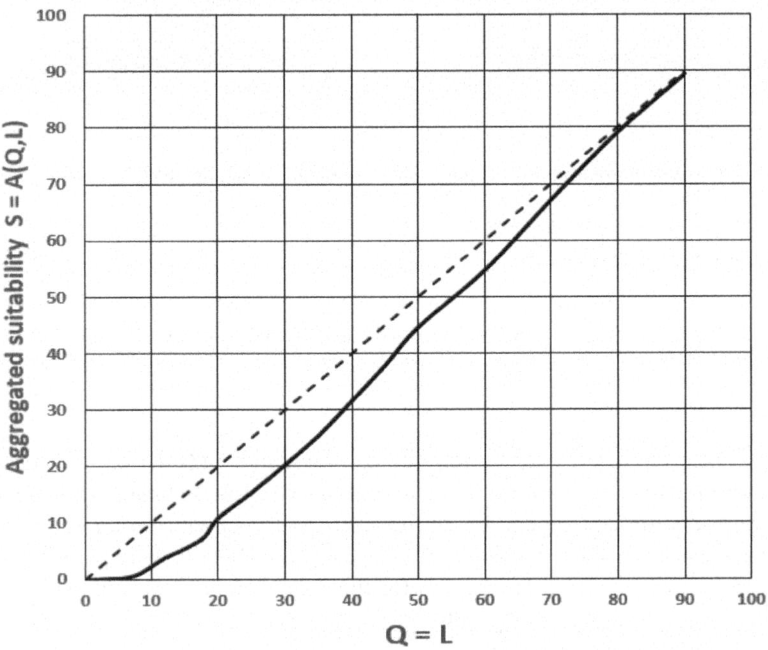

Fig. 14.8 Average values of responses of 28 respondents (the diagonal section function)

analyzed $Q = L$ values we have the mean aggregated suitability $S < Q = L$. Therefore, in this experiment, the mental aggregator $S = A(Q, L)$ is hyperconjunctive. The local andness has a typical decreasing shape shown in Fig. 14.9.

In the special case where $Q = L$, the hyperconjunctive aggregator has the diagonal section function in the following interpolative global andness-directed GCD form [DUJ18]:

$$S = \delta(Q, \alpha) = \begin{cases} (5 - 4\alpha)Q + 4(\alpha - 1)Q^2, & 1 < \alpha < 1.25 \\ Q^{2(\sqrt{3/(2-\alpha)}-1)} & 1.25 < \alpha < 2 \end{cases}$$

From Table 14.5 we know the values $Q_i, S_i, i = 1, \ldots, 20$. So, the best value of global andness α is one that minimizes the following mean absolute error function:

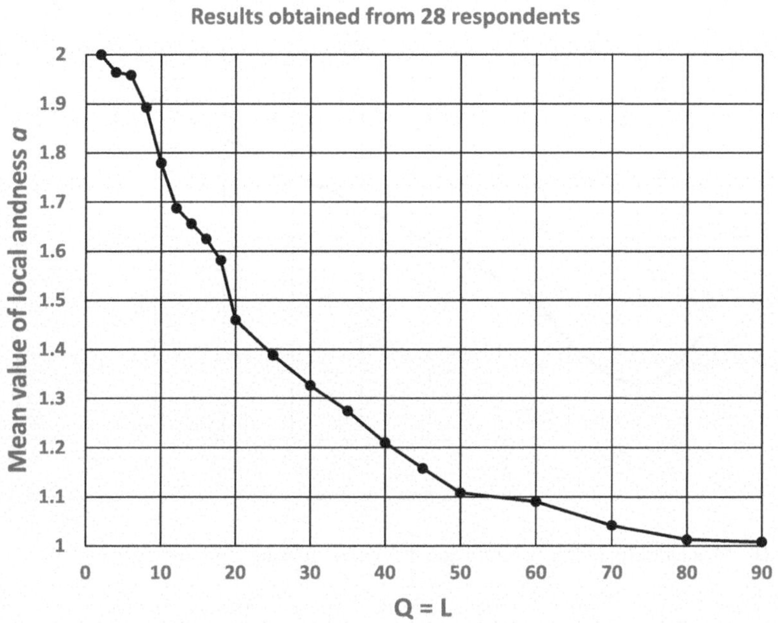

Fig. 14.9 Mean local andness a as a function of equal Q, L values

$$E(\alpha) = \frac{1}{20}\sum_{i=1}^{20}|S_i - \delta(Q_i, \alpha)|.$$

The function $E(\alpha)$ is shown in Fig. 14.10. The optimum value of global andness is $\alpha = 1.1$ and the corresponding mean absolute error is $E(1.1) = 3.324\%$.

The conclusion of this homebuyer experiment is that human mental aggregation uses hyperconjunction. The properties of commonsense logical reasoning can be investigated using the diagonal section function. The hyperconjunctive reasoning of our group of 28 respondents can be characterized using both the local andness and the global andness. The local andness is a strictly decreasing function of the value of identical inputs. The mean global andness of 1.1 is between the pure conjunction and the product t-norm.

Andness Variability for Extreme Inputs

Experiments with simultaneity of stakeholder's goals frequently show the phenomenon of variability of andness as a function of the values of inputs. The basic shape of that phenomenon is shown in Fig. 14.11. Generally, the andness of subjective conjunction is expected to depend on the type and importance of stakeholder's goals. Higher levels of andness reflect a high importance of attaining stakeholder's goals, and theoretically, that should have a stable value independent of specific input values. For example, in the case of a respondent who plays the role of a homebuyer

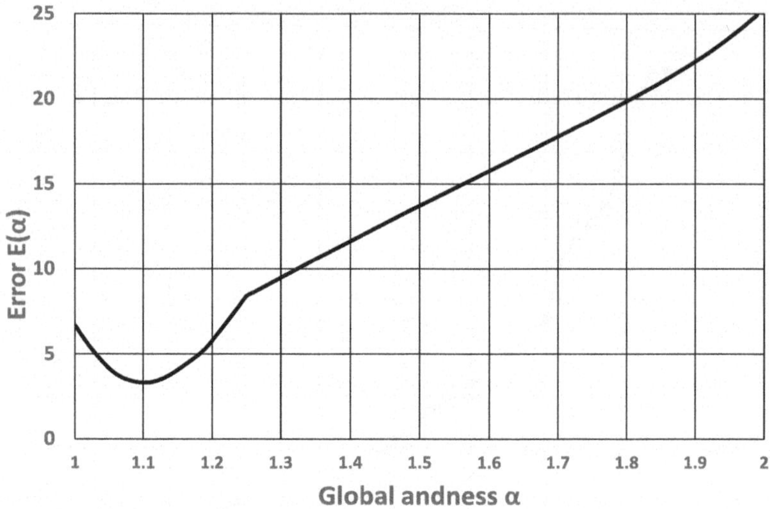

Fig. 14.10 Mean absolute error function and the optimum value of global andness

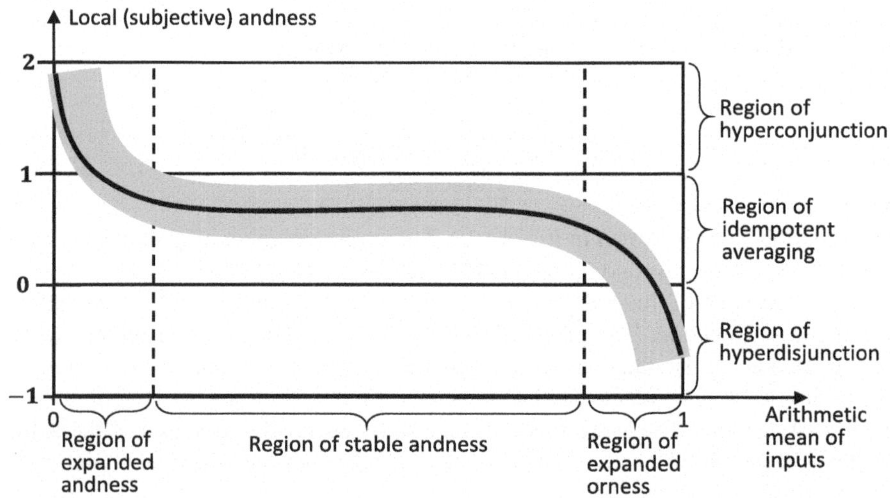

Fig. 14.11 Regions of expanded local andness and expanded local orness

who has the goals to find a good home in a highly suitable location, we might compute the following local andness:

Input: $(80\%, 40\%)$. Respondent's output: 50%. Local andness: $a = \frac{80-50}{80-40} = \frac{30}{40} = 0.75$.

Input: $(95\%, 35\%)$. Respondent's output: 50%. Local andness: $a = \frac{95-50}{95-35} = \frac{45}{60} = 0.75$.

Input: (30%, 70%). Respondent's output: 40%. Local andness: $a = \frac{70-40}{70-30} = \frac{30}{40} = 0.75$.

Input: (30%, 50%). Respondent's output: 35%. Local andness: $a = \frac{50-35}{50-30} = \frac{15}{20} = 0.75$.

In all these examples, the respondent's output suitability yields a stable value of local andness $a = 0.75$. However, some experiments show the instability of andness for extreme (very low or very high) values of inputs. In Fig. 14.11, these are two regions denoted as the region of expanded andness and the region of expanded orness.

The values of inputs can be collectively characterized by the arithmetic mean of inputs. That is used as abscissa in Fig. 14.11. For small values of the arithmetic mean of inputs (or for small values of the maximum of inputs) our experiments frequently show a significant increase of andness. The explanation of this phenomenon is simple: suppose that the stakeholder decided to request a high andness (e.g., to simultaneously have a very suitable home location and a high-quality home). So, if the inputs are very low (e.g., below 10–20%), then the respondent who plays the role of such a stakeholder can frequently consider that such a combination is unacceptable, and it must be rejected. That usually yields the overall suitability 0 or a hyperconjunctive value which is less than the minimum input. The result is a sharp increase of local andness in the region od expanded andness.

A similar situation is encountered in the region of very high input suitability (a high value of the minimum input). E.g., if a home is located in an ideal location (location suitability = 100%), and the quality of home is close to perfect (e.g. 90%), the respondent can be impressed with this exceptional opportunity and decide to neglect small imperfections assigning the highest overall suitability of 100%, which is a pure disjunction. Likewise, if two very high inputs (e.g. 95% and 95%) produce respondent's suitability of 100%, then the resulting orness is increased and enters the region of hyperdisjunction. Such respondent's reasoning creates a sharp decline of andness (and the corresponding increase of orness) in the region of very high input suitability. Therefore, the resulting values of local andness frequently have the shape of an integral and belong to the shaded area, as the sample function shown in Fig. 14.11.

The main result of presented experiments in the goal realization area is that the *local andness of subjective conjunction aggregators is not constant*. For small values of arguments, many respondents justifiably gravitate towards the use of hyperconjunctive aggregators. For large values of arguments, regardless of the conjunctive decision problem, some respondents gravitate towards the use of hyperdisjunctive aggregators. This behavior shows the duality of hyperconjunctive and hyperconjunctive reasoning. For arguments in the area between extremes, respondents typically use conjunctive averaging aggregators and steady values of local andness.

The main conclusion of this chapter is that human mental aggregation is observable, measurable, and can be modeled in GL with rather good accuracy. That opens a wide space for creating decision-support systems that can generate results that are provably similar to human reasoning. Therefore, this area of research deserves serious attention.

We experimentally investigated logic aggregators in the area of modeling simultaneity (the subjective conjunction). We showed that it is useful to differentiate two types of human mental engagement: (1) as a neutral observer of selected properties of objects, and (2) as a stakeholder/decision-maker interested in goal attainment. These are two different types of simultaneity and each of them has characteristic properties that can be experimentally investigated.

In decision engineering, mathematical models can be validated in only one way: by making experiments. Without experimental validation, questionable theoretical assumptions can easily penetrate in decision-support systems. One of such assumptions is that the product norm and the pure conjunction can always be used as models of simultaneity (or intersection of fuzzy sets). It is safe to claim that the credible models of simultaneity of attributes are always located in the range of andness visibly below 1, rejecting both the conjunction and the product norm. Five independent experiments with more than 250 respondents convincingly verify this conclusion.

Experiments with the simultaneity of goal realizations are significantly affected by the variability of individual interests and experiences of responding subjects. Unsurprisingly, that generates a spectrum of aggregators which justifiably include hyperconjunction and hyperdisjunction. While the experiments with simultaneity of attributes generated conclusions that might be considered definitive, the experiments with simultaneity of goals open various questions/problems that remain for the future work. Such problems include respondent preparation techniques, response filtering algorithms, empirical studies of the commonsense use of hyperconjunction and hyperdisjunction, and the development of nonstationary GCD aggregators where andness is not a constant parameter, but a function of input arguments.

References

ALS83 Alsina, C., E. Trillas, and L. Valverde. 1983. On Some Logical Connectives for Fuzzy Sets Theory. *Journal of Mathematical Analysis and Applications* 93: 15–26.

DUJ79 Dujmović, J. 1979. Partial Absorption Function. *Journal of the University of Belgrade, Series Mathematics and Physics* 659: 156–163.

DUJ18 Dujmović, J. 2018. *Soft Computing Evaluation Logic*. Wiley.

DUJ04a Dujmović, J., and W.Y. Fang. 2004. An Empirical Analysis of Assessment Errors for Weights and Andness in LSP Criteria. In *Modeling Decisions for Artificial Intelligence*, ed. V. Torra and Y. Narukawa, 139–150. Springer LNAI 3131.

DUJ04b Dujmović, J., and W.Y. Fang. 2004. Reliability of LSP Criteria. In *Modeling Decisions for Artificial Intelligence*, edited by V. Torra and Y. Narukawa, 151–162. Springer LNAI 3131.

DUJ22 Dujmović, J., and D. Tomasevich. 2022. Experimental Analysis and Modeling of Human Conjunctive Logic Aggregation. In *2022 IEEE International Conference on Fuzzy Systems (FUZZ-IEEE)*, 1–8. Padua, Italy. https://doi.org/10.1109/FUZZ-IEEE55066.2022.9882665.

GOG69 Goguen, J.A. 1969. The Logic of Inexact Concepts. *Synthese* 19: 325–373.

HIS88 Hisdal, E. 1988. Are Grades of Membership Probabilities? *Fuzzy Sets and Systems* 25: 325–348.

KLE99 Klement, E.P., and M.M. Navara. 1999. Propositional Fuzzy Logics Based on Frank T-Norms: A Comparison. In: Fuzzy Sets, Logics and Reasoning about Knowledge. Applied Logic Series, vol 15, ed. D. Dubois, H. Prade, and E.P. Klement. Dordrecht: Springer. https://doi.org/10.1007/978-94-017-1652-9_3.

KLI95 Klir, G.J., and B. Yuan. 1995. *Fuzzy Sets and Fuzzy Logic*. Prentice-Hall.

KOV92 Kovalerchuk, B., and V. Taliansky. 1992. Comparison of Empirical and Computed Values of Fuzzy Conjunction. *Fuzzy Sets and Systems* 46: 49–53.

MAY91 Mayor, G., and J. Torrens. 1991. On a Family of T-Norms. *Fuzzy Sets and Systems* 41: 161–166.

MIL56 Miller, G.A. 1956. The Magical Number Seven, Plus or Minus Two: Some Limits on Our Capacity for Processing Information. *The Psychological Review* 63: 81–97.

ODE77 Oden, G.C. 1977. Integration of Fuzzy Logical Information. *Journal of Experimental Psychology: Human Perception and Performance* 3 (4): 565–575. https://doi.org/10.1037/0096-1523.3.4.565.

RAD00 Radojević, D.G. 2000. [0,1]-Valued Logic: A Natural Generalization of Boolean Logic. *Yugoslav Journal of Operations Research* 10 (2): 185–216.

SAA03 Saaty, T.L., and M.S. Ozdemir. 2003. Why the Magic Number Seven Plus or Minus Two. *Mathematical and Computer Modelling* 38: 233–244.

THO79 Thole, U., H.-J. Zimmermann, and P. Zysno. 1979. On the suitability of minimum and product operators for the intersection of fuzzy sets. *Fuzzy Sets and Systems* 2: 167–180.

WIK25 Wikipedia. *Fuzzy Logic*. https://en.wikipedia.org/wiki/Fuzzy_logic. Retrieved 18 Jan 2025.

WIK22 Wikipedia. 2022. *The T-Norm Fuzzy Logics*. https://en.wikipedia.org/wiki/T-norm_fuzzy_logics.

ZIM87 Zimmermann, H.-J. 1987. *Fuzzy Sets, Decision-Making, and Expert Systems*. Kluwer Academic Publishers.

ZIM96 Zimmermann, H.-J. *Fuzzy Set Theory and Its Applications*. New York: Springer Science+Business Media.

ZIM80 Zimmermann, H.-J., and P. Zysno. 1980. Latent Connectives in Human Decision-making. *Fuzzy Sets and Systems* 4: 37–51.

ZYS79 Zysno, P. 1979. *One Class of Operators for the Aggregation of Fuzzy Sets*. Amsterdam: EURO III Congress.

Chapter 15
A Fourth Look at GCD: Continuum-Valued, Andness-Directed, Importance-Weighted, Idempotence-Selectable, Annihilator-Selectable, Dualized, Interpolative, and Conditional GL Aggregators

Mathematical models of aggregators are a very wide area and a topic of many papers and a significant or exclusive topic of several books. In [YAG94, BEL07, GRA09, KAP15, BEL16] aggregators are not interpreted as logic functions. In this book, as well as in [DUJ18], aggregators are interpreted as andness-directed logic functions, based on observable properties of the commonsense human logic reasoning. Our main goal in this chapter is to create analytic models of the GCD aggregator. The GCD models must be consistent with observable properties of natural human reasoning and convenient for applications in decision-support systems. Thus, we will primarily focus on those topics that are related to *andness-directed importance-weighted GCD aggregators* and the Segmented Interpolative Method (SIM) for their design. Included are the following topics:

- Andness-directedness.
- GCD models with fixed and implicit andness.
- GCD models with adjustable parameters.
- GCD models with adjustable conjunction degrees.
- Andness-directed GCD models based on parameter transformation.
- Andness-directed GCD aggregators based on the Segmented Interpolative Method.
- Conditional logic aggregation.

Andness-Directedness

In graded propositional calculus we create formulas based on superposition of GCD aggregators and the standard negation. Aggregation properties of each GCD aggregator of n arguments depend on $n+1$ adjustable parameters: n importance weights and a desired andness, shown in Fig. 13.1. However, the logic properties of GCD aggregators are adjustable only by selecting a desired andness, because andness is defined for equal importance weights. Therefore, andness/orness is the crucial adjustable parameter in Graded Logic, and it must be *explicitly visible and easily*

J. Dujmović, *Graded Logic*, Intelligent Systems Reference Library 273,
https://doi.org/10.1007/978-3-031-88558-7_15

adjustable in analytic models of GCD aggregators.[1] Such aggregators are classified as *andness-directed* and denoted $GCD(\boldsymbol{x}; \boldsymbol{W}, \alpha)$, where input arguments are $\boldsymbol{x} = (x_1, \ldots, x_n)$, $x_i \in [0, 1]$, $i = 1, \ldots, n$, $n \geq 2$, and the importance weights are $\boldsymbol{W} = (W_1, \ldots, W_n)$, $0 < W_i \leq 1$, $i = 1, \ldots, n$. In a general case, the andness α is assumed to cover the full range of logic aggregators from the drastic conjunction to the drastic disjunction: $-1/(n-1) \leq \alpha \leq n/(n-1)$. In graded propositional calculus, we define andness-directedness of GCD aggregators as follows:

Definition 15.1 Graded logic aggregator $GCD(\boldsymbol{x}; \boldsymbol{W}, \alpha)$ is *fully andness-directed* if and only if it satisfies the following condition:

$$\forall \alpha \in \left[\frac{-1}{n-1}, \frac{n}{n-1}\right] \Rightarrow \frac{n}{n-1} - \frac{n+1}{n-1} \int\limits_{[0,1]^n} GCD\big(\boldsymbol{X}; \underline{\boldsymbol{W}}, \alpha\big) dx_1 \cdots dx_n = \alpha,$$

$\underline{\boldsymbol{W}} = \left(\frac{1}{n}, \ldots, \frac{1}{n}\right)$ (sum-normalized), or $\underline{\boldsymbol{W}} = (1, \ldots, 1)$ (max-normalized or count-normalized), $n \geq 2$.

Considering that there are no natural GCD functions that have andness as an explicitly visible parameter, all andness-directed importance-weighted GCD functions use numeric approximations (usually based on polynomials). The quality of approximation can be assessed and controlled using the following error function:

$$E_n(\alpha) = \left| \alpha - \frac{n}{n-1} + \frac{n+1}{n-1} \int\limits_{[0,1]^n} GCD\big(\boldsymbol{X}; \underline{\boldsymbol{W}}, \alpha\big) dx_1 \cdots dx_n \right|,$$

$$-\frac{1}{n-1} \leq \alpha \leq \frac{n}{n-1}, n \geq 2.$$

Error analysis in [DUJ21d] shows that such approximations can have sufficient accuracy where the maximum values of $E_n(\alpha)$ can be easily kept below 1%.

Definition 15.1 specifies GCD aggregators that are *fully andness-directed* because the andness-directedness is supported in the full (maximum) range of andness, including both idempotent and nonidempotent aggregators. On the other hand, GCD aggregators can also be *partially andness-directed* if they support andness directedness in a reduced range of andness, typically in the range of idempotent aggregators $0 \leq \alpha \leq 1$.

[1] This requirement reflects all professional decision-making applications of graded logic functions where decision analysts specify, verify, and modify desired values of andness. In such applications andness is an adjustable input parameter. On the other hand, there are infrequent situations where we can use a specific aggregator that has a fixed andness (e.g., the harmonic mean $z = 2xy/(x+y)$, where $\alpha = 0.773$). In such cases, andness-directedness is not an indispensable property.

GCD Aggregators with Fixed and Implicit Andness

In Fig. 13.8 we presented 7 families of logic aggregators used for building graded propositional calculus formulas. These families include 113 GCD functions, including those that occur multiple times. In Chap. 13, we focused on granularity, precision, importance of inputs and techniques for selecting the most appropriate GCD function. Now is the time to focus on another fundamental question: how to make all these GCD logic aggregators?

The theory of means and aggregation functions [GIN58, MIT69, MIT77, BUL88, MIT89, CAL02, BUL03, CAL03, CAL04, BEL07, GRA09, MES15, BEL16] offers a wide spectrum of functions that are applicable as aggregators. Generally, all functions that are nondecreasing in all arguments and idempotent in extreme points $(0, \ldots, 0)$ and $(1, \ldots, 1)$ are aggregators and many of them are special cases of logic aggregators. Means and t-norms/conorms are the most popular families of such functions. The andness of popular aggregators can be easily computed, but it is never explicitly visible in the original function.

Theoretically, any function that is an aggregator could serve as GCD. However, if we approach GCD aggregators from the practical standpoint of GL applications, it is easy to see that GCD logic aggregators must aggregate any number of logic arguments, be weighted, applicable in the whole range of andness (from the drastic conjunction to the drastic disjunction), and andness-directed, i.e., selectable using the desired value of andness. That yields the need for the continuous transition through the whole range of andness.

All parameters that define the logic properties of an aggregator must be explicitly visible. There is no traditional aggregation function that satisfies these conditions, because the most important logic parameter is andness. So, it is necessary to develop methodology for creating general forms of GCD logic aggregators that are universal, weighted, andness-directed, and easy to use. In the remainder of this chapter, we will move in that direction.

A survey of conjunctive aggregation functions of two variables and their disjunctive duals is presented in Table 10.2. These functions have a spectrum of fixed values of andness taken from the full range of andness. In special cases where GCD can be commutative and aggregate two variables, these functions can be useful. In cases of idempotent aggregators where we need weights and more than two variables, we can use a binary aggregation tree method [DUJ15c, BEL16a, DUJ17a] to generalize bivariate idempotent symmetric aggregators and create multivariate weighted GCD aggregators. This computational technique uses a binary tree of bivariate aggregators and with L levels and 2^L inputs. Then, each input argument is connected to as many inputs of the binary tree as needed to express its importance weight. Computational aspects of this procedure are presented in [BEL16, DUJ18].

In some cases of GCD aggregators the andness is adjustable, but not explicitly visible. The most popular among such aggregators is the *ordered weighted averaging*, OWA, introduced in 1988 by Yager [YAG88, YAG97, TOR07, GRA09, KAC15, BEL16]. The idea of this aggregator is that conjunctive aggregation is sensitive to small values of arguments, and disjunctive aggregation is sensitive to large values of arguments. Therefore, if the arguments are sorted, it is possible to provide conjunctive or disjunctive "attitudinal character of aggregator" [YAG04b] by using weights that emphasize either small or large arguments.

Let the array of arguments be $\boldsymbol{x} = (x_1, \ldots, x_n)$ and let $\sigma(i)$ be the index of the ith largest element. So, the sorted arguments are $x_{\sigma(1)} \geq \ldots \geq x_{\sigma(n)}$, where $x_{\sigma(1)} = \max(x_1, \ldots, x_n) = \vee_{i=1}^{n} x_i$ and $x_{\sigma(n)} = \min(x_1, \ldots, x_n) = \wedge_{i=1}^{n} x_i$. The original OWA aggregator is defined as the scalar product of the vector of sorted arguments and the vector of normalized "logical weights" $\boldsymbol{v} = (v_1, \ldots, v_n)$, $0 \leq v_i \leq 1, i = 1, \ldots, n$, $v_1 + \cdots + v_n = 1$, as follows:

$$OWA(\boldsymbol{x}; \boldsymbol{v}) = v_1 x_{\sigma(1)} + \ldots + v_n x_{\sigma(n)}.$$

Therefore, OWA is a commutative, strictly soft and idempotent aggregator. If $\boldsymbol{v} = (1, 0, \ldots, 0)$, then $OWA(\boldsymbol{x}; \boldsymbol{v})$ becomes disjunction, and if $\boldsymbol{v} = (0, \ldots, 0, 1)$ then $OWA(\boldsymbol{x}; \boldsymbol{v})$ becomes conjunction. If $\boldsymbol{v} = (1/n, \ldots, 1/n)$ then $OWA(\boldsymbol{x}; \boldsymbol{v})$ is the arithmetic mean. The andness and orness of $OWA(\boldsymbol{x}; \boldsymbol{v})$ depend only on selected weights, as follows:

$$\begin{aligned}
\alpha &= 1 - \omega = 1 - \left[(n-1)v_1 + (n-2)v_2 + \cdots + v_{n-1}\right]/(n-1) \\
&= 1 - v_1, \quad n = 2 \\
&= 1 - \left(v_1 + \frac{v_2}{2}\right), \quad n = 3 \\
&= 1 - \left(v_1 + \frac{2v_2}{3} + \frac{v_3}{3}\right), \quad n = 4 \\
&= 1 - \left(v_1 + \frac{3v_2}{4} + \frac{2v_3}{4} + \frac{v_4}{4}\right), \quad n = 5, \text{etc.}
\end{aligned}$$

The weights $v_1, \ldots, v_n$ express logic properties and not the importance of arguments. This is the reason why we distinguish *logic weights* and *importance weights*. In the case of logic weights the andness and orness are "implicit" because they are not explicitly visible but hidden in a specific distribution of weights.

In all aggregators we studied so far, a desired value of andness produces a single aggregator and a corresponding single aggregated value. So, by selecting the value of andness of GCD, we uniquely select the properties of the aggregator. For example, according to Chap. 9, the andness of geometric mean $y = (x_1 \cdots x_n)^{1/n}$ is $\alpha_n^{geo} = \frac{n}{n-1} - \frac{n+1}{n-1}\left(\frac{n}{n+1}\right)^n$. For example, if $n = 3$, then $\alpha_{n3}^{geo} = \frac{21}{32}$. So, if we model GCD using the power mean $y = \left(\frac{x_1^r + x_2^r + x_3^r}{3}\right)^{1/r}$ and want the aggregator that has the andness $\frac{21}{32}$, the unique solution is $r = 0$ and GCD is the hard aggregator $y = (x_1 x_2 x_3)^{1/3}$. On the other hand, if we want OWA aggregator that has the andness $\frac{21}{32}$, then we get the conditions $v_1 + \frac{v_2}{2} = \frac{11}{32}$ and $v_1 + v_2 < 1$ (because $v_1 + v_2 + v_3 = 1$). Of course, we have an infinite number of triplets v_1, v_2, v_3 that satisfy these conditions and each of them gives a different OWA aggregator and different aggregated result. This property can be called the OWA uncertainty: an infinite number of OWA aggregators create the same andness, and the selection of desired andness is not sufficient to uniquely specify the properties of this aggregator. In other words, the original OWA aggregator cannot be andness-directed in a way that is possible with other aggregators.

From the graded logic standpoint, the original OWA aggregator has 4 fundamental problems: (1) uncertainty, (2) commutativity (no importance weights), (3) no support for annihilators (OWA is additive, i.e., a strictly soft aggregator), and (4) no andness-directedness (andness is implicit, and not an explicitly visible input parameter). Elimination of these drawbacks attracted numerous research contributions. For example, Torra introduced importance weights in [TOR97] and Yager in [YAG98]. Other improvements have been proposed in [YAG94, TOR07, TOR21]. The problem of uncertainty can be solved by introducing an additional condition that logic weights must satisfy. In [OHA87] the maximum entropy ME-OWA was proposed and in [YAG94] the S-OWA was proposed as the interpolative form between the conjunction/disjunction of arguments and the arithmetic mean of arguments[2]:

$$y = p\max(x_1, \ldots, x_n) + \frac{(1-p)(x_1 + \ldots + x_n)}{n}, \quad 0 \le p \le 1(\text{disjunctive form})$$

$$y = q\min(x_1, \ldots, x_n) + \frac{(1-q)(x_1 + \ldots + x_n)}{n}, \quad 0 \le q \le 1(\text{conjunctive form}).$$

The disjunctive form can be directly interpreted as OWA with the following logic weights:

$$v_1 = p + \frac{(1-p)}{n}, v_2 = \ldots = v_n = \frac{(1-p)}{n}.$$

If we apply the mean orness theorem to the disjunctive form, then we have the following:

$$\omega = p + \frac{1-p}{2} = \frac{p+1}{2} \Rightarrow p = 2\omega - 1, 1 - p = 2(1-\omega),$$

$$y = (2\omega - 1)\max(x_1, \ldots, x_n) + \frac{2(1-\omega)(x_1 + \ldots + x_n)}{n}, 1/2 \le \omega \le 1.$$

By applying the mean orness theorem to the conjunctive form, we have the following:

[2] It is useful to note that the idea of interpolative aggregation is visible already in the Heronian mean of two variables which is located between the arithmetic mean and the geometric mean: $H(x, y) = \frac{2}{3}\frac{x+y}{2} + \frac{1}{3}\sqrt{xy} = (x + \sqrt{xy} + y)/3$. This form can be generalized using a parameter p as follows: $H(x, y; p) = p\frac{x+y}{2} + (1-p)\sqrt{xy}, 0 \le p \le 1$. Using the mean andness theorem, we get for $H(x, y; p)$ the following andness $= p\frac{1}{2} + (1-p)\frac{2}{3} = \frac{4-p}{6}$. Consequently, $p = 4 - 6\alpha$, and we can have the andness-directed aggregator $GCD(x, y; \alpha) = (4 - 6\alpha)\frac{x+y}{2} + (6\alpha - 3)\sqrt{xy}$, which is applicable in the range $\frac{1}{2} \le \alpha \le \frac{2}{3}$.

$$\omega = \frac{1-q}{2} \Rightarrow q = 1-2\omega, \quad 1-q = 2\omega,$$

$$y = (1-2\omega)\min(x_1, \ldots, x_n) + \frac{2\omega(x_1 + \ldots + x_n)}{n}, \quad 0 \le \omega \le 1/2.$$

This aggregator can be interpreted as OWA, and it is orness-directed. The corresponding andness-directed OWA is obtained using $\omega = 1 - \alpha$:

$$OWA(\boldsymbol{x}; \alpha) = \begin{cases} (1-2\alpha)\max(x_1, \ldots, x_n) + \frac{2\alpha(x_1+\ldots+x_n)}{n}, & 0 \le \alpha \le 1/2 \\ (2\alpha-1)\min(x_1, \ldots, x_n) + \frac{2(1-\alpha)(x_1+\ldots+x_n)}{n}, & 1/2 \le \alpha \le 1 \end{cases}$$

This aggregator is strictly soft, and not importance-weighted. Since the arithmetic mean is explicitly visible, we can make an importance-weighted andness-directed OWA-based GCD by simply using the importance-weighted arithmetic mean:

$$GCD(\boldsymbol{x}; \boldsymbol{W}, \alpha) = \begin{cases} (1-2\alpha)\max(x_1, \ldots, x_n) + \frac{2\alpha(W_1x_1+\ldots+W_nx_n)}{n}, & 0 \le \alpha \le 1/2 \\ (2\alpha-1)\min(x_1, \ldots, x_n) + \frac{2(1-\alpha)(W_1x_1+\ldots+W_n)}{n}, & 1/2 \le \alpha \le 1 \end{cases}$$

To fix the problem of strictly soft aggregation, the arithmetic mean in OWA definition can be replaced with the geometric mean [RAL97] (where the aggregator is strictly hard) or with weighted power mean [YAG04b], and quasi-arithmetic means [BEL16]. A survey of OWA modifications can be found in [DUJ23]. There is also an iterative version of OWA, proposed in [DUJ91] and its andness-directed version [DUJ21d]. Flexible selectability of annihilators is not available in most versions of the OWA family of aggregators.

GCD Models with Adjustable Parameters

The simplest ways to introduce importance weights and andness in aggregation functions are based on those means that naturally include weights and adjustable parameters that affect andness. The most important such means are the descendants of the very general Bajraktarević mean [BUL03]:

$$\begin{aligned} &y = F^{-1}\left(\frac{\sum_{i=1}^{n} w_i(x_i)F(x_i)}{\sum_{i=1}^{n} w_i(x_i)}\right), \\ &F : [0, 1] \to]-\infty, +\infty[\quad (\textit{strictly monotone}), \\ &x_i \in [0, 1], w_i : [0, 1] \to]0, +\infty[, i = 1, \ldots, n. \end{aligned}$$

The primary descendant of the Bajraktarević mean is obtained if the weight functions are positive constants $w_i, i = 1 \ldots, n$. Then we get the sum-one normalized quasi-arithmetic means:

$$y = F^{-1}\left(\frac{\sum_{i=1}^{n} w_i F(x_i)}{\sum_{i=1}^{n} w_i}\right) = F^{-1}\left(\sum_{i=1}^{n} W_i F(x_i)\right), W_i = \frac{w_i}{\sum_{i=1}^{n} w_i}, \sum_{i=1}^{n} W_i = 1.$$

Various important means can be obtained as descendants of the quasi-arithmetic mean, obtained for various specific functions F. Two most popular special cases are the parameterized exponential function and the parameterized power functions, yielding the following weighted exponential means and the weighted power means:

$$F(x) = e^{rx} \Rightarrow y = \begin{cases} \frac{1}{r} ln \sum_{i=1}^{n} W_i e^{rx_i}, & r \neq 0 \\ \sum_{i=1}^{n} W_i x_i, & r = 0 \\ \wedge_{i=1}^{n} x_i, & r = -\infty \\ \vee_{i=1}^{n} x_i, & r = +\infty \end{cases}$$

$$F(x) = x^r \Rightarrow y = \begin{cases} \left(\sum_{i=1}^{n} W_i x_i^r \right)^{1/r}, & r \neq 0 \\ \prod_{i=1}^{n} x_i^{W_i}, & r = 0 \\ \wedge_{i=1}^{n} x_i, & r = -\infty \\ \vee_{i=1}^{n} x_i, & r = +\infty \end{cases}$$

In both cases, the parameter r is freely adjustable in the whole range $-\infty \leq r \leq +\infty$, and that provides a continuous transition from the pure conjunction to the pure disjunction, i.e., for the andness range $0 \leq \alpha \leq 1$. For any desired value of andness we can compute the corresponding value of the exponent $r(\alpha)$. This technique is discussed later in this chapter.

GCD Models with Adjustable Conjunction Degrees

The need for andness-like parameters and continuous transition in the range of conjunctive and disjunctive aggregators was clear to several authors. The simplest idea in this area consists of taking a strong conjunctive aggregator $Con(\boldsymbol{x})$ and its dual $1 - Con(1 - \boldsymbol{x})$ and then use an andness-like parameter $c \in [0, 1]$ to perform the following continuous additive or multiplicative transition between the selected (extreme) aggregators:

$$A_{add}(\boldsymbol{x}) = cCon(\boldsymbol{x}) + (1 - c)[1 - Con(\mathbf{1} - \boldsymbol{x})], \quad \boldsymbol{x} = (x_1, \ldots, x_n)$$

$$A_{mul}(\boldsymbol{x}) = Con(\boldsymbol{x})^c[1 - Con(\mathbf{1} - \boldsymbol{x})]^{1-c}, \quad 0 \leq c \leq 1$$

So, for $c = 1$ both $A_{add}(\boldsymbol{x})$ and $A_{mul}(\boldsymbol{x})$ become a strong conjunctive aggregator and for $c = 0$ they become a strong disjunctive aggregator. Consequently, the parameter c is a form of *conjunction degree*, similar to the andness α. An obvious advantage of these aggregators is that the parameter c can be directly computed from desired aggregated values, as follows:

$$c_{add} = \frac{1 - Con(\mathbf{1} - \boldsymbol{x}) - A_{add}(\boldsymbol{x})}{1 - Con(\mathbf{1} - \boldsymbol{x}) - Con(\boldsymbol{x})}$$

$$c_{mul} = \frac{\log(1 - Con(\mathbf{1} - \boldsymbol{x})) - \log(A_{mul}(\boldsymbol{x}))}{\log(1 - Con(\mathbf{1} - \boldsymbol{x})) - \log(Con(\boldsymbol{x}))}$$

Therefore, if we have N observations (or measurements) of inputs $\boldsymbol{x}$, and corresponding (actual or desired) aggregated values $A_{add}(\boldsymbol{x})$ or $A_{mul}(\boldsymbol{x})$, then we can get a sequence of conjunction degrees $c_{1,add}, \ldots, c_{N,add}$, or $c_{1,mul}, \ldots, c_{N,mul}$. Their mean values $\overline{c}_{add} = \left(c_{1,add}, \ldots, c_{N,add}\right)/N$ or $\overline{c}_{mul} = \left(c_{1,mul}, \ldots, c_{N,mul}\right)/N$ are now giving the best approximations of measured/desired mapping:

$$A_{add}(\boldsymbol{x}) = \overline{c}_{add} Con(\boldsymbol{x}) + (1 - \overline{c}_{add})[1 - Con(\mathbf{1} - \boldsymbol{x})],$$

$$A_{mul}(\boldsymbol{x}) = Con(\boldsymbol{x})^{\overline{c}_{add}}[1 - Con(\mathbf{1} - \boldsymbol{x})]^{1-\overline{c}_{add}}.$$

This process is similar to the training of preferential neurons studied in Chap. 13. Usually, N observations of mappings $\boldsymbol{x} \to A_{add}(\boldsymbol{x})$ or $\boldsymbol{x} \to A_{mul}(\boldsymbol{x})$ are obtained from experts, and the resulting functions $A_{add}(\boldsymbol{x})$ or $A_{mul}(\boldsymbol{x})$ are adjusted analytic models that fit experimental data.

Unfortunately, the applicability of these aggregators in Graded Logic is rather limited because of the following five problems related to this form of aggregation:

(1) How to select a convenient form of strong conjunction $Con(\boldsymbol{x})$ of n variables?
(2) How to achieve idempotency?
(3) What happens for $c = 1/2$, when both $A_{add}(\boldsymbol{x})$ and $A_{mul}(\boldsymbol{x})$ should become the arithmetic mean?
(4) How to make the pure conjunction and the pure disjunction using $A_{add}(\boldsymbol{x})$ or $A_{mul}(\boldsymbol{x})$?
(5) How to make selectable annihilators using $A_{add}(\boldsymbol{x})$ and $A_{mul}(\boldsymbol{x})$?

Let us consider these questions in the context of the popular gamma aggregators studied in [ZYS79, ZIM79, ZIM80, DOM82, ZIM87, ZIM96]. The idea of gamma aggregators is to interpolate aggregators between the models of intersection and union of fuzzy sets. Thus, the conjunctive aggregation is performed by the simple product $\prod_{i=1}^{n} x_i$ (one of basic t-norms) and the disjunctive aggregation by its De Morgan dual $1 - \prod_{i=1}^{n}(1 - x_i)$. The gamma aggregators are then defined as follows:

$$y_a(\boldsymbol{x}, \gamma, n) = (1 - \gamma)\left(\prod_{i=1}^{n} x_i\right) + \gamma\left(1 - \prod_{i=1}^{n}(1 - x_i)\right), \quad 0 \le \gamma \le 1$$

$$y_g(\boldsymbol{x}, \gamma, n) = \left(\prod_{i=1}^{n} x_i\right)^{(1-\gamma)}\left(1 - \prod_{i=1}^{n}(1 - x_i)\right)^{\gamma}.$$

The parameter γ is a *disjunction degree* and its role is similar to orness. These aggregators use products of arguments and can be weighted using exponential importance weights [ZYS79, ZIM80, DOM82, ZIM87]:

$$y_a(\boldsymbol{x}, \boldsymbol{p}, \gamma, n) = (1-\gamma)\left(\prod_{i=1}^{n} x_i^{p_i}\right) + \gamma\left(1 - \prod_{i=1}^{n}(1-x_i)^{p_i}\right),$$

$$y_g(\boldsymbol{x}, \boldsymbol{p}, \gamma, n) = \left(\prod_{i=1}^{n} x_i^{p_i}\right)^{(1-\gamma)}\left(1 - \prod_{i=1}^{n}(1-x_i)^{p_i}\right)^{\gamma},$$

$$\boldsymbol{p} = (p_1, \ldots, p_n), \quad \sum_{i=1}^{n} p_i = n.$$

The importance weights $p_1, \ldots, p_n$ are exponential, and they can be greater than or less than 1.

The gamma aggregators offer compensative aggregation in the range from the product t-norm to the product t-conorm. That is the answer to the above question (1). The answer to question (2) is negative: gamma aggregators are based on products of arguments, and they are not idempotent. Regarding question (3), the arithmetic mean is not a special case of gamma aggregator for any n, and cannot be obtained for the disjunction degree $\gamma = 1/2$. The answer to question (4) is also negative: the fundamental logic operators of pure conjunction and pure disjunction are not special cases of the gamma aggregator. Regarding the question (5), it is easy to see that the additive version of gamma aggregator is strictly soft, and the multiplicative version is strictly hard. Therefore, hard and soft aggregators cannot be selected by adjusting the value of the parameter γ. In addition, the gamma aggregator cannot model the necessary hard graded disjunction. Therefore, the conclusion is that the gamma aggregators can be useful for fitting empirical data provided by domain experts, but they do not support necessary properties of graded logic aggregators. They use parameters γ and $p_1, \ldots, p_n$ that cause nonlinear effects of aggregation and cannot be independently intuitively selected by decision-makers.

We could select $Con(\boldsymbol{x}) = x_1 \wedge \cdots \wedge x_n$ and try to restrict the aggregation between the full conjunction and its dual, the full disjunction $x_1 \vee \cdots \vee x_n$, as follows:

$$A_{add}(\boldsymbol{x}) = c(x_1 \wedge \cdots \wedge x_n) + (1-c)(x_1 \vee \cdots \vee x_n), \quad \boldsymbol{x} = (x_1, \ldots, x_n)$$

$$A_{mul}(\boldsymbol{x}) = (x_1 \wedge \cdots \wedge x_n)^c (x_1 \vee \cdots \vee x_n)^{1-c}, \quad 0 \le c \le 1.$$

If n > 2, these aggregators eliminate the impact of all inputs except the largest and the smallest. That is not an acceptable property for logic applications. However, if $n = 2$, then all arguments are equally influential, and logic properties are controlled by the conjunction degree c:

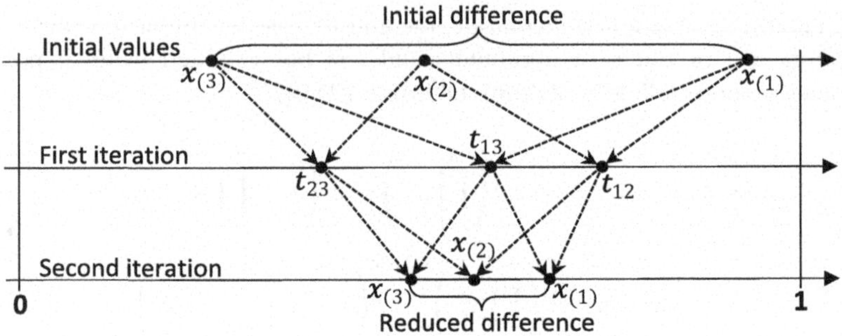

Fig. 15.1 Converging iterative procedure for reducing difference between arguments for $n = 3$

$$A_{add}(\boldsymbol{x}) = c(x_1 \wedge x_2) + (1 - c)(x_1 \vee x_2),$$

$$A_{mul}(\boldsymbol{x}) = (x_1 \wedge x_2)^c (x_1 \vee x_2)^{1-c}, \quad 0 \le c \le 1.$$

The iterative OWA aggregator (ItOWA), proposed in [DUJ91] and studied in [DUJ07b, DUJ21d], is based on an idea that in the case of n variables we can apply the basic 2-variable aggregation to each of $n(n-1)/2$ pairs of variables. Figure 15.1 illustrates this process in the case $n = 3$ and using $A_{add}(\boldsymbol{x})$ for aggregating each of three groups of two variables.

$A_{add}(\boldsymbol{x})$ has no annihilators for $0 < c < 1$ and $A_{mul}(\boldsymbol{x})$ supports the annihilator 0 for all $c > 0$. Our goal is to create a soft ItOWA aggregator (called SItOWA) $y_3 = A_3(x_1, x_2, x_3; c)$ that aggregates three arguments using the conjunction degree c. Initially, we sort the arguments: $x_{(1)} \ge x_{(2)} \ge x_{(3)}$. In the first iteration we aggregate three pairs of inputs: (1,2), (1,3), and (2,3). That creates three new values ($t_{12} \ge t_{13} \ge t_{23}$) that have reduced difference between the largest and the smallest value: $x_{(1)} - x_{(3)} \ge t_{12} - t_{23}$. The process continues according to the following algorithm:

while $x_{(1)} - x_{(3)} \ge \varepsilon$ ***do*** $\quad //\varepsilon$ = small positive error $\left(\text{e.g.}10^{-6}\right)$

$\quad t_{12} = (1-c)x_{(1)} + cx_{(2)}; \quad //x_{(1)} \ge t_{12} \ge x_{(2)}$

$\quad t_{13} = (1-c)x_{(1)} + cx_{(3)}; \quad //x_{(1)} \ge t_{13} \ge x_{(3)}\ ; x_{(1)} \ge t_{12} \ge t_{13}$

$\quad t_{23} = (1-c)x_{(2)} + cx_{(3)}; \quad //x_{(2)} \ge t_{23} \ge x_{(3)}\ ; x_{(1)} \ge t_{12} \ge t_{13} \ge t_{23} \ge x_{(3)}$

$\quad x_{(1)} = t_{12}; \quad x_{(2)} = t_{13}; \quad x_{(3)} = t_{23};$

end

$y_3 = x_{(2)}$; //the middle value, $x_{(2)}$ is the desired value of function $A_3(x_1, x_2, x_3; c)$

Obviously, this process converges because in each iteration $t_{12} - t_{23} \le x_{(1)} - x_{(3)}$. In each iteration we compute the new set of arguments as follows:

$$\begin{bmatrix} x_{(1)} \\ x_{(2)} \\ x_{(3)} \end{bmatrix}_{i+1} = \begin{bmatrix} t_{12} \\ t_{13} \\ t_{23} \end{bmatrix} = \begin{bmatrix} 1-c & c & 0 \\ 1-c & 0 & c \\ 0 & 1-c & c \end{bmatrix} \begin{bmatrix} x_{(1)} \\ x_{(2)} \\ x_{(3)} \end{bmatrix}_i, \quad i = 0, 1, 2, \ldots$$

In each iteration the difference $x_{(1)} - x_{(3)}$ becomes smaller and smaller and rather fast we get the resulting value $x_{(1)} \cong x_{(2)} \cong x_{(3)} = y_3 = A_3(x_1, x_2, x_3; c)$. Therefore, for n = 3, the iterative aggregation process can be written in the following matrix notation:

$$\begin{bmatrix} y_3 \\ y_3 \\ y_3 \end{bmatrix} = \lim_{k\to\infty} \begin{bmatrix} 1-c & c & 0 \\ 1-c & 0 & c \\ 0 & 1-c & c \end{bmatrix}^k \begin{bmatrix} x_{(1)} \\ x_{(2)} \\ x_{(3)} \end{bmatrix} = \begin{bmatrix} v_{31}(c) & v_{32}(c) & v_{33}(c) \\ v_{31}(c) & v_{32}(c) & v_{33}(c) \\ v_{31}(c) & v_{32}(c) & v_{33}(c) \end{bmatrix} \begin{bmatrix} x_{(1)} \\ x_{(2)} \\ x_{(3)} \end{bmatrix},$$
$$x_{(1)} \geq x_{(2)} \geq x_{(3)}.$$

This is a convergent process that returns the following SItOWA aggregator [DUJ91]:

$$y_3 = A_3(x_1, x_2, x_3; c) = \frac{(1-c)^2 x_{(1)} + c(1-c)x_{(2)} + c^2 x_{(3)}}{1-c+c^2}.$$

Consequently, SItOWA of three variables can be written as follows:

$$A_3(x_1, x_2, x_3; c) = v_{31}(c)x_{(1)} + v_{32}(c)x_{(2)} + v_{33}(c)x_{(3)},$$
$$v_{31}(c) = \frac{(1-c)^2}{1-c+c^2}, \; v_{32}(c) = \frac{c(1-c)}{1-c+c^2}, \; v_{33}(c) = \frac{c^2}{1-c+c^2}.$$

Expectedly, $A_3(x_1, x_2, x_3; c)$ satisfies the following conditions:

$$v_{31}(c) + v_{32}(c) + v_{33}(c) = 1,$$
$$A_3(x_1, x_2, x_3; 0) = x_{(1)} = x_1 \vee x_2 \vee x_3,$$
$$A_3(x_1, x_2, x_3; 1) = x_{(3)} = x_1 \wedge x_2 \wedge x_3,$$
$$A_3(x_1, x_2, x_3; {}^1/_2) = \frac{x_{(1)} + x_{(2)} + x_{(3)}}{3} = \frac{x_1 + x_2 + x_3}{3}.$$

In the case of 3 variables this process is efficient but in the general case of n variables the number of pairs is $n(n-1)/2$ and the number of resulting pairs would grow very fast resulting in very inefficient procedure. A better idea is to use weights $v_{31}(c), v_{32}(c), v_{33}(c)$ to compute the weights $v_{41}(c), v_{42}(c), v_{43}(c), v_{44}(c)$ of the SItOWA aggregator of 4 variables, as follows:

$$y_4 = A_4(x_1, x_2, x_3, x_4; c) = v_{41}(c)x_{(1)} + v_{42}(c)x_{(2)} + v_{43}(c)x_{(3)} + v_{44}(c)x_{(4)},$$

$$\begin{bmatrix} y_4 \\ y_4 \\ y_4 \\ y_4 \end{bmatrix} = \lim_{k\to\infty} \begin{bmatrix} v_{31} & v_{32} & v_{33} & 0 \\ v_{31} & v_{32} & 0 & v_{33} \\ v_{31} & 0 & v_{32} & v_{33} \\ 0 & v_{31} & v_{32} & v_{33} \end{bmatrix}^k \begin{bmatrix} x_{(1)} \\ x_{(2)} \\ x_{(3)} \\ x_{(4)} \end{bmatrix} = \begin{bmatrix} v_{41} & v_{42} & v_{43} & v_{44} \\ v_{41} & v_{42} & v_{43} & v_{44} \\ v_{41} & v_{42} & v_{43} & v_{44} \\ v_{41} & v_{42} & v_{43} & v_{44} \end{bmatrix} \begin{bmatrix} x_{(1)} \\ x_{(2)} \\ x_{(3)} \\ x_{(4)} \end{bmatrix},$$

$$x_{(1)} \geq x_{(2)} \geq x_{(3)} \geq x_{(4)}.$$

Obviously, this procedure can be applied to any number of variables, using the following general matrix form:

$$\begin{bmatrix} y_n \\ y_n \\ \cdots \\ y_n \end{bmatrix} = \lim_{k\to\infty} \begin{bmatrix} v_{(n-1)1} & \cdots & v_{(n-1)(n-1)} & 0 \\ v_{(n-1)1} & \cdots & 0 & v_{(n-1)(n-1)} \\ \cdots & 0 & \cdots & \cdots \\ 0 & v_{(n-1)1} & \cdots & v_{(n-1)(n-1)} \end{bmatrix}^k \begin{bmatrix} x_{(1)} \\ x_{(2)} \\ x_{(3)} \\ x_{(4)} \end{bmatrix}$$

$$= \begin{bmatrix} v_{n1} & \cdots\cdots & v_{nn} \\ v_{n1} & \cdots\cdots & v_{nn} \\ \cdots & \cdots & \cdots \\ v_{n1} & \cdots\cdots & v_{nn} \end{bmatrix} \begin{bmatrix} x_{(1)} \\ x_{(2)} \\ \cdots \\ x_{(n)} \end{bmatrix}, \quad x_{(1)} \geq x_{(2)} \geq \cdots \geq x_{(n)}.$$

The coefficients $v_{ij}(c)$ computed using the above procedure for up to 5 variables can be found in [DUJ07b]. These results are verified and expanded to any number of variables in [TRO16, AHN18]. Troiano and Díaz [TRO16] proved that $SItOWA(\boldsymbol{X}; c)$ can generally be written as follows:

$$SItOWA(\boldsymbol{X}; c) = v_{n1}(c)x_{(1)} + \cdots + v_{nn}(c)x_{(n)} = \frac{\sum_{i=1}^{n}(1-c)^{n-i}c^{i-1}x_{(i)}}{\sum_{i=1}^{n}(1-c)^{n-i}c^{i-1}}, \quad n \geq 2.$$

The conjunction degree c is similar to andness, but it is not same as the global andness. So, for each value of c we can numerically compute the corresponding andness $\alpha = \varphi_n(c)$. According to [DUJ22], the inverse functions $c = \varphi_n^{-1}(\alpha)$ can be precisely approximated using simple polynomials. That gives the following andness-directed SItOWA for $n \geq 2$ and $\alpha \in [0, 1]$:

$$SItOWA(\boldsymbol{X}; \alpha) = v_{n1}(\alpha)x_{(1)} + \cdots + v_{nn}(\alpha)x_{(n)}$$

$$= \frac{\sum_{i=1}^{n}(1-\varphi_n^{-1}(\alpha))^{n-i}\left(\varphi_n^{-1}(\alpha)\right)^{i-1}x_{(i)}}{\sum_{i=1}^{n}(1-\varphi_n^{-1}(\alpha))^{n-i}\left(\varphi_n^{-1}(\alpha)\right)^{i-1}}.$$

The polynomials that approximate $v_{ni}(\alpha)$ for most frequently used values $n = 2, 3, 4, 5$ can be found in [DUJ22], together with the same procedure that uses $A_{mul}(\boldsymbol{x})$ to develop a hard iterative version (HItOWA) of the ItOWA aggregator and a technique for using importance weights in the iterative aggregators of ItOWA family.

The important concept of andness-directedness for importance-weighted aggregators, is considered a fundamental property of GCD aggregators. This concept is used in AIWA aggregators, proposed by H.L. Larsen in [LAR03] and further developed it in [LAR09, LAR12]. Larsen's AIWA aggregator (Andness-directed Importance-weighted Averaging Operator) uses a specific conjunction degree, different from global andness, called AIWA andness and denoted a a, and max-one-normalized implicative weights, as follows:

$$h(\boldsymbol{x};\boldsymbol{w},a)=\begin{cases}\lim\limits_{r\to(1-a)/a}\left(\frac{\sum_{i=1}^{n}(w_i x_i)^r}{\sum_{i=1}^{n}w_i^r}\right)^{1/r}, & 0\le a\le\frac{1}{2}\\ 1-h(\boldsymbol{1}-\boldsymbol{x};\boldsymbol{w},1-a), & \frac{1}{2}\le a\le 1\end{cases}$$

$$=\begin{cases}\max(w_1x_1,\dots,w_nx_n), & a=0,\ r=+\infty\\ \left(\frac{\sum\limits_{i=1}^{n}(w_ix_i)^r}{\sum\limits_{i=1}^{n}w_i^r}\right)^{1/r}, & 0<a\le\frac{1}{2},\quad 1\le r<+\infty\\ \frac{\sum\limits_{i=1}^{n}w_ix_i}{\sum\limits_{i=1}^{n}w_i}, & a=\frac{1}{2},\ r=1\\ \left(1-\frac{\sum\limits_{i=1}^{n}(w_i(1-x_i))^{1/r}}{\sum\limits_{i=1}^{n}w_i^{1/r}}\right)^r, & \frac{1}{2}\le a<1,\quad 0\le r<1\\ \min[1-w_1(1-x_1),\dots,1-w_n(1-x_n)], & a=1,\ r=0\end{cases}$$

$$r=\frac{1}{a}-1=\frac{1-a}{a}=\frac{\bar{a}}{a};\quad a=\frac{1}{r+1};\ \max(w_1,\dots,w_n)=1,$$
$$\min[1-w_1(1-x_1),\dots,1-w_n(1-x_n)]=1-\max[w_1(1-x_1),\dots,w_n(1-x_n)]$$

The definition of AIWA aggregator shows De Morgan duality and the disjunctive segment for $0<a<1/2$ is a soft aggregator. Consequently, the dual partial conjunction is also predominantly soft, and AIWA cannot be used in cases where it is necessary to have a hard graded conjunction/disjunction aggregator.

AIWA aggregator is related to power means. Assuming equal weights $w_1=\cdots=w_n=1$ and $a=1/(r+1)$ the AIWA aggregator becomes the power mean:

$$h(\boldsymbol{x};(1,\dots,1),a)=\begin{cases}\left(\frac{1}{n}\sum\limits_{i=1}^{n}x_i^r\right)^{1/r}, & a\le\frac{1}{2},\ r\ge 1,\text{(soft graded disjunction)}\\ \left(1-\frac{1}{n}\sum\limits_{i=1}^{n}(1-x_i)^{1/r}\right)^r, & a\ge\frac{1}{2},\ r\le 1,\ \text{(soft graded conjunction).}\end{cases}$$

AIWA aggregator has unique property that its extreme values are weighted conjunction and weighted disjunction. In the case of implicative weights, it is not acceptable that an argument is important and not highly satisfied $w_i\to x_i=\overline{w_i\wedge\overline{x_i}}=\overline{w_i}\vee x_i$.

So, the graded conjunction with implicative weights is $y = \wedge_{i=1}^{n}(w_i \rightarrow x_i) = \wedge_{i=1}^{n}(\overline{w_i} \vee x_i)$. In this model we can use various forms of implication, and AIWA uses the Reichenbach implication [REI34] where the conjunction is expressed as the algebraic product: $w_i \rightarrow x_i = 1 - w_i(1 - x_i)$. This form of weighted conjunction yields $h(\boldsymbol{x}; \boldsymbol{w}, 1) = \wedge_{i=1}^{n}(1 - w_i(1 - x_i))$.

Let us recall that GL supports De Morgan laws and the logic aggregator $A(\boldsymbol{x}; \boldsymbol{w}, \alpha)$ can be implemented as follows: $A(\boldsymbol{x}; \boldsymbol{w}, \alpha) = \overline{\overline{A(\boldsymbol{x}; \boldsymbol{w}, \alpha)}} = 1 - \overline{A(\boldsymbol{x}; \boldsymbol{w}, \alpha)} = 1 - A(\boldsymbol{1} - \boldsymbol{x}; \boldsymbol{w}, 1 - \alpha)$. The dual of aggregator $A(\boldsymbol{x}; \boldsymbol{w}, \alpha)$ is the aggregator $A(\boldsymbol{x}; \boldsymbol{w}, 1 - \alpha) = 1 - A(\boldsymbol{1} - \boldsymbol{x}; \boldsymbol{w}, \alpha)$. Consequently, a dual of $h(\boldsymbol{x}; \boldsymbol{w}, 1)$ is $h(\boldsymbol{x}; \boldsymbol{w}, 0)$ as follows:

$$\begin{aligned} h(\boldsymbol{x}; \boldsymbol{w}, 0) &= 1 - h(\boldsymbol{1} - \boldsymbol{x}; \boldsymbol{w}, 1) = 1 - \wedge_{i=1}^{n}(1 - w_i x_i) \\ &= 1 - (1 - \max_i w_i x_i) = \vee_{i=1}^{n} w_i x_i. \end{aligned}$$

Since $0 < w_i \leq 1$, we can note the following inequality:

$$\wedge_{i=1}^{n} x_i \leq \min_i (1 - w_i(1 - x_i)) \leq h(\boldsymbol{x}; \boldsymbol{w}, a) \leq \max_i w_i x_i \leq \vee_{i=1}^{n} x_i; \ \max_i w_i = 1.$$

In this expression, the equality holds for $x_1 = \ldots = x_n = x$ and AIWA is an idempotent aggregator.

For AIWA andness $a \in \{0, 1\}$, implicative weights yield the weighted conjunction/disjunction, making AIWA not compatible with conjunction/disjunction in classic Boolean logic. Some aggregation properties of implicative weights can be exposed in the following simple case: $n = 2$, $1 = w_1 > w_2$, $x_2 = 1$, and $a = 0$. In this case, $h(\boldsymbol{x}; \boldsymbol{w}, 0) = \max(x_1, w_2 x_2) = \max(x_1, w_2)$. If $0 \leq x_1 \leq w_2$, i.e., the output value equals the weight of the less important input, insensitive to the changes of more important input x_1 in the range $[0, w_2]$. In addition, in some cases AIWA does not support nonincreasing monotonicity in andness [DUJ18]. In many applications these properties are not desirable, regardless of the benefits of andness-directedness. A detailed analysis of sometimes questionable properties of the weighted conjunction/disjunction, and the use of AIWA as a GCD aggregator can be found in [DUJ18].

Andness-Directed GCD Models Based on Parameter Transformation

Andness is the most important input parameter of GCD. It is selected by decision-makers in all applications of the graded propositional calculus. Consequently, andness must be explicitly visible and easily adjustable input parameter of all GCD aggregators; that property is called *andness-directedness*. Based on applications of Graded Logic, analytic models of GCD must be andness-directed.

The first technique for making andness-directed GCD is called the *parameter transformation technique*. The idea of this technique is to use aggregators that have an adjustable parameter, and then to replace that parameter with a function of andness in such a way that the resulting aggregator (which is now andness-directed) has that specific andness, as requested in Definition 15.1.

Suppose that we have a function $F(\boldsymbol{x}; \boldsymbol{W}, r)$ which has logic arguments $\boldsymbol{x}$, normalized importance weights $\boldsymbol{W}$, and a parameter r that affects the logic properties of F. Since the parameter r is not the andness, we first compute the andness of F:

$$\alpha = \frac{n - (n+1)\int_0^1 \cdots \int_0^1 F\left(\boldsymbol{x}; \underline{\boldsymbol{W}}, r\right) dx_1 \cdots dx_n}{n-1} = \varphi_n(r), \underline{\boldsymbol{W}} = (1/n, \ldots, 1/n).$$

Using the function $\alpha = \varphi_n(r)$, we can now numerically compute and analytically approximate the inverse function $r = \varphi_n^{-1}(\alpha)$ and use it to define the andness-directed GCD:

$$GCD(\boldsymbol{x}; \boldsymbol{W}, \alpha) = F\left(\boldsymbol{x}; \boldsymbol{W}, \varphi_n^{-1}(\alpha)\right), 0 \le \alpha \le 1.$$

Therefore, the function $\varphi_n^{-1} : [0, 1] \to \mathbb{R}$ is the *parameter transformation function* which maps the desired andness to the original parameter r. In this way the original parameter r becomes a hidden internal parameter and the aggregator becomes andness-directed.

The function $r = r_n = \varphi_n^{-1}(\alpha)$ depends on n, and it is necessary to compute a separate function approximation for each value of n. Fortunately, in practice, $2 \le n \le 5$ and this effort is moderate. The parameter transformation technique is used for creating andness-directed aggregators in [TOR21, DUJ21d, DUJ21e, TOR22].

Let us now exemplify the use of the parameter transformation technique for weighted power means [GIN58, BUL03, BEL16] which use the parameter $r \in [-\infty, +\infty]$. It is important to note that the computation of $F(\boldsymbol{x}; \boldsymbol{W}, r) = \lim_{p \to r}\left(\sum_{i=1}^{n} W_i x_i^p\right)^{1/p}$ cannot be done in a naïve brute force way because for very small values of $|r|$ (close to 0), as well as for very large values of $|r|$, there are sensitive numerical phenomena that should be properly controlled. A solution to these problems can be found in [SLA74]. If the computation of $F(\boldsymbol{x}; \boldsymbol{W}, r)$ is performed in a numerically safe way for all values of r, then for each r and n we can numerically compute the andness $\alpha = \varphi_n(r)$. The next step is to create the inverse function $r_n = \varphi_n^{-1}(\alpha)$. The only way to get this function is to use a convenient numeric approximation. A simple analytic approximation of this function, which gives exact results for $\alpha = 0, {}^1/_2, 1$, is the following:

$$r_n = \varphi_n^{-1}(\alpha) = \frac{0.25 + c_{n1}(^1/2 - \alpha) + c_{n2}(^1/2 - \alpha)^2 + c_{n3}(^1/2 - \alpha)^3 + c_{n4}(^1/2 - \alpha)^4}{\alpha(1-\alpha)}.$$

Precise values of coefficients $c_{n1}, c_{n2}, c_{n3}, c_{n4}$ for $n = 2, 3, 4, 5$ can be found in [DUJ18]; to increase precision, the resulting coefficients are different for $\alpha > 0.5$ and for $\alpha < 0.5$. In many practical cases, however, we can be satisfied with lower precision, and in such cases the following approximate exponent value $r = \rho$ can be used for all values of α, and for any $n \in \{2, 3, 4, 5\}$:

Table 15.1 Differences between the desired andness and the obtained andness for fixed values of the power mean exponent

Desired andness α	Approximate fixed value of exponent ρ	Obtained andness α for $n = 2, 3, 4, 5$			
		$\varphi_2(\rho)$	$\varphi_3(\rho)$	$\varphi_4(\rho)$	$\varphi_5(\rho)$
0.05	32.07	0.041	0.048	0.054	0.057
0.1	14.87	0.085	0.097	0.11	0.11
0.15	9.127	0.13	0.15	0.16	0.17
0.2	6.25	0.18	0.20	0.21	0.22
0.25	4.52	0.23	0.25	0.26	0.27
0.3	3.365	0.28	0.30	0.31	0.32
0.35	2.535	0.33	0.35	0.36	0.36
0.4	1.908	0.39	0.40	0.41	0.41
0.45	1.411	0.44	0.45	0.45	0.46
0.5	1	0.50	0.50	0.50	0.50
0.55	0.6439	0.56	0.55	0.55	0.55
0.6	0.3167	0.62	0.60	0.60	0.59
0.65	−0.006546	0.67	0.66	0.65	0.65
0.7	−0.356	0.71	0.71	0.71	0.7
0.75	−0.7768	0.76	0.75	0.76	0.76
0.8	−1.352	0.80	0.80	0.81	0.81
0.85	−2.265	0.84	0.85	0.85	0.86
0.9	−4.066	0.89	0.90	0.90	0.91
0.95	−9.477	0.94	0.95	0.95	0.96

$$\rho = \frac{0.25 + 1.89425(^1/_2 - \alpha) + 1.7044(^1/_2 - \alpha)^2 + 1.47532(^1/_2 - \alpha)^3 - 1.42532(^1/_2 - \alpha)^4}{\alpha(1 - \alpha)}.$$

The expected precision of the approximate exponent ρ and the corresponding andness α for various values of desired andness are presented in Table 15.1. We intentionally show this model where simplicity yields a reduced level of precision. The errors of resulting values of andness in the conjunctive segment $\alpha \geq {}^1/_2$ are mostly less than 2%. The reduced precision can be justified by the fact that the input degrees of truth $x_1, \ldots, x_n$ are generated by humans and frequently their precision can be modest (e.g., according to "magical number seven" [MIL56], human imprecision of the order of 10–15% is rather common). Generally, the precision of a mathematical model should be selected according to the precision of input data. If the precision of a mathematical model is for an order of magnitude better than the precision of input data, that is acceptable in applications: for processing imprecise input data we don't need an extreme precision of the model.

Precision of parameters of aggregation functions and the precision of the degrees of truth directly affect the confidence in the validity of results generated by graded propositional calculus [DUJ20, DET21, DET22, DET23]. As shown in Chap. 13, the parameters of GCD aggregators can be selected in two ways: (1) by independent direct selection of the desired andness and the relative importance of inputs, or (2) using analytic techniques and software tools for joint selection of weights and andness [DUJ91, SEA11, DUJ18]. The direct selection of parameters is the most frequent approach used by analysts after a basic training that exposes geometric properties of aggregators. Thus, the precision of andness is also limited by the granularity of andness rating scales.

Dualized GCD Aggregators Based on Andness-Directed Segmented Interpolative Method

In this section we present our recommended solution to the logic aggregation problems. Proposed are andness-directed interpolative GCD aggregators (ADIGCD) that make Graded Logic fully continuum-valued: according to GL postulates (Chap. 3) our ADIGCD aggregators are continuum-valued, andness-directed, importance-weighted, idempotence-selectable, and annihilator-selectable. Aggregators that are presented in previous sections of this chapters do not support all these important properties. The reasons for their presentation in this chapter are (1) they are frequent in the research literature, (2) some of them are popular in applications, and (3) by comparing their missing properties with the properties of ADIGCD, we can properly assess the suitability of each logic aggregator for applicability in various graded propositional calculus problems. Note that we use the ADIGCD notation only to denote the GCD aggregator construction method, and not its logic properties.

Interpolation of aggregators is a technique that inserts an aggregator $A(\boldsymbol{x})$ between two aggregators $A_1(\boldsymbol{x})$ and $A_2(\boldsymbol{x})$ using a parameter p as follows:

$$A(\boldsymbol{x}) = (1-p)A_1(\boldsymbol{x}) + pA_2(\boldsymbol{x}), \quad 0 \le p \le 1.$$

The parameter p specifies the degree of similarity between the aggregator $A(\boldsymbol{x})$ and the aggregators $A_1(\boldsymbol{x})$ and $A_2(\boldsymbol{x})$. If $p = 0$, then $A(\boldsymbol{x}) = A_1(\boldsymbol{x})$, and if $p = 1$, then $A(\boldsymbol{x}) = A_2(\boldsymbol{x})$. For $0 < p < 1$ the properties of aggregator $A(\boldsymbol{x})$ are a mixture of properties of aggregators $A_1(\boldsymbol{x})$ and $A_2(\boldsymbol{x})$. More generally, if we want that $A(\boldsymbol{x}) = A_1(\boldsymbol{x})$ for $p = p_1$ and $A(\boldsymbol{x}) = A_2(\boldsymbol{x})$ for $p = p_2$, then the corresponding interpolative aggregator is the following:

$$A(\boldsymbol{x}) = \frac{p_2 - p}{p_2 - p_1} A_1(\boldsymbol{x}) + \frac{p - p_1}{p_2 - p_1} A_2(\boldsymbol{x}), \quad p_1 \le p \le p_2.$$

To exemplify properties of interpolated aggregators, let us interpolate an aggregator between the geometric and harmonic means, using andness as the parameter. According to [DUJ18], the andness of harmonic mean $H(x, y) = 2xy/(x + y)$ is $\alpha_h = \ln 16 - 2$, and the andness of geometric mean $G(x, y) = \sqrt{xy}$ is $\alpha_g = 2/3$. So, the interpolated aggregator is

$$A(x, y; \alpha) = \frac{\alpha_h - \alpha}{\alpha_h - \alpha_g} G(x, y) + \frac{\alpha - \alpha_g}{\alpha_h - \alpha_g} H(x, y), \quad \alpha_g \le \alpha \le \alpha_h.$$

Table 15.2 Average errors of interpolative aggregator between the geometric and harmonic means

r	0	-0.1	-0.2	-0.3	-0.4	-0.5	-0.6	-0.7	-0.8	-0.9	-1
$\alpha(r)$	0.667	0.681	0.695	0.707	0.719	0.729	0.740	0.748	0.757	0.764	0.773
$100\overline{E}(r)$	0	0.092	0.161	0.197	0.211	0.203	0.192	0.152	0.112	0.058	0

Table 15.3 Interpolation between soft and hard aggregators

$A_1(x)$	$A_2(x)$	$A(x) = (1-p)A_1(x) + pA_2(x),\quad 0 < p < 1$
soft	soft	soft
soft	hard	soft
hard	soft	soft
hard	hard	hard

More precisely, we have

$$A(x, y; \alpha) = \frac{\ln 16 - 2 - \alpha}{\ln 16 - 8/3}\sqrt{xy} + \frac{\alpha - 2/3}{\ln 16 - 8/3}\frac{2xy}{x+y}, \quad \frac{2}{3} \leq \alpha \leq \ln 16 - 2.$$

On the other hand, both the geometric mean and the harmonic mean are special cases of the power mean:

$$M(x, y; r) = \left(0.5x^r + 0.5y^r\right)^{1/r}, \quad M(x, y; -1) = \frac{2xy}{x+y}, \quad M(x, y; 0) = \sqrt{xy}.$$

Now, we can estimate the usefulness of interpolated aggregator by comparing $A(x, y; \alpha)$ in the range $\frac{2}{3} \leq \alpha \leq \ln 16 - 2$ and the native mean $M(x, y; r)$ in the range $-1 \leq r \leq 0$. For values of r from this interval we can compute numerically the andness $\alpha(r)$ using program from Fig. 10.1 and use it to compute the absolute error $E(x, y; r) = |M(x, y; r) - A(x, y; \alpha(r)|$. Then, the mean error is $\overline{E}(r) = \int_0^1 \int_0^1 E(x, y; r)dxdy$. The corresponding results as shown in Table 15.2. Unsurprisingly, the maximum value of the mean error $\overline{E}(-0.4) = 0.00211$ occurs in the middle region of exponent r. This value is sufficiently small to show that the interpolative aggregator is a very good approximation of the original power mean and preserves well the properties of geometric and harmonic means. So, this concept can be used to create a general method for design of andness-directed aggregators.

Interpolative aggregators can be soft (not supporting annihilators) or hard (supporting annihilators), depending on the type of boundary aggregators they interpolate. According to additive form of interpolation, Table 15.3 shows that hard interpolative aggregators are obtained only between two hard anchor aggregators. So, these properties can be used to adjust a threshold aggregator, which is an aggregator where soft aggregation transforms to hard aggregation.

Suppose that we need the following andness-directed logic aggregator: $A(\boldsymbol{x}; \alpha), \alpha_1 \leq \alpha \leq \alpha_3$. In the increasing sequence of andness $\alpha_1 < \alpha_2 < \alpha_3$ we want that in the range $\alpha_1 \leq \alpha < \alpha_2$ the aggregator $A(\boldsymbol{x}; \alpha)$ is soft, and in the

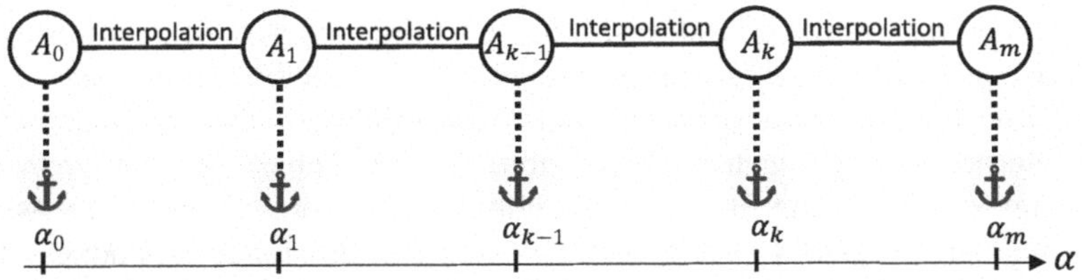

$$ADIGCD(\boldsymbol{x}; \boldsymbol{W}, \alpha) = \frac{\alpha_k - \alpha}{\alpha_k - \alpha_{k-1}} A_{k-1}(\boldsymbol{x}; \boldsymbol{W}, \alpha_{k-1}) + \frac{\alpha - \alpha_{k-1}}{\alpha_k - \alpha_{k-1}} A_k(\boldsymbol{x}; \boldsymbol{W}, \alpha_k), \quad \alpha_{k-1} \le \alpha \le \alpha_k, \quad k = 1, \ldots, m.$$

Fig. 15.2 Segmented interpolative method for creating andness-directed GCD aggregators

range $\alpha_2 \le \alpha \le \alpha_3$ the aggregator $A(\boldsymbol{x}; \alpha)$ is hard. So, we have the threshold andness $\alpha = \alpha_2$. That is the andness where $A(\boldsymbol{x}; \alpha)$ changes properties from soft to hard aggregation. To create this aggregator, we need a soft aggregator $S(\boldsymbol{x}; \alpha)$ and two hard aggregators, $H_1(\boldsymbol{x}; \alpha)$ and $H_2(\boldsymbol{x}; \alpha) \le H_1(\boldsymbol{x}; \alpha) \le S(\boldsymbol{x}; \alpha)$. The desired aggregator $A(\boldsymbol{x}; \alpha)$ can now be defined as follows:

$$A(\boldsymbol{x}; \alpha) = \begin{cases} \frac{\alpha_2 - \alpha}{\alpha_2 - \alpha_1} S(\boldsymbol{x}; \alpha) + \frac{\alpha - \alpha_1}{\alpha_2 - \alpha_1} H_1(\boldsymbol{x}; \alpha), & \alpha_1 \le \alpha < \alpha_2 \\ \frac{\alpha_3 - \alpha}{\alpha_3 - \alpha_2} H_1(\boldsymbol{x}; \alpha) + \frac{\alpha - \alpha_2}{\alpha_3 - \alpha_2} H_2(\boldsymbol{x}; \alpha), & \alpha_2 \le \alpha \le \alpha_3. \end{cases}$$

So, in the range $\alpha_1 \le \alpha < \alpha_2$ the aggregator $A(\boldsymbol{x}; \alpha)$ is soft, and in the range $\alpha_2 \le \alpha \le \alpha_3$ the aggregator $A(\boldsymbol{x}; \alpha)$ is hard. The threshold andness $\alpha = \alpha_2$ is the lowest andness for which $A(\boldsymbol{x}; \alpha)$ has hard aggregation properties. In a special case, we can use only one suitable hard aggregator $H(\boldsymbol{x}; \alpha)$ and create the aggregator $A(\boldsymbol{x}; \alpha)$ with the same threshold andness as follows:

$$A(\boldsymbol{x}; \alpha) = \begin{cases} \frac{\alpha_2 - \alpha}{\alpha_2 - \alpha_1} S(\boldsymbol{x}; \alpha) + \frac{\alpha - \alpha_1}{\alpha_2 - \alpha_1} H(\boldsymbol{x}; \alpha), & \alpha_1 \le \alpha < \alpha_2 \\ H(\boldsymbol{x}; \alpha), & \alpha_2 \le \alpha \le \alpha_3. \end{cases}$$

The concept of interpolative aggregators can be used to create IGCD aggregators in the whole range from the drastic conjunction to the drastic disjunction. A general method for creating *segmented interpolative andness-directed aggregators* in the andness range $\alpha_0 \le \alpha \le \alpha_m$ is shown in Fig. 15.2 [DUJ08, DUJ15a, DUJ18]. We first select a sequence of logic "*anchor aggregators*" $A_0, A_1, \ldots, A_m$ that have increasing andness $\alpha_0 < \alpha_1 < \ldots < \alpha_m$, and other desirable or necessary properties. The intermediate (transitional) aggregation between two adjacent anchor aggregators is realized using linear interpolation based on formula shown in Fig. 15.2, or by using an aggregator that naturally covers that range of andness.

By selecting a sufficient number of anchor aggregators, we can adjust the precision of resulting *andness-directed interpolative* GCD aggregator $ADIGCD(\boldsymbol{x}; \boldsymbol{W}, \alpha)$, as well as a desired range of andness. Since each interpolative segment of aggregation can have different aggregation properties, this method eliminates restrictive properties (e.g., the undesirable support for annihilators) that are characteristic for non-interpolative aggregators. In addition, the segmented interpolative method provides aggregation in the full range of andness. Therefore, we generally assume that the segmented *ADIGCD* aggregator covers the full range of andness from $\alpha_0 = -1/(n-1)$, to $\alpha_m = n/(n-1)$. In other words, we assume the outermost aggregators $A_0 = 1 - \lfloor \prod_{i=1}^{n}(1-x_i) \rfloor$ and $A_m = \lfloor \prod_{i=1}^{n} x_i \rfloor$.

The central anchor aggregator must be the logic neutrality, i.e., the weighted arithmetic mean. The anchor aggregators can have a variety of properties. However, if the chain of anchor aggregators is not symmetric with respect to the arithmetic mean, then De Morgan duality is not supported. Since the duality reflects the equal status of conjunctive and disjunctive logic aggregation, we assume that duality is a mandatory property of GCD. This requirement yields a useful consequence: it is necessary to design only conjunctive (or only disjunctive) part of GCD aggregator and define a general form of GCD as a conjunctive-dualized aggregator.

To create the conjunctive part of the GCD aggregator we propose the *conjunctive andness-directed interpolative GCD aggregator CADIGCD*($\boldsymbol{x}$; $\boldsymbol{W}$, α) that is based on the following five anchor aggregators:

1. **Logic Neutrality**: The weighted arithmetic mean.
2. **Threshold Conjunction**: A hard aggregator with threshold andness $\alpha_\theta = 3/4$ which yields unform soft and hard properties of GCD in the idempotent range.
3. **Pure conjunction**: the end of idempotent range and the beginning of the segment of t-norms.
4. **Product t-norm**: The most frequently used nonidempotent hyperconjunctive aggregator.
5. **Drastic conjunction**: The strongest form of simultaneity.

Based on these anchor aggregators the resulting *CADIGCD* is shown in Fig. 15.3. The analytic form that covers the full range of andness $\alpha_{\min} \le \alpha \le \alpha_{\max}$ is presented in Table 15.4. For simplicity, this uniform aggregator is denoted UGCD, and it is assumed that UGCD is andness-directed, interpolative, and dual. The conjunctive part of this aggregator in Fig. 15.3 includes four intermediate segments: soft graded conjunction, hard graded conjunction, low hyperconjunction, and high hyperconjunction. Note that for $n = 3$ all segments have the same size $1/4$. That contributes to the decision to use the product t-norm as a central hyperconjunctive aggregator.

The interpolative UGCD aggregator shown in Table 15.4 is a conjunctive-dualized interpolative logic aggregator: the disjunctive part of the GCD aggregator is realized recursively as a De Morgan dual of the conjunctive part. The result is the following general interpolative *continuum-valued, andness-directed, importance-weighted, idempotence-selectable, annihilator-selectable, and conjunctive-dualized* GCD aggregator that covers the complete area of andness/orness, from drastic conjunction to drastic disjunction:

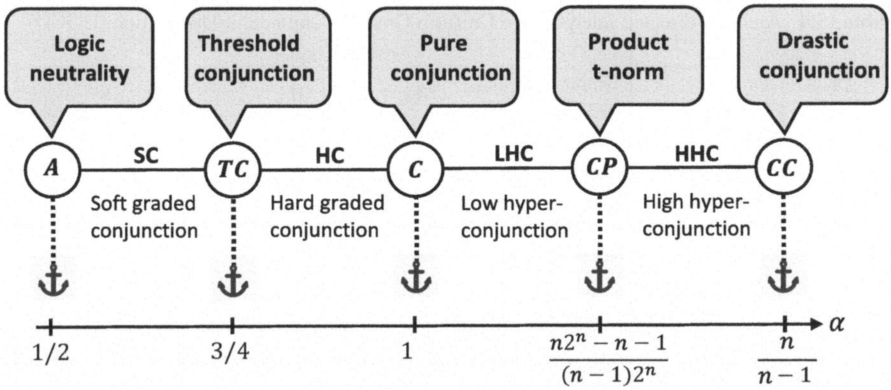

Fig. 15.3 The conjunctive part of andness-directed interpolative UGCD aggregator

$$UGCD(\boldsymbol{x}; \boldsymbol{W}, \alpha) = \begin{cases} CADIGCD(\boldsymbol{x}; \boldsymbol{W}, \alpha), & \frac{1}{2} \le \alpha \le \frac{n}{n-1} \\ 1 - UGCD(\boldsymbol{1} - \boldsymbol{x}; \boldsymbol{W}, 1 - \alpha), & -\frac{1}{n-1} \le \alpha < \frac{1}{2}. \end{cases}$$

There are no known reasons why the default size of the soft idempotent aggregation andness range should be different from the size of the hard idempotent aggregation andness range [DUJ15a]. Consequently, the interpolative GCD is assumed to be uniform, so that the andness range of soft idempotent aggregators has the same size as the range of hard idempotent aggregators, yielding the default threshold andness $\alpha_\theta = 0.75$. This uniform distribution is a logic property and the corresponding GCD aggregator is denoted UGCD. In Table 15.4 we model hard idempotent UGCD using the weighted power mean with negative exponent $r_n(\alpha)$ which depends on the number of variables. According to [DUJ18], for the default threshold andness $\alpha_\theta = 0.75$, the corresponding exponents are $r_2(0.75) = -0.7201$, $r_3(0.75) = -0.7317$, $r_4(0.75) = -0.7205$, $r_5(0.75) = -0.7054$. Nine characteristic special cases of the UGCD function of two variables (hard, soft, idempotent, nonidempotent, conjunctive and disjunctive) are shown in Fig. 15.4.

The uniform aggregator $UGCD(\boldsymbol{x}; \boldsymbol{W}, \alpha)$ covers the whole range of andness, $\alpha_{\min} \le \alpha \le \alpha_{\max}$ and idempotent aggregators in the range $0 < \alpha < 1$ are weighted. The selection of desired value of andness can be based on using software tools such as AnSy (Chap. 13) or by using an appropriate andness rating scale presented in Figs. 13.3–13.7. The precision of UGCD is adjustable and Fig. 15.4 (comparable with Fig. 9.2) shows a standard medium precision interpolative UGCD with 15 idempotent aggregators, expanded with 2 hyperconjunctive anchor aggregators and 2 hyperdisjunctive anchor aggregators. The line of andness in Fig. 15.5 is denoted with α and the complementary line of orness is denoted with ω. Note that in Table 15.4 the hard graded conjunction $HC(\boldsymbol{x}; \boldsymbol{W}, \alpha)$ is not interpolative but implemented as the weighted power mean. Similarly, the high hyperconjunction is not interpolative,

Table 15.4 Andness-directed interpolative Uniform Graded Conjunction/Disjunction (UGCD)

Classification				GCD name	Analytic model	α
GCD	Conjunctive aggregators	Non-idempotent conjunctive aggregation (Hyperconjunction)	HARD	Drastic conjunction	$CC(\mathbf{x};\alpha_{max}) = \lfloor \Pi_{i=1}^{n} x_i \rfloor$	$\alpha_{max} = \frac{n}{n-1}$
				High hyperconjunction	$HHC(\mathbf{x};\alpha) = \Pi_{i=1}^{n} x_i^{p(\alpha)}$ $p(\alpha) = \left(\frac{n+1}{n-(n-1)\alpha}\right)^{1/n} - 1$	$\alpha_t \le \alpha \le \alpha_{max}$
				Product t-norm	$CP(\mathbf{x};\alpha_t) = \Pi_{i=1}^{n} x_i$	$\alpha_t = \frac{n2^n - n - 1}{(n-1)2^n}$
				Low hyperconjunction	$LHC(\mathbf{x};\alpha) = \frac{\alpha_t - \alpha}{\alpha_t - 1} C(\mathbf{x}) + \frac{\alpha - 1}{\alpha_t - 1} CP(\mathbf{x};\alpha)$	$1 \le \alpha \le \alpha_t$
		Idempotent logic aggregation		Pure conjunction	$C(\mathbf{x}) = min(x_1, \ldots, x_n)$	$\alpha = 1$
				Hard graded conjunction	$HC(\mathbf{x};\mathbf{W},\alpha) = \left(\Sigma_{i=1}^{n} W_i x_i^{r_n(\alpha)}\right)^{1/r_n(\alpha)}$	$\alpha_\theta \le \alpha \le 1$
				Threshold conjunction	$TC(\mathbf{x};\mathbf{W},\alpha_\theta) = \left(\Sigma_{i=1}^{n} W_i x_i^{r_n(\alpha_\theta)}\right)^{1/r_n(\alpha_\theta)}$	$\alpha = \alpha_\theta = \frac{3}{4}$
			SOFT	Soft graded conjunction	$SC(\mathbf{x};\mathbf{W},\alpha) = \frac{\alpha_\theta - \alpha}{\alpha_\theta - 1/2} A(\mathbf{x};\mathbf{W},\alpha) + \frac{\alpha - 1/2}{\alpha_\theta - 1/2} TC(\mathbf{x};\mathbf{W},\alpha_\theta)$	$\frac{1}{2} \le \alpha \le \alpha_\theta$
	Neutrality			Logic neutrality	$A(\mathbf{x};\mathbf{W},\alpha) = \Sigma_{i=1}^{n} W_i x_i$	$\alpha = \frac{1}{2}$
	Disjunctive aggregators			Soft graded disjunction	$SD(\mathbf{x};\mathbf{W},\alpha) = 1 - SC(\mathbf{1} - \mathbf{x};\mathbf{W}, 1 - \alpha)$	$1 - \alpha_\theta \le \alpha \le \frac{1}{2}$
			HARD	Threshold disjunction	$TD(\mathbf{x};\mathbf{W},\alpha) = 1 - TC(\mathbf{1} - \mathbf{x};\mathbf{W}, 1 - \alpha)$	$\alpha = 1 - \alpha_\theta = \frac{1}{4}$
				Hard graded disjunction	$HD(\mathbf{x};\mathbf{W},\alpha) = 1 - HC(\mathbf{1} - \mathbf{x};\mathbf{W}, 1 - \alpha)$	$0 \le \alpha \le 1 - \alpha_\theta$
				Pure disjunction	$D(\mathbf{x}) = max(x_1, \ldots, x_n)$	$\alpha = 0$
		Non-idempotent disjunctive aggregation (Hyperdisjunction)	HARD	Low hyperdisjunction	$LHD(\mathbf{x};\alpha) = 1 - LHC(\mathbf{1} - \mathbf{x}; 1 - \alpha)$	$\alpha_c \le \alpha \le 0$
				Product t-conorm	$DP(\mathbf{x};\alpha_c) = 1 - CP(\mathbf{1} - \mathbf{x};\alpha_t) = 1 - \Pi_{i=1}^{n}(1 - x_i)$	$\alpha_c = 1 - \alpha_t = \frac{n + 1 - 2^n}{(n-1)2^n}$
				High hyperdisjunction	$HHD(\mathbf{x};\alpha) = 1 - HHC(\mathbf{1} - \mathbf{x}; 1 - \alpha) = 1 - \Pi_{i=1}^{n}(1 - x_i)^{p(1-\alpha)}$	$\alpha_{min} \le \alpha \le \alpha_c$
				Drastic disjunction	$DD(\mathbf{x};\alpha_{min}) = 1 - CC(\mathbf{1} - \mathbf{x};\alpha_{max})$	$\alpha_{min} = \frac{-1}{(n-1)}$

Note: shaded cells denote anchor aggregators; intermediate aggregators are either natural or interpolated

but implemented as the product-power function $HHC(\mathbf{x};\alpha) = \prod_{i=1}^{n} x_i^{p(\alpha)}$, $p(\alpha) \ge 1$. Obviously, for $p(\alpha) \gg 1$, we have $HHC(\mathbf{x};\alpha) = 1$ only if $x_1 = \ldots = x_n = 1$, and in all other cases $HHC(\mathbf{x};\alpha) \cong 0$. So, for $p(\alpha) \to +\infty$ we have that $HHC(\mathbf{x};\alpha)$ becomes the drastic conjunction.

In Table 15.4, the pure conjunction ($\alpha = 1$) and the pure disjunction ($\omega = 1$) are not weighted and consequently, for these values, the interpolative GCD is consistent with classic Boolean logic. That is compatible with the concept of the balance of

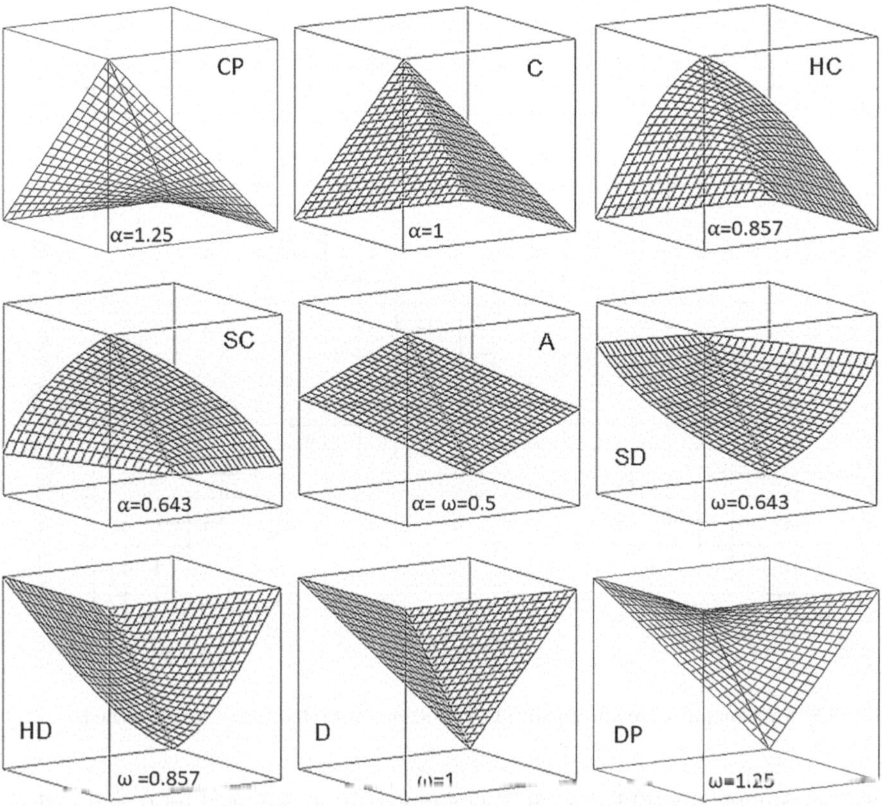

Fig. 15.4 Nine characteristic special cases of the GCD function of two variables

impact of logic importance (*LI*) and relative importance (*RI*), shown in Fig. 12.2, which is expanded outside of the range of idempotent aggregators as follows:

$$LI = \min(|\alpha - \omega|, 1), \quad RI = \max(1 - |\alpha - \omega|, 0).$$

If the relative importance is positive, then GCD uses importance weights, and it is not commutative. However, if the relative importance is $RI = 0$, then GCD is not weighted, and that means it is a commutative aggregator. For $|\alpha - \omega| \in [0, 1]$ we have the situation shown in Fig. 12.2, and outside this range we assume $LI = 1$ and $RI = 0$. In other words, for $|\alpha - \omega| < 1$ we assume that GCD must be noncommutative because $LI < 1$ and $RI = 1 - LI > 0$. However, for $|\alpha - \omega| \geq 1$, GCD must be commutative because we assume $LI = 1$ and $RI = 0$. For $|\alpha - \omega| = 1$ we have that GCD becomes commutative classic conjunction or disjunction, and GL behaves as the classical Boolean logic. Thus, for $\alpha = 1$, and for $\omega = 1$, the logic importance is so strong that all input arguments already have equal power of impact on the aggregated output; they become commutative, and the impact of different weights

GCD	ANDNESS (α) $\alpha < 0$				0	$\frac{1}{14}$	$\frac{2}{14}$	$\frac{3}{14}$	$\frac{4}{14}$	$\frac{5}{14}$	$\frac{6}{14}$	$\frac{1}{2}$	$\frac{8}{14}$	$\frac{9}{14}$	$\frac{10}{14}$	$\frac{11}{14}$	$\frac{12}{14}$	$\frac{13}{14}$	1	$\alpha > 1$				GCD
CC	ω																▲						α	CC
HHC		ω															▲					α		HHC
CP			ω														▲				α			CP
LHC				ω													▲			α				LHC
C					ω												▲		α					C
HC+						ω											▲	α						HC+
HC							ω										α	◄	◄	◄	◄	◄	◄	HC
HC-								ω								α								HC-
SC+									ω						α									SC+
SC										ω				α										SC
SC-											ω		α											SC-
A												α												A
SD-											α		ω											SD-
SD										α				ω										SD
SD+									α						ω									SD+
HD-								α								ω								HD-
HD							α										ω	◄	◄	◄	◄	◄	◄	HD
HD+						α											▼	ω						HD+
D					α												▼		ω					D
LHD				α													▼			ω				LHD
DP			α														▼				ω			DP
HHD		α															▼					ω		HHD
DD	α																▼						ω	DD
GCD	$\omega < 0$				0	$\frac{1}{14}$	$\frac{2}{14}$	$\frac{3}{14}$	$\frac{4}{14}$	$\frac{5}{14}$	$\frac{6}{14}$	$\frac{1}{2}$	$\frac{8}{14}$	$\frac{9}{14}$	$\frac{10}{14}$	$\frac{11}{14}$	$\frac{12}{14}$	$\frac{13}{14}$	1	$\omega > 1$ ORNESS (ω)				GCD

Fig. 15.5 Medium precision interpolative UGCD aggregator (idempotent area is shaded)

(and noncommutativity) fades out. If commutativity is accepted for $\alpha = 1$ and for $\omega = 1$ then it is reasonable to use commutative aggregators also for $\alpha > 1$ and for $\omega > 1$. That is implemented in Table 15.4.

The product t-norm is commutative: $CP(\boldsymbol{x}; \alpha) = \prod_{i=1}^{n} x_i$. In a special case where there are strong arguments for using weights in hyperconjunctive aggregation, it is possible to make a count-normalized weighted version of this aggregator (which is no longer a t-norm) $CP(\boldsymbol{x}; \boldsymbol{p}, \alpha) = \prod_{i=1}^{n} x_i^{p_i}, \sum_{i=1}^{n} p_i = n$. An additional analysis of this product-power version of GCD aggregator can be found in the weight normalization section of Chap. 12 and Fig. 12.1.

Theoretically, instead of product t-norm $CP(\boldsymbol{x}; \alpha)$ we could use the weighted version of the product aggregator $CP(\boldsymbol{x}; \boldsymbol{p}, \alpha)$, and then provide a weighted interpolative GCD aggregator in the whole range of andness. For example, if $n = 2$, then $CP(\boldsymbol{x}; \boldsymbol{p}, \alpha) = \left(x_1^{p_1} x_2^{p_2}\right)^{\sqrt{3/(2-\alpha)}-1}$, $p_1 + p_2 = 2$. For computing andness we need $p_1 = p_2 = 1$ and now $CP(\boldsymbol{x}; \boldsymbol{p}, \alpha) = (x_1 x_2)^{\sqrt{3/(2-\alpha)}-1}$. Therefore, we have the weighted product $CP(\boldsymbol{x}; \boldsymbol{p}, 5/4) = x_1^{p_1} x_2^{p_2}$, and the weighted pseudo conjunction $CP(\boldsymbol{x}; \boldsymbol{p}, 1) = \left(x_1^{p_1} x_2^{p_2}\right)^{\sqrt{3}-1}$. For weighted product we have the volume $V = \int_0^1 \int_0^1 x_1 x_2 dx_1 dx_2 = 1/4$ and $\alpha = 2 - 3V = 5/4$. Similarly, for pseudo conjunction we have $V = \int_0^1 \int_0^1 (x_1 x_2)^{\sqrt{3}-1} dx_1 dx_2 = 1/3$ and $\alpha = 2 - 3V = 1$. The pseudo conjunction has the same andness $\alpha = 1$ as the real conjunction because they have the same volume under the surface inside the unit hypercube. Using the technique of interpolative aggregation shown in Fig. 15.3 and Table 15.4 it is theoretically possible to have weighted logic aggregation in the whole range of andness from drastic conjunction to drastic disjunction. However, that could be meaningful if and only if it is possible to prove that such a form of aggregation is used in human commonsense reasoning.

In Chap. 9 we have shown that the volumes under the conjunction and disjunction functions depend on the number of variables n: $V_{con}(n) = 1/(n+1)$ and $V_{dis}(n) = n/(n+1)$. Since all idempotent functions are located between conjunction and disjunction, it follows that the volume of the unit hypercube that contains idempotent functions is $V_{id}(n) = V_{dis}(n) - V_{con}(n) = (n-1)/(n+1)$. Consequently, the volume that contains nonidempotent functions is $V_{nid}(n) = 1 - V_{id}(n) = 2/(n+1)$. The geometric interpretation of logic properties shows that for larger values of n n, idempotent aggregators become dominant and the volume of the unit hypercube that contains nonidempotent aggregators diminishes. This situation is illustrated in Fig. 15.6 (see also Figs. 9.4 and 9.5). There are four distinct regions of idempotent GCD aggregator (hard and soft partial conjunction and disjunction) and four regions of nonidempotent GCD aggregators (above and below the product t-norm and t-conorm). Idempotent GCD aggregators are denoted as "partial" because they are combinations of complementary conjunctive and disjunctive properties. Note that the regions above the product t-norms and below the product t-conorm very quickly reduce with the increase of n.

The segmented interpolative method for construction of logic aggregators is the only method that provides logic GCD aggregators in the whole range of andness. It can be compared with other logic aggregation models as shown in Table 15.5. Applications convincingly show that most logic aggregators operate in the range of andness/orness [0, 1]. A wider range, between the product norm and conorm, is used by the Gamma aggregator [ZYS79, ZIM80]. Some logic aggregators operate only in selected segments of the range of andness. The most frequent such a restriction is the use of t-norms as models of conjunction and t-conorms as models of disjunction, as in the case of fuzzy logic and IBA [RAD00]. The motivations for such a restriction are the mathematical benefits of t-norms and t-conorms [FOD94]: commutativity, monotonicity, associativity, neutral elements, and fixed annihilators. It should be noted, however, that such a specific set of restrictions cannot be observed in human commonsense logic.

Verbalizing GCD Aggregation

The GCD function is derived from observing human commonsense logical reasoning. Of course, the human commonsense reasoning is not numerical. It is verbalized, i.e. it is based on expressing logic requirements using descriptive constructs of stakeholder's natural language. The basic ranges of andness of GCD are verbalized as follows (see also Table 17.7):

"Must have all" Hard conjunctive GCD $\alpha \geq 0.75$

"Nice to have most"	Soft conjunctive GCD	$0.5 < \alpha < 0.75$
"Nice to have"	Soft neutral GCD	$\alpha = 0.5$
"Nice to have some"	Soft disjunctive GCD	$0.25 < \alpha < 0.5$
"Enough to have any"	Hard disjunctive GCD	$\alpha \leq 0.25$

The first step in creating a GCD function is the selection of the most appropriate of the above groups (all > most > some > any). The next step is to select an appropriate andness intensity in the selected group/range, and the weights of inputs. It is rather easy to select one of the available major groups (andness intervals). However, we also need to provide precision that is appropriate for specific evaluation problems. So, after selecting the verbalized group, we can refine the selection by asking whether the intensity is medium (in the middle of the interval), or high (stronger than medium), or low (weaker than medium). In Table 15.5, the medium intensity aggregators are CP, HC, SC, SD, HD, and DP. The high intensity aggregators are HHC, HC+, SC+, SD+, HD+, and HHD. The low intensity aggregators are LHC, HC-, SC-, SD-, HD-, and LHD. According to Table 15.5, the GCD aggregators can be selected by answering four simple questions:

1. Conjunctive, or neutral, or disjunctive?
2. Idempotent, or nonidempotent?
3. Hard or soft? (nonidempotent GCD is always hard).
4. High, or medium, or low?

Obviously, answering these questions is very simple because the choice is between only two or three options. For example, let the stakeholder be a homebuyer, and let the inputs be the quality of home (Q), the suitability of home location (L) and the affordability of home (A). Let us consider the following question: would the homebuyer buy a home that has an unacceptable quality ($Q = 0$), or an unsuitable location ($L = 0$), or an unacceptably high price ($A = 0$)? Obviously, the answer is "no." Thus, the homebuyer *must have all* (Q and L and A) simultaneously sufficiently satisfied. In the case of average precision (Tables 15.4, 15.5 and 17.4) the candidate hard aggregators are C, HC + , HC, and HC- because it is regularly not reasonable to expect that the overall home suitability H satisfies $H < \min(Q, L, A)$. We should also eliminate the conjunction C, because it is not reasonable to use $H = \min(Q, L, A)$. Why? It is easy to explain the reason: it is not reasonable to consider that $H = \min(0.5, 0.5, 0.5) = \min(1, 0.5, 1) = 0.5$. Indeed, nobody would consider an extremely high quality and very affordable home as equivalent to another home that has very modest quality and an exaggerated price, while they are located in the same/equivalent location. On the other hand, the function $H = GCD(Q, L, A)$ is obviously the most important aggregator in the final stage of evaluation, and consequently it is reasonable to eliminate the weakest aggregator HC- in the "must have all" group. That brings us to select between HC + and HC. If the available homes

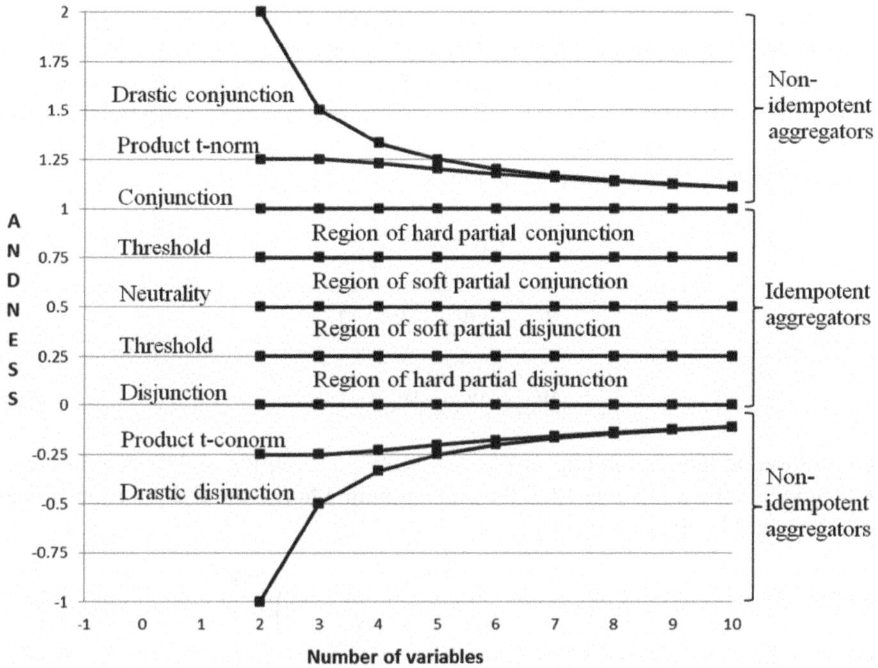

Fig. 15.6 Idempotent and nonidempotent regions of GCD aggregator for various numbers of variables

Table 15.5 Types, classification, and comparison of selected logic aggregators

<table>
<tr><th colspan="23">Type of logic aggregator</th></tr>
<tr><td colspan="11">Conjunctive (simultaneity)</td><td rowspan="4">Neutral</td><td colspan="11">Disjunctive (substitutability)</td></tr>
<tr><td colspan="4">Not Idempotent</td><td colspan="7">Idempotent</td><td colspan="7">Idempotent</td><td colspan="4">Not Idempotent</td></tr>
<tr><td colspan="4">Hyper Con.</td><td colspan="4">Hard</td><td colspan="3">Soft</td><td colspan="3">Soft</td><td colspan="4">Hard</td><td colspan="4">Hyper Dis.</td></tr>
<tr><td>Dras. Con.</td><td colspan="3">Level</td><td>∧</td><td colspan="3">Level</td><td colspan="3">Level</td><td colspan="3">Level</td><td colspan="3">Level</td><td>∨</td><td colspan="3">Level</td><td>Dras. Dis.</td></tr>
<tr><td>CC</td><td>HHC</td><td>CP</td><td>LHC</td><td>C</td><td>HC+</td><td>HC</td><td>HC−</td><td>SC+</td><td>SC</td><td>SC−</td><td>A</td><td>SD−</td><td>SD</td><td>SD+</td><td>HD−</td><td>HD</td><td>HD+</td><td>D</td><td>LHD</td><td>DP</td><td>HHD</td><td>DD</td></tr>
<tr><td></td><td></td><td colspan="3">Fuzzy Logic</td><td></td><td></td><td></td><td></td><td></td><td></td><td></td><td></td><td></td><td></td><td></td><td></td><td></td><td colspan="3">Fuzzy Logic</td><td></td><td></td></tr>
<tr><td></td><td></td><td></td><td></td><td colspan="15">← – – – OWA (soft), AIWA, ItOWA (soft) – – –→</td><td></td><td></td><td></td><td></td></tr>
<tr><td></td><td></td><td colspan="19">← – – – Gamma multiplicative (hard) and additive (soft) – – –→</td><td></td><td></td></tr>
<tr><td colspan="23">← – – – – – – – – – – Graded Conjunction/Disjunction (GCD) – – – – – – – – – –→</td></tr>
<tr><td colspan="23">RANGE OF LOGIC AGGREGATORS (shaded area is not supported)</td></tr>
</table>

are in an area where good homes are rather frequent, the stakeholder might decide to use the high intensity of hard simultaneity, i.e., HC + .

The final step is the selection of weights. Suppose that the homebuyer considers that the most important component is Q and the least important component is L, so that Q is two times more important than L and the importance od A is in the middle between Q and L. That yields the equations

$$W_Q = 2W_L,$$

$$W_A = (W_Q + W_L)/2,$$

$$W_Q + W_A + W_L = 1.$$

The solution od these equations is $W_Q = 4/9$, $W_A = 3/9$, $W_L = 2/9$. According to GCD models shown in Table 17.3, the corresponding home suitability propositional formula would be the following:

$$H = \left(\frac{4}{9}Q^{-6.53} + \frac{3}{9}A^{-6.53} + \frac{2}{9}L^{-6.53}\right)^{-1/6.53}.$$

If the stakeholder/decision-maker prefers to use the HC aggregator, then the corresponding GCD would be the following:

$$H = \left(\frac{4}{9}Q^{-2.539} + \frac{3}{9}A^{-2.539} + \frac{2}{9}L^{-2.539}\right)^{-1/2.539}.$$

Conditional Logic Aggregation

In some cases, propositional logic formulas must be combined with logic conditions. The most frequent such a condition is the simple *if–then-else* control structure. For example, a home location can be fully acceptable if it is close to a shopping area; otherwise, it is acceptable only if it is close to good public transportation. Such a criterion can be created using the if–then-else control structure shown in Fig. 15.7. We use a selector function which compares the input degree of truth x (e.g. the distance from home to a shopping area) with a threshold value T (maximum acceptable distance) as follows:

$$b = SEL(x, T) = \begin{cases} 1, \ x \geq T \\ 0 \ \ x < T \end{cases}.$$

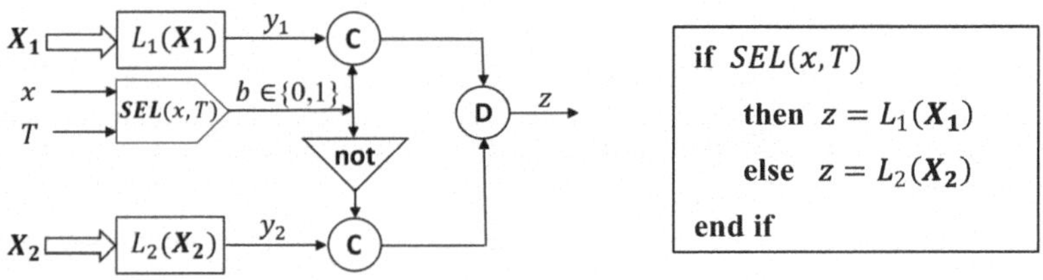

Fig. 15.7 A general if–then-else construct implemented in GL

The selector function can be combined with the GL conjunction (C), disjunction (D), and negation (not) as shown in Fig. 15.7 to achieve the following general if–then-else construct:

$$z = \begin{cases} L_1(X_1), & x \geq T \\ L_2(X_2), & x < T \end{cases}.$$

Here $y_1 = L_1(X_1) \in [0, 1]$ denotes a graded propositional calculus formula based on an array of input degrees of truth X_1. Likewise, $y_2 = L_2(X_2) \in [0, 1]$denotes a graded propositional calculus formula based on an array of input degrees of truth X_2. Generally, $X_1 \neq X_2$, but frequently we can have $X_1 = X_2 = X$. Similar reasoning can be applied to L_1 and L_2; e.g., these can be the same propositional formulas that differ only in weights or only in selected inputs. The selector variables x x and T T can be independent inputs or selected components of arrays X_1 and X_2. Using the standard conditional expression notation, the output value is `z = (b ? y1: y2)`.

Multiple if–then-else constructs can be combined and provide a high flexibility for the development of sophisticated graded propositional calculus formulas. Two specific nested if–then-else structures are if-elsif-else and if-thenif-else, shown in Fig. 15.8. The if-elsif-else construct has k conditions ($C1, \ldots, Ck$) used for selecting one of k propositional calculus formulas ($L1, \ldots, Lk$). If none of k conditions are satisfied the default propositional formula D (possibly empty) is executed. In the case of the if-thenif-else construct, the conditions $C1, \ldots, Ck$, activated in sequence, are permissions to continue and execute propositional formulas $L1, \ldots, Lk$. If one of the conditions is not satisfied, the default propositional formula D (possibly empty) is executed.

Conditional logic aggregators can be based on selector functions that activate various aggregation structures, as exemplified in Fig. 15.9. The idea of such compound logic aggregation is that we can define $k > 1$ component aggregators $A_i(X; W_i, \alpha_i), i = 1, \ldots, k$, and automatically (dynamically) select the most suitable among them using appropriate selector functions. Thus, the component aggregator $A_i(X; W_i, \alpha_i)$ is related to a specific selector function $B_i = C_i(X; W_i, \alpha_i) \in \{0, 1\}$. Most frequently, a specific combination of input arguments requires a specific type of aggregation, and in such a case the selector block automatically activates a corresponding component aggregator, which can have any structure and complexity. We

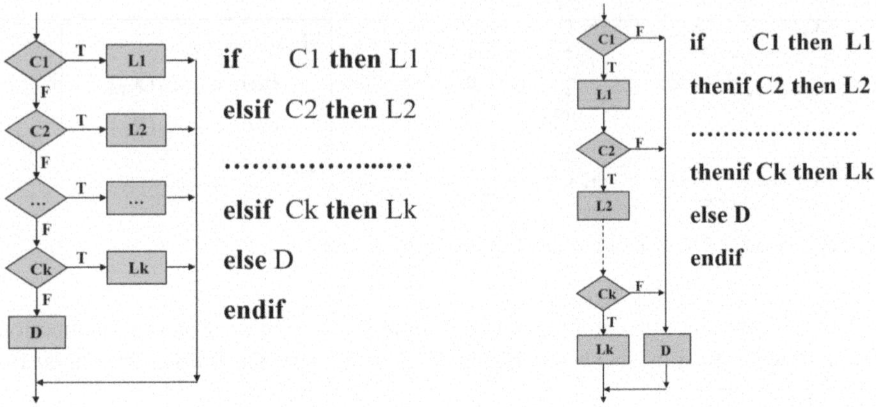

Fig. 15.8 If-elsif-else and if-thenif-else control structures for conditional GL functions

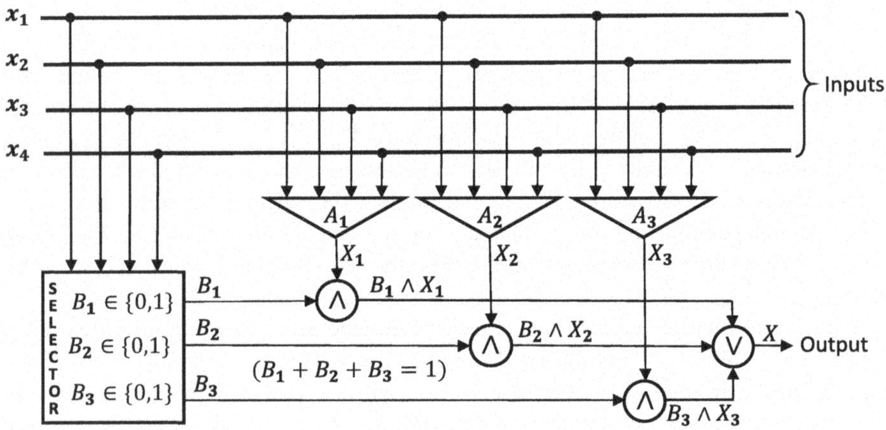

Fig. 15.9 A compound logic aggregator with 4 inputs and 3 component aggregators

assume that conditions for activating component aggregators are *mutually exclusive*, so that $B_1 + \ldots + B_k = 1$, and the selector block activates only one of existing component aggregators. The sample structure in Fig. 15.9 generates the output suitability

$$X = (B_1 \wedge X_1) \vee (B_2 \wedge X_2) \vee (B_3 \wedge X_3) = \vee_{i=1}^{3}[C_i(\mathbf{X}; \mathbf{W}_i, \alpha_i) \wedge A_i(\mathbf{X}; \mathbf{W}_i, \alpha_i)],$$

$$\mathbf{X} = (x_1, x_2, x_3, x_4).$$

Logic aggregators have various parameters (like andness and importance weights) that are used to adjust desirable aggregator properties. Parameter-directed aggregators are those aggregators that have explicitly visible and easily adjustable parameters. Parameter-implicit aggregators have logic parameters that are not explicitly visible but can be indirectly computed, like in the cases of t-norms [KLI95], OWA [YAG98, TOR21, TOR07], and iterative OWA [DUJ91].

A useful property of conditional aggregators is that they can be parameter-directed, if they are a combination of parameter-directed component aggregators $A_i(\boldsymbol{X}; \boldsymbol{W}_i, \alpha_i), i = 1, \ldots, k$ Therefore, an interesting research topic is to find conditional logic aggregators that can successfully replace some of the parameter-implicit aggregators, such as the iterative OWA, t-norms, and fuzzy integrals. Such solutions are presented in [DUJ21d, DUJ21e] and in the recent work of Torra [TOR21, TOR22, TOR24]. The replacement of parameter-implicit aggregators with parameter-directed aggregators can significantly increase the applicability of graded logic aggregators.

Graded propositional calculus formulas with fixed structure and constant parameters (weights and andness/orness) are called *stationary* GL models. They are most frequently used in decision-making models that provide the overall suitability of various competitive objects/alternatives. Much less frequently, we can use *nonstationary* GL models [DUJ18], where the parameters (weights, andness, orness) can be functions of input attributes. Of course, while the design of stationary decision models is simple and requires very modest effort, the design of nonstationary models and conditional aggregators needs significantly higher effort and a thorough justification.

Modeling Hyperconjunction Using T-Norms

In decision-making applications, both hyperconjunction and hyperdisjunction are infrequently used because they model extreme cases of simultaneity and substitutability (compare Fig. 11.1 and Table 11.2). In human commonsense logic, idempotent aggregators (andness in the range [0, 1]) are much more frequently used. The hyperconjunction and hyperdisjunction are dual, and consequently it is sufficient to investigate the role and the properties of hyperconjunction. According to Table 15.4, in the basic case of 2 variables, we use the following andness-directed interpolative GCD model of hyperconjunction:

$$\mathrm{Hcon}(x_1, x_2; \alpha) = \begin{cases} \min(x_1, x_2), & \alpha = 1 \\ (5 - 4\alpha)\min(x_1, x_2) + 4(\alpha - 1)x_1 x_2, & 1 \leq \alpha \leq 5/4 \\ (x_1 x_2)^{\sqrt{3/(2-\alpha)}-1}, & 5/4 \leq \alpha < 2 \\ \lfloor x_1 x_2 \rfloor, & \alpha = 2 \end{cases}$$

This GCD model is extremely simple, explicitly andness-directed, related to commonsense logical reasoning, and it covers the whole range of hyperconjunction, from the andness 1 to the andness 2. In addition, the hyperdisjunction is $Hdis(x_1, x_2; \alpha) = 1 - Hcon(1 - x_1, 1 - x_2; 1 - \alpha), -1 \leq \alpha \leq 0$. If desirable, for $1 < \alpha < 2$, the components of the product $x_1 x_2$ can be importance-weighted $(x_1^p x_2^q, p + q = 2)$. *Hcon* is not an associative function, but it is interpolated between three associative anchor aggregators (Fig. 15.3) and has properties sufficiently close to associativity. The presented interpolative model can easily be extended for more than two variables by inserting them in the product expression (and adjusting the exponent) and expanding the list of arguments in the minimum function, as shown in Table 15.4.

There are alternative ways to create hyperconjunction and hyperdisjunction, and it is useful to compare them with the above GCD version. In mathematical literature, the functionality like hyperconjunction is provided by triangular norms (t-norms) [KLE00, MEN01, FOD94, KLI95, YAG04, NGU06, BEL07, GRA09]. T-norms have extensively been studied in the context of the theory of fuzzy sets [WIK25, WIK22, WIK24b, WIK24c]. They are used as mathematical models of intersection of fuzzy sets. T-norms are commutative, associative, nondecreasing binary functions $T : [0, 1]^2 \rightarrow [0, 1]$ with the neutral element 1 (strictly hard with annihilator 0), defined as follows:

Commutativity: $T(x, y) = T(y, x), \forall x, y \in [0, 1]$.
Monotonicity: $T(x, y) \leq T(X, Y), 0 \leq x \leq X \leq 1, 0 \leq y \leq Y \leq 1$.
Associativity: $T(x, T(y, z)) = T(T(x, y), z), \forall x, y, z \in [0, 1]$.
Neutral element: $T(x, 1) = x, \forall x \in [0, 1]$ (Consequences: $T(0, 1) = T(1, 0) = 0$.
Annihilator: $T(x, 0) = 0, \forall x \in [0, 1]$. (Consequence of monotonicity and $T(1, 0) = 0$.

Since $T(0, 0) = T(1, 0) = T(0, 1) = 0$, $T(1, 1) = 1$, and t-conorms are defined as duals to t-norms, it follows that t-norms/conorms, similarly to GCD, include bivalent conjunction and disjunction as special cases, and can be considered a generalization of classical Boolean logic. T-norms are primarily a part of the fuzzy set theory and usually appear in the literature under the title of fuzzy logic.

It is easy to check that the pure conjunction $\min(x, y)$ and the product $T(x, y) = xy$ are t-norms. If $T(x, x) < x, 0 < x < 1$, as in the case of product, then the t-norm is Archimedean. A proof of the important theorem that *the pure conjunction is the only idempotent t-norm* can be found in [KLI95]. So, other means cannot be t-norms, and consequently the andness of t-norms cannot be below 1. Therefore, except for the pure conjunction, the andness of t-norms is greater than 1 and t-norms and t-conorms are models of hyperconjunction and hyperdisjunction and could be of interest for modeling GCD. Unfortunately, T-norms/conorms are not andness-directed, not weighted, strictly hard, and not frequently related to human commonsense logical reasoning and decision making. These drawbacks attracted research attention and Yager developed weighted t-norms and conorms [YAG04], and Torra introduced andness directedness for t-norms and conorms [TOR22].

Observations and experiments prove that simultaneity and substitutability are human graded percepts. Consequently, graded propositional logic models must be andness-directed, i.e., andness and orness must be explicitly visible and easily adjustable. Andness-directedness is the prerequisite that a GCD aggregator must satisfy to get the status of the graded logic function and the corresponding applicability [DUJ21b, DUJ21c, DUJ21d, TOR21, TOR22]. Therefore, the andness-directedness is necessary for all hyperconjunction/hyperdisjunction aggregation functions. The andness-directed t-norms and t-conorms were introduced in the recent work of Torra [TOR22] which will be presented later in this section. Andness-directed t-norms and t-conorms have the status of graded logic functions.

From the standpoint of modeling hyperconjunction, it is convenient to classify t-norms in three groups:

(1) Fixed andness ($\alpha = const.$),
(2) Subrange of andness ($\alpha_{\min} \leq \alpha \leq \alpha_{\max}, \alpha_{\max} - \alpha_{\min} < 2$), and
(3) Full range of andness ($1 \leq \alpha \leq 2$).

The t-norms in groups 2 and 3 must be parameterized: they must contain an adjustable parameter (r). By adjusting the parameter r we can change/adjust the andness of the t-norm $T(x, y; r)$ and obtain desired logic properties of that aggregation function, so that we can use it as an alternative to GCD.

As an example of fixed andness let us use the Hamacher t-norm $h(x, y) = xy/(x + y - xy)$, where the andness is $\alpha_h = 2 - 3\int_0^1\int_0^1 h(x, y)dxdy = 1.1309$. So, $Hcon(x_1, x_2; \alpha_h)$ should be close to $h(x, y)$. The corresponding mean difference and maximum difference are the following:

$$D_{mean} = \int_0^1\int_0^1 |Hcon(x_1, x_2; \alpha_h) - h(x_1, x_2)|dx_1dx_2 = 0.011,$$

$$D_{\max} = \max_{\substack{0 \leq x_1 \leq 1 \\ 0 \leq x_2 \leq 1}} |Hcon(x_1, x_2; \alpha_h) - h(x_1, x_2)| = 0.052.$$

Obviously, with the mean difference around 1% and the maximum difference around 5% of the unit range, for all practical purposes, $Hcon(x_1, x_2; \alpha_h)$ is an acceptable approximation of $h(x_1, x_2)$.

For a parameterized t-norm $T(x, y; r)\ r \in \mathbb{P} \subset \mathbb{R}$ the value of andness is a function of the parameter r:

$$\alpha = \varphi(r) = 2 - 3\int_0^1\int_0^1 T(x, y; r)dxdy.$$

For each desired value of andness α, we can now compute the corresponding value of parameter $r = \varphi^{-1}(\alpha)$. The andness of analyzed t-norm satisfies the following relation (see Definition 15.1):

$$2 - 3\int_0^1\int_0^1 T\big(x, y; \varphi^{-1}(\alpha)\big)dxdy = \alpha.$$

Consequently, $T\big(x, y; \varphi^{-1}(\alpha)\big)$ is an andness-directed hyperconjunctive aggregator. Of course, the function $\varphi^{-1} : [0, 1] \rightarrow \mathbb{R}$ is usually obtained only as an interpolative numerical approximation [TOR22].

As an example of the t-norm that covers a subrange of andness let us use the following Frank t-norm [KLI95, KLE99] which uses the parameter $r \in [0, +\infty]$:

$$T_F(x, y; r) = \begin{cases} \min(x, y) & \text{if } r = 0; \quad \alpha = 1 \\ xy & \text{if } r = 1; \quad \alpha = 1.25 \\ \max(0, x + y - 1) & \text{if } r = +\infty; \ \alpha = 1.5 \\ \log_r\left(1 + \frac{(r^{1-x}-1)(r^{1-y}-1)}{r-1}\right) & \text{otherwise}; \quad 1 < \alpha < 1.5. \end{cases}$$

The Frank t-norm covers the andness in the range $\alpha \in [1, 1.5]$, i.e., from the pure conjunction to the Łukasiewicz t-norm. Three most popular t-norms (minimum, algebraic product, and Łukasiewicz bounded difference) are special cases connected by the Frank t-norm. That is a half of the hyperconjunction range $\alpha \in [1, 2]$ for two variables.

The andness of the Frank t-norm function $\varphi_F : [0, +\infty] \rightarrow [1, 1.5]$ for all values of the parameter r r in the range [0,4] is shown in Fig. 15.10. It is easy to note the inconvenient shape of the function $\alpha = \varphi_F(r)$: for small values of r r, the microscopic changes of r r cause large variations of andness. Consequently, it is rather difficult to adjust this aggregator to produce desired value of andness in the vicinity of the pure conjunction.

The simple andness-directed hyperconjunction $Hcon(x, y; \alpha)$ is rather close to $T_F\big(x, y; \varphi_F^{-1}(\alpha)\big)$. That is visible in Fig. 15.11: the mean difference for $1 \leq \alpha \leq 1.25$ is below 1% and for $1.25 \leq \alpha \leq 2$ it is below 5%. The increase of difference for $\alpha = 1.5$ is understandable from a geometrical interpretation: Łukasiewicz t-norm $\max(0, x + y - 1)$ has a discontinuity of the first derivative for $x + y = 1$ and is flat in the triangle $0 < x + y < 1$. That is not an observable property in the commonsense human reasoning. On the other hand, the hyperconjunction $Hcon(x, y; 1.5) = (xy)^{\sqrt{6}-1} = (xy)^{1.44949}$, which has the same andness, is a continuous function that has similar properties but is closer to the continuous nature of human reasoning (see Fig. 14.8). The idea that natural properties change gradually, rather than suddenly ("natura non facit saltus") is frequently violated in the fuzzy logic theory and in the mathematical literature devoted to t-norms. As opposed to that, in Graded Logic, we support the natural concept of continuity as much as possible.

Frank's t-norm cannot cover the whole range of hyperconjunction. However, there are several popular parameterized t-norms that provide hyperconjunctive aggregation in the whole range of andness from minimum to drastic conjunction (for 2 variables, $1 \leq \alpha \leq 2$). According to [KLI95] and the study of Torra [TOR22], they include the Dombi t-norm $T_D(x, y; r)$, three t-norms developed by Schweizer and Sklar, denoted

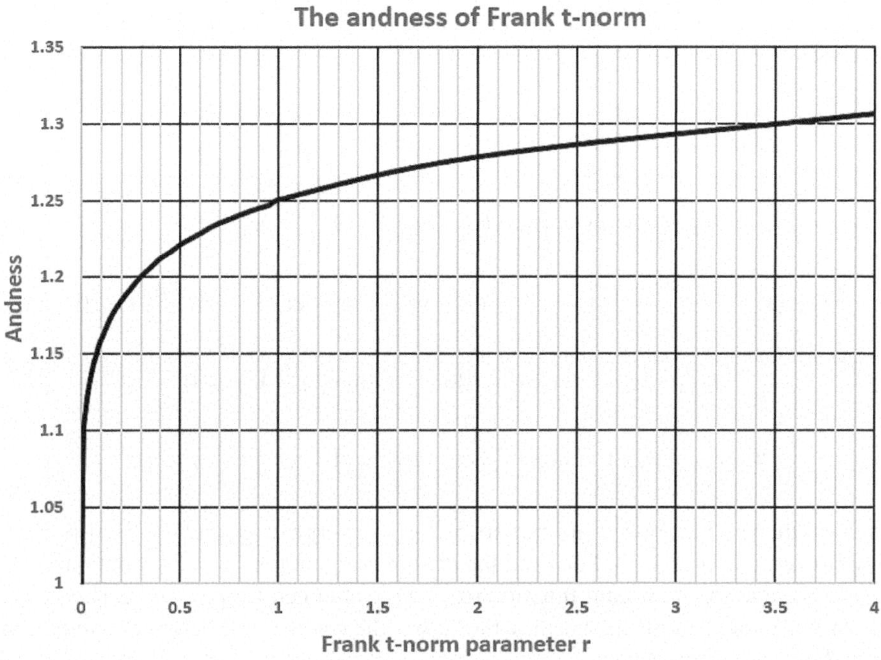

Fig. 15.10 The andness of Frank t-norm as a function of parameter r

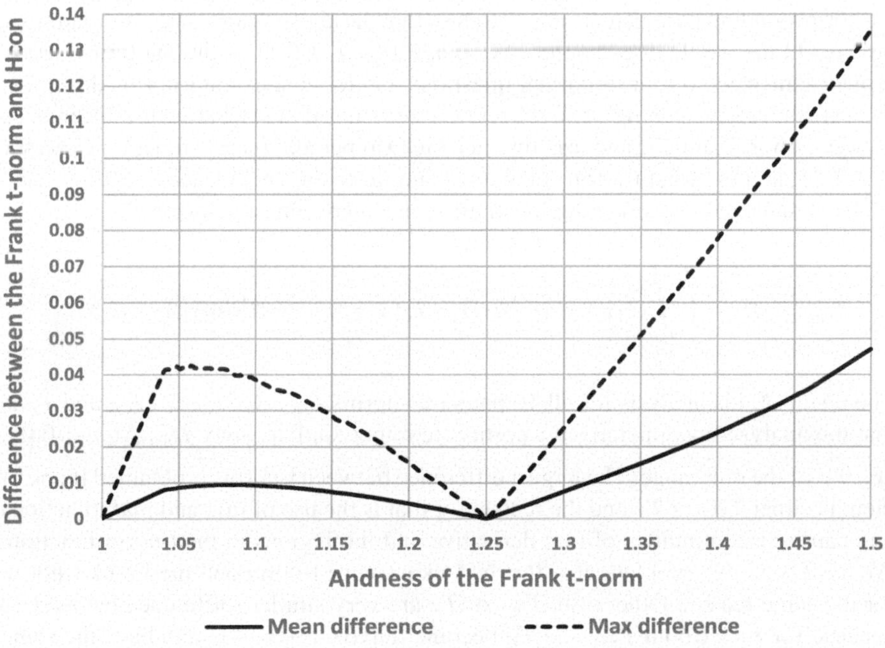

Fig. 15.11 Mean difference and maximum difference between Hcon and the Frank t-norm

$T_{SS1}(x, y; r)$, $T_{SS2}(x, y; r)$ and $T_{SS3}(x, y; r)$, and the Yager t-norm $T_Y(x, y; r)$, as follows:

$$T_D(x, y; r) = \left\{1 + \left[\left(\frac{1}{x} - 1\right)^r + \left(\frac{1}{y} - 1\right)^r\right]^{1/r}\right\}^{-1}, r > 0,$$

$$T_{SS1}(x, y; r) = \left[\max\left(0, x^r + y^r - 1\right)\right]^{1/r}, |r| > 0,$$

$$T_{SS2}(x, y; r) = 1 - \left[(1 - x)^r + (1 - y)^r - (1 - x)^r(1 - y)^r\right]^{1/r}, r > 0,$$

$$T_{SS3}(x, y; r) = \exp\left(-\left(|\ln x|^r + |\ln y|^r\right)^{1/r}\right), r > 0,$$

$$T_Y(x, y; r) = 1 - \min\left\{1, \left[(1 - x)^r + (1 - y)^r\right]^{1/r}\right\}, r > 0.$$

If $r \to 0$ then all the above t-norms except T_{SS1} converge to xy, and if $r \to +\infty$ then they converge to $\min(x, y)$. For T_{SS1}, the drastic conjunction ($\alpha = 2$) is obtained for $r \to +\infty$, and the minimum function ($\alpha = 1$) is obtained for $r \to -\infty$.Therefore, these t-norms are significant, because they cover the whole range of hyperconjunction $xy \leq T(x, y; r) \leq \min(x, y)$.

The logic properties of these t-norms are analyzed in [TOR22]. Torra's goal was to make the above t-norms andness-directed, i.e., to compute for each value of andness the corresponding parameter $r = \varphi^{-1}(\alpha)$, so that the above t-norms become andness-directed in the whole hyperconjunctive range $1 \leq \alpha \leq 2$. In addition, Torra's study provides answers to four important questions: (1) to what extent t-norms differ, (2) what are the reasons for high differences between t-norms, (3) what are the differences between t-norms and the andness-directed GCD hyperconjunction function *Hcon* for $\alpha > 1.25$, and (4) what t-norm is the most similar to the GCD *Hcon*.

Torra analyzed the following mean difference between two t-norms:

$$D_{mean}^{T_1,T_2}(\alpha) = \int_0^1 \int_0^1 \left|T_1\left(x, y; \varphi_1^{-1}(\alpha)\right) - T_2\left(x, y; \varphi_2^{-1}(\alpha)\right)\right| dxdy, \ 1 \leq \alpha \leq 2.$$

The result of this analysis for all 10 pairs of t-norms T_D T_{SS1}, T_{SS2} , T_{SS3}, and T_Y is that the analyzed parameterized t-norms are rather similar: $\max\limits_{1 \leq \alpha \leq 2} D_{mean}(\alpha) < 0.09$, i.e., 9% of the unit range. The largest difference between t-norms is obtained if one of them is either T_{SS1} or T_Y and the reason for that is the use of min and max functions that cause discontinuities of first derivative and flat regions in the t-norm function. We have seen this problem also for the Łukasiewicz t-norm and the Frank t-norm, for the same reason. Otherwise, T_{SS1} or T_Y are very similar (difference below 4%) because for $r = 1$ both T_{SS1} and T_Y become $\max(0, x + y - 1)$ and have the same andness $\alpha = 1.5$.

The mean difference between *Hcon* and the analyzed t-norms is the following:

$$D_{mean}^{Hcon,T}(\alpha) = \int_0^1 \int_0^1 \left| (xy)^{\sqrt{3/(2-\alpha)}-1} - T\left(x, y; \varphi_T^{-1}(\alpha)\right) \right| dxdy, \quad 5/4 \le \alpha \le 2.$$

The results can be summarized as follows:

$$\max_{1.25 \le \alpha \le 2} D_{mean}^{Hcon,SS2}(\alpha) < 3\%, \quad \max_{1.25 \le \alpha \le 2} D_{mean}^{Hcon,SS3}(\alpha) < 4\%,$$

$$\max_{1.25 \le \alpha \le 2} D_{mean}^{Hcon,D}(\alpha) < 4\%, \quad \max_{1.25 \le \alpha \le 2} D_{mean}^{Hcon,Y}(\alpha) < 5\%, \quad \max_{1.25 \le \alpha \le 2} D_{mean}^{Hcon,SS1}(\alpha) < 6\%.$$

In addition, in the range $1.25 \le \alpha \le 1.75$ the mean difference between *Hcon* and *SS*3 is below 3% and the smallest of all analyzed t-norms. According to [TOR22], in the analyzed set of five t-norms, the central location has T_{SS3} which is very similar to *Hcon*. Thus, if we need $\alpha \ge 1.25$ and want to use a t-norm, then the suggested t-norm is T_{SS3}, and if we need the benefits of GCD, then the suggested hyperconjunction is *Hcon*.

It is useful to note that t-norms can be developed using additive generator functions [KLI95, KLE00, TOR07, BEL07, GRA09, YAG04]. An additive generator function is a strictly decreasing continuous function $f : [0, 1] \to [0, +\infty]$, where $f(1) = 0$. E.g., such a generator is $f(x) = y = -\ln x$ and its inverse function is $x = f^{-1}(y) = e^{-y}$. There is also an important related concept of pseudoinverse function (denoted $f^{(-1)}$) which is defined as $f^{(-1)}(y) = f^{-1}(y)$ if $y \le f(0)$ and otherwise $f^{(-1)}(y) = 0$. In our example, $f^{(-1)}(y) = e^{-y}$.

An important group of t-norms $T(x, y)$ includes those t-norms that satisfy the subidempotency condition $\forall x \in]0, 1[\Rightarrow T(x, x) < x$. Such t-norms are called the Archimedean t-norms, and they can be created using the following additive generator functions:

$$T(x, y) = f^{(-1)}(f(x) + f(y)).$$

In our example, we have the following:

$$f^{(-1)}(f(x) + f(y)) = f^{(-1)}(-\ln x - \ln y) = e^{-(-\ln x - \ln y)} = e^{\ln xy} = xy = T(x, y).$$

Therefore, the generator function $f(x) = -\ln x$ generates the product t-norm. Of course, other generator functions create many other Archimedean t-norms which can be used as models of hyperconjunction.

T-norms can be importance weighted. In a series of papers summarized in [YAG04], Yager proposed the use of weighted linear generators, as follows:

$$T((w_1, x_1), \ldots, (w_n, x_n)) = f^{(-1)}\left(\sum_{i=1}^{n} w_i f(x_i)\right).$$

Here $w_1, \ldots, w_n$ denote adjustable importance weights. For the generator $f(x) = -\ln x$ we have the following exponentially importance-weighted product t-norm:

$$T((w_1, x_1), \ldots, (w_n, x_n)) = e^{-(-w_1 \ln x_1 - \ldots - w_n \ln x_n)} = \prod_{i=1}^{n} x_i^{w_i}.$$

If $w_1 = \ldots = w_n = 1$, then we have the standard commutative multiple variable product t-norm $T(x_1, \ldots, x_n) = \prod_{i=1}^{n} x_i, n \geq 2$. Note that in this case $w_1 + \ldots + w_n = n$, i.e., we have the count-normalized weights.

Applicability of t-norms in Decision-Making

In decision-making applications of GL, the centroid of all logic aggregation operations is the logic neutrality, i.e., the most frequently used aggregator is the arithmetic mean and idempotent conjunctive and disjunctive soft and hard GCD aggregators in its vicinity. On the other hand, the strictly hard and nonidempotent hyperconjunction and hyperdisjunction are the lateral segments of GCD (see Fig. 11.1 and Table 11.2). They have extreme properties and are not frequent in decision-making applications. So, it is useful to discuss the applicability of t-norms in this area.

T-norms are strictly hard aggregators: they strictly support the annihilator 0 and all their input arguments reflect only mandatory requirements. So, t-norms and conorms cannot be used in situations where we need soft aggregators. On the other hand, there are families of aggregators that are strictly soft (e.g., the exponential means, or the original soft version of OWA [YAG88]). The strictly soft aggregators can model only optional inputs, and no input argument can reflect a mandatory requirement. With negligible and primarily theoretical exceptions, there are no real evaluation-based decision-making problems that can be solved using only t-norms/conorms or using only soft aggregators.[3]

Excluding the idempotent minimum and maximum aggregators, all other t-norms/conorms are strictly nonidempotent. Experiments with commonsense reasoning and decision making, reported in Chapter 14 and [DUJ15a], as well as experiences in decision engineering practice, show beyond any doubt that the real-life propositional calculus formulas must simultaneously include both hard and soft aggregators. In almost all human evaluation criteria, some inputs are mandatory, and some are optional. In addition, in most evaluation problems we use idempotent aggregators, and that is not possible with t-norms/conorms.

The key aggregator in the whole area of hyperconjunction is the product t-norm. The reason for that is the probabilistic interpretation of that aggregator. The andness of the product function[4] is moderately higher than the andness of the pure conjunction

[3] The analysis of general applicability of t-norms is beyond the scope of this book. However, it is worth noting that in the book with characteristic title "Fuzzy Logic with Engineering Applications" by T.J. Ross [ROS10] the presence of t-norms/conorms is negligible, except the use of minimum and maximum as the standard (i.e., most suitable) operation for modeling the intersection and union of fuzzy sets.

[4] For n variables, $\alpha_{prod}(n) = \left[n - (n+1)2^{-n}\right]/(n-1)$. See Fig. 12.1.

and if two or more inputs can be interpreted as probabilities of attaining a positive result, then the use of that specific special case of hyperconjunction is fully justifiable. For example, if a candidate is expected to have skills in two independent areas, where x and y are the probabilities of achieving successful results in these areas, then the overall suitability can easily be defined as xy, which is the probability of simultaneous success in both areas. So, the basic reason for using hyperconjunction in decision models is the aggregation of arguments which can have a probabilistic (or similar) interpretation. The product of n variables is the central anchor aggregator in our interpolative hyperconjunction aggregator *Hcon*, as well as in the hyperconjunctive segment of the UGCD aggregator shown in Fig. 15.3.

T-norms can be weighted, but the need for importance weights in hyperconjunctive GCD is a delicate problem because the answer depends on andness. Generally, the likelihood of using importance weights decreases when the andness (or orness) increases, and it can easily reach zero in the range $[1, n/(n-1)]$. As we discussed in Chap. 12, for andness above 1, the logic importance increases to high levels where all GCD inputs can be highly and equally important. In such cases, the relative importance becomes less influential, commutativity can be acceptable, and unweighted t-norms can be used as models of hyperconjunctive GCD.

It is worth noting a significant degree of contradiction between the request for very high andness and the request for different weights of arguments. The very high andness means that all inputs must be simultaneously highly satisfied, and consequently there are no unimportant inputs. As opposed to that, different weights are necessary to differentiate between important and unimportant inputs. So, high andness is a claim that there are no unimportant inputs, and variable weights indicate the possible presence of low weights and unimportant inputs. Considering the high andness of t-norms, it follows that the use of weighted t-norms in decision models is regularly not justified. Indeed, in the high range of andness, the percept of the overall importance of arguments is primarily affected by the logic importance, i.e., by the high andness or orness. For example, if the suitability of a candidate is defined using the probabilistic model $S = xy$, then the obvious assumption is that x and y are equally important.

Even in the case of a high andness, it is theoretically possible to find the situation where one of inputs might be considered more important than the other one. For example, if x is declared to be two times more important than y, then some analysts would be tempted to modify the probabilistic model using weighted Archimedean t-norm $S = x^{4/3}y^{2/3}$. Both models are based on weighted linear generators. Another viewpoint would be to claim that if the andness is sufficiently above 1, then all arguments are by definition very important and must be highly satisfied, so that it is not possible that one argument can be two times more important than another argument. Of course, it would be more difficult to reject the request for different weights if the proposed differences between weights are sufficiently small. Such problems can be avoided by creating a training set of desired input–output mapping and adjusting GCD parameters to best fit the training set. The experimental validation of weighted GCD hyperconjunction is one of topics for the future work.

T-norms are strictly associative aggregators, and it is useful to identify the benefits of associativity. Of course, associativity is a mathematical property that is necessary for the development of some mathematical theories. In GL applications, the strict associativity is not a significant advantage because all GCD aggregators are approximately associative. Our segmented interpolative aggregation is based on anchor aggregators which are either associative, or close to associativity. In the vicinity of associative anchor aggregators, the associativity theoretically does not hold, but the corresponding practical consequences are usually negligible. For example, the analysis in [DUJ75a] shows that mean differences between various associativity structures are approximately 1%. So, all GCD aggregators are *approximately associative*, in the sense that, e.g., the differences between aggregators $GCD(a, GCD(b, c, d))$, $GCD(GCD(a, b), GCD(c, d)), GCD(a, GCD(b, GCD(c, d)))$, and $GCD(a, b, c, d)$ are sufficiently small and have negligible impact on the results of GL models used in decision-making applications. In addition, in the process of building GL aggregation structures, it is easier to simultaneously see all inputs of aggregators and their weights, compared to multiple binary aggregations.

Various forms of hyperconjunction create the need for their comparison. Theoretically, it is not difficult to compute the mean difference and the maximum difference between various existing models of hyperconjunction in the range $1 \le \alpha \le n/(n-1)$. However, practical decision engineering aspects of this problem are more difficult. The human-centric approach to Graded Logic assumes that the primary criterion for the validity of GCD models is their consistency with observable and measurable properties of human reasoning. Therefore, the correct way for validation of various models of hyperconjunction and hyperdisjunction must include experiments with human subjects based on requirements that should simultaneously come from both decision engineering and cognitive psychology.

Fully Importance-Weighted GCD Aggregators

In the andness range $0 \le \alpha \le 1$, most models of the *GCD* aggregator provide idempotency and consistency with classical Boolean logic. As the andness approaches the value 1, the impact of weights diminishes, and for the full conjunction ($\alpha = 1$) it fully disappears, assuming that all inputs, which are mandatory and must be positive, have equal and strong impact. This is consistent with observable properties of human commonsense reasoning. For the hyperconjunction in the range $1 < \alpha \le \frac{\mathrm{n}}{\mathrm{n}-1}$, we assume that very high simultaneity implies high and equal importance, so that the weights of all inputs are constant and equal.

In AI applications where GCD aggregators model natural phenomena beyond the observable human commonsense logic, there is no need to restrict the use of weights and to provide a strict idempotency. Weighted aggregation in the whole range from drastic conjunction to drastic disjunction can be realized by introducing weights in the count-normalized product-power aggregators $y = \left(\prod_{i=1}^{n} x_i\right)^p$, where the parameter p is used to adjust the desired andness. The volume under such a surface is $V = 1/(p+1)^n$ yielding the following relationships between the andness and the exponent p:

$$\alpha(n,p) = \frac{n-(n+1)/(p+1)^n}{n-1},$$

$$p(n,\alpha) = \left(\frac{n+1}{n-(n-1)\alpha}\right)^{1/n} - 1, \quad p(n,1) = (n+1)^{1/n} - 1.$$

The nonidempotent aggregator $y = \left(\prod_{i=1}^{n} x_i\right)^{(n+1)^{1/n}-1}$ has the andness 1 and is called the pseudoconjunction [DUJ18]. The product-power aggregators are count-normalized. Consequently, they can be weighted using weights of the sum-normalized aggregators ($w_1 + \ldots + w_n = 1$) as follows:

$$y = \left(\prod_{i=1}^{n} x_i^{nw_i}\right)^{p(n,\alpha)}, \quad 0.75 \le \alpha \le n/(n-1).$$

This range of andness is selected to create uniform aggregators where for $0.5 \le \alpha < 0.75$ we have soft aggregators and for $\alpha \ge 0.75$ hard aggregators.

The fully importance-weighted GCD aggregators can be created by interpolating soft aggregators between the arithmetic mean $y = \sum_{i=1}^{n} w_i x_i$ and the weighted product-power hard border function $y = \left(\prod_{i=1}^{n} x_i^{nw_i}\right)^{p(n,0.75)}$. Then, we can use the weighted product-power function in the range $0.75 \le \alpha \le n/(n-1)$. For simplicity, let us use the case $n = 2$ creating the aggregator $GCD_2(x, y; w, \alpha)$ where $w =$ *the importance degree of* x, and $1 - w =$ *the importance degree of* y, assuming the weight range $0 < w < 1$. The corresponding andness-power relationships and the fully importance-weighted recursive aggregator are the following:

$$\alpha(2,p) = 2 - \frac{3}{(p+1)^2}, \quad p(2,\alpha) = \sqrt{\frac{3}{2-\alpha}} - 1 = \begin{cases} \sqrt{12/5} - 1, & \alpha = 3/4 \\ \sqrt{3} - 1, & \alpha = 1 \\ 1, & \alpha = 5/4 \end{cases}$$

Therefore, the following is the fully importance-weighted recursive aggregator of two degrees of truth:

$$z = \begin{cases} GCD_2(x,y;w,\alpha) = \begin{cases} \lfloor xy \rfloor, & \alpha = 2, \\ \left(x^{2w} y^{2(1-w)}\right)^{\sqrt{\frac{3}{2-\alpha}}-1}, & ¾ \le \alpha < 2, \\ (3-4\alpha)(wx + (1-w)y) + \\ (4\alpha - 2)\left(x^{2w} y^{2(1-w)}\right)^{\sqrt{\frac{3}{2-\alpha}}-1}, & ½ < \alpha < ¾, \\ wx + (1-w)y, & \alpha = \omega = ½, \end{cases} \\ 1 - GCD_2(1-x, 1-y;\, w, 1-\alpha), & -1 \le \alpha < 0.5 \end{cases}$$

This aggregator is weighted in the whole range $-1 < \alpha < 2$ and idempotent only for $\alpha = 0.5$. The following are the important special cases of this weighted aggregator:

$$GCD_2(x, y; w, 0.75) = \left(x^{2w}y^{2(1-w)}\right)^{\sqrt{\frac{12}{5}}-1} \quad \text{(lowest andness of hard graded conjunction)}$$

$$GCD_2(x, y; w, 1) = \left(x^{2w}y^{2(1-w)}\right)^{\sqrt{3}-1} \quad \text{(nonidempotent weighted pseudoconjunction)}$$

$$GCD_2(x, y; w, 1.25) = \left(x^{2w}y^{2(1-w)}\right) \quad \text{(weighted product).}$$

In the range $^3/_4 \le \alpha \le 1$, the differences between idempotent hard aggregators (Tables 15.4 and 17.4) and the fully weighted nonidempotent aggregators are small. For example, in the case of 2 variables, the mean absolute difference between the pure conjunction $\min(x, y)$ and the pseudoconjunction $(xy)^{\sqrt{3}-1}$ inside $[0, 1]^2$ is 4.77% and the maximum difference is 13.94%.

Logic Aggregation: Summary of Seven GCD Properties and Conclusions

This chapter terminates the sequence of chapters devoted to fundamental GL topics: necessary properties of GL functions and methods for their design. Our basic conclusion is that a human-centric graded logic must be based on GCD aggregators that satisfy postulates of Graded Logic (Chapter 3). Thus, GCD must be *continuum-valued in arguments, andness-directed, importance-weighted, idempotence-selectable, annihilator-selectable, based on unified and complementary conjunction/disjunction, and dualized*. These seven properties are necessary because they are explicitly visible in natural human commonsense reasoning. Without these properties, it is not possible to model natural human reasoning, and without the full compatibility with natural human commonsense reasoning, all propositional logic-based decision models become questionable. Following is the summary of seven fundamental GCD properties.

(1) The arguments of GCD aggregators must be *continuum-valued* because, in decision-making, human percepts are graded: the value statements that claim the full satisfaction of various specific stakeholder's requirements are only partially true.
(2) GCD must be *andness-directed* because humans use both simultaneity and substitutability as complementary graded percepts. The necessary penalizing of insufficient simultaneity or substitutability is not fixed (constant) but adjustable in a wide range from very weak to drastic, reflecting the goals and requirements of a specific stakeholder/decision-maker.
(3) GCD must be *importance-weighted* in most cases where truth has semantic identity. If a statement claims a specific property of an object that is evaluated by a stakeholder, then each such property can have a different importance that reflects the goals and interests of the decision-maker. Most human decisions are importance-driven, and their mathematical models must include importance weights as explicitly visible and adjustable parameters.
(4) GCD must be *idempotence-selectable* because in natural human reasoning the results of logic aggregation of two or more degrees of truth can be either inside or outside the min–max range. If the expected results are inside the min–max

range, and the GCD aggregator must be idempotent, then it is a mean. If the expected results are outside the min–max range, then the GCD aggregator must be nonidempotent, and it can be either a t-norm/conorm or another nonidempotent function. In exceptional cases, nonidempotent aggregators can be used in the whole range of andness.

(5) GCD must be *annihilator-selectable* because in natural human reasoning both conjunction and disjunction can be either hard (supporting annihilators) or soft (not supporting annihilators). If a stakeholder wants to reject objects that do not satisfy one of attributes that represent mandatory requirements, then the conjunctive annihilator 0 must be supported. If the simultaneity is desirable but can tolerate some zero inputs, then the annihilator 0 must not be supported. Similar selectability of annihilator 1 holds also for disjunction. The annihilator 1 must be supported if any input has the power to fully satisfy the stakeholder's requirements; if that is not desirable, the annihilator 1 must not be supported.

(6) GCD must be *conjunction/disjunction unified and complementary* in the sense that GCD simultaneously includes complementary conjunctive and disjunctive properties as models of simultaneity and substitutability. Any increase of conjunctive properties means a decrease of disjunctive properties, and vice versa. GCD unifies conjunction and disjunction in a single logic operator. That is an observable property of human reasoning, and the fundamental property of GL.

(7) GCD must be *dualized*, i.e., analytically dualized because naturally dual aggregators with all necessary properties are not available. That is necessary because, in human commonsense logic, duality is present as two equivalent ways of expressing the same logical relationship.

The presented seven properties of the GCD logic function are fundamental prerequisites for human-centric graded propositional logic. All functions that are candidates for logic aggregators must satisfy these prerequisites because without them most applications of graded logic lose credibility. To create a GCD function that satisfies all necessary properties, we proposed an implementation based on the andness-directed segmented interpolative method (Tables 15.4 and 17.4). Our implementations of $GCD(\boldsymbol{x};\boldsymbol{W},\alpha)$ fully support all postulates of Graded Logic, and the fundamental property of nondecreasing monotonicity in all arguments: $\boldsymbol{x} \le \boldsymbol{z} \Rightarrow GCD(\boldsymbol{x};\boldsymbol{W},\alpha) \le GCD(\boldsymbol{z};\boldsymbol{W},\alpha)$. In addition, $GCD(\boldsymbol{x};\boldsymbol{W},\alpha)$ satisfies the nonincreasing monotonicity in andness (and the nondecreasing monotonicity in orness): $\alpha_1 \le \alpha_2 \Rightarrow GCD(\boldsymbol{x};\boldsymbol{W},\alpha_1) \ge GCD(\boldsymbol{x};\boldsymbol{W},\alpha_2)$.

The segmented interpolative method for creating logic aggregators, combined with conditional logic aggregation provides all properties that are necessary in logic-based decision making. Using GCD as the basic logic aggregation model, we can now focus on the graded propositional calculus and its applications.

References

AHN18 Ahn, B.S. 2018. An Alternative Approach to Obtaining an Analytical Solution to Dujmovic's Iterative OWA. *International Journal of Uncertainty, Fuzziness and Knowledge-Based Systems* 26 (2): 261–267.

BEL07 Beliakov, G., A. Pradera, and T. Calvo. 2007. *Aggregation Functions: A Guide for Practitioners*. Springer.

BEL16a Beliakov, G., and J. Dujmović. 2016. Extension of Bivariate Means to Weighted Means of Several Arguments by Using Binary Trees. *Information Sciences* 331: 137–147. https://doi.org/10.1016/j.ins.2015.10.040.

BEL16 Beliakov, G., H. Bustince Sola, and T. Calvo Sanchez. 2016. A Practical Guide to Averaging Functions. In *Studies in Fuzziness and Soft Computing*, vol. 329. Springer.

BUL03 Bullen, P.S. 2003, 2010. *Handbook of Means and Their Inequalities*. Kluwer.

BUL88 Bullen, P.S., D.S. Mitrinović, and P.M. Vasić. 1988. *Means and Their Inequalities*. Dordrecht: D. Reidel.

CAL02 Calvo, T., A. Kolesarova, M. Komornikova, and R. Mesiar. 2002. Aggregation Operators: Properties, Classes and Construction Methods. In *Aggregation Operators*, ed. T. Calvo, G. Mayor, and R. Mesiar. New Trends and Applications, 3–104. Heildeberg: Physica-Verlag.

CAL03 Calvo, T., and R. Mesiar. 2003. Fusion with Quantitative Weights. In *EUSFLAT 2003 Proceedings*, ed. M. Wagenknecht, and R. Hampel, 312–317. http://www.eusflat.org/publications_proceedings_EUSFLAT_2003.php.

CAL04 Calvo, T., R. Mesiar, and R. Yager. 2004. Quantitative Weights and Aggregation. *IEEE Transactions on Fuzzy Systems* 12 (1): 62–69.

DET21 De Tré, G., and J. Dujmović. 2021. Dealing with Data Veracity in Multiple Criteria Handling: An LSP-Based Sibling Approach. In *Proceedings of the International Conference on Flexible Query Answering Systems, FQAS 2021, Bratislava, Slovakia*, ed. T. Andreasen et al. Flexible Query Answering Systems, 82–96. Springer LNAI 12871.

DET22 De Tré, G., M. Peelman, and J. Dujmović. 2022. Logic Operators and Sibling Aggregators for Z-Grades. *Proceedings of the IPMU 2022*. In *Information Processing and Management of Uncertainty in Knowledge-Based Systems, 19th International Conference, IPMU 2022 Milan, Italy, July 11–15, 2022 Proceedings, Part II*, ed. Davide Ciucci, Inés Couso, Jesús Medina, Dominik Ślęzak, Davide Petturiti, Bernadette Bouchon-Meunier, and Ronald R. Yager. Communications in Computer and Information Science, vol. 1602, 572–583. Springer.

DET23 De Tré, G., M. Peelman, and J. Dujmović. 2023. Logic Reasoning Under Data Veracity Concerns. International Journal of Approximate Reasoning 161: 108977. https://doi.org/10.1016/j.ijar.2023.108977.

DOM82 Dombi, J., and P. Zysno. 1982. Comments on the Gamma-Model. In Cybernetics and Systems Research, ed. R. Trappl, 711–714. North-Holland.

DUJ75a Dujmović, J. 1975. Extended Continuous Logic and the Theory of Complex Criteria. *Journal of the University of Belgrade, EE Dept., Series Mathematics and Physics* 537: 197–216.

DUJ91 Dujmović, J. 1991. Preferential Neural Networks. Chapter 7 in *Neural Networks—Concepts, Applications, and Implementations*, Vol. II, ed. P. Antognetti, and V. Milutinović. Prentice-Hall Advanced Reference Series, 155–206. Prentice-Hall.

DUJ07b Dujmović, J. 2007. Continuous Preference Logic for System Evaluation. *IEEE Transactions on Fuzzy Systems* 15 (6): 1082–1099.

DUJ08 Dujmović, J. 2008. Characteristic Forms of Generalized Conjunction/Disjunction. In *Proceedings of the IEEE World Congress on Computational Intelligence*, 1075–1080b, Hong Kong: IEEE Press, ISBN 978-1-4244-1819-0.

DUJ15a Dujmović, J. 2015. Weighted Compensative Logic with Adjustable Threshold Andness and Orness. *IEEE Transactions on Fuzzy Systems* 23 (2): 270–290. https://doi.org/10.1109/TFUZZ.2014.2312018.

DUJ15c Dujmović, J. 2015. An Efficient Algorithm for General Weighted Aggregation. In *Proceedings of the 8th International Summer School on Aggregation Operators (AGOP 2015)*, ed. M. Baczynski, B. De Baets, and R. Mesiar, 115–120. University of Silesia, ISBN: 978-83-8012-519-3.

DUJ18 Dujmović, J. 2018. *Soft Computing Evaluation Logic*. Wiley.

DUJ21d Dujmović, J. 2021. Andness-Directed Iterative OWA Aggregators. In *Modeling Decisions for Artificial Intelligence. MDAI 2021*, ed. V. Torra, and Y. Narukawa. Lecture Notes in Computer Science, vol. 12898, 3–16. Cham: Springer. https://doi.org/10.1007/978-3-030-85529-1_1.

DUJ20 Dujmović, J. 2020. Interpretability and Explainability of LSP Evaluation Criteria. In *Proceedings of the 2020 IEEE World Congress on Computational Intelligence*, paper F-22042, ISBN: 978-1-7281-6932-3/20.

DUJ21e Dujmović, J. 2021. Numerical Comparison of Idempotent Andness-Directed Aggregators. In *Modeling Decisions for Artificial Intelligence. MDAI 2021*, ed. V. Torra, and Y. Narukawa. Lecture Notes in Computer Science, vol. 12898, 67–77. Cham: Springer. https://doi.org/10.1007/978-3-030-85529-1_6.

DUJ17a Dujmović, J., and G. Beliakov. 2017. Idempotent Weighted Aggregation Based on Binary Aggregation Trees. *International Journal of Intelligent Systems* 32 (1): 31–50. https://doi.org/10.1002/int.21828.

DUJ22 Dujmović, J., and D. Tomasevich. 2022. Experimental Analysis and Modeling of Human Conjunctive Logic Aggregation. In *2022 IEEE International Conference on Fuzzy Systems (FUZZ-IEEE)*, 1–8. Padua, Italy. https://doi.org/10.1109/FUZZ-IEEE55066.2022.9882665.

DUJ21c Dujmović, J., and V. Torra. 2021. Properties and Comparison of Andness-Characterized Aggregators. *International Journal of Intelligent Systems* 36 (3): 1366–1385.

DUJ21b Dujmović, J., and V. Torra. 2022. Aggregation Functions in Decision Engineering: Ten Necessary Properties and Parameter-Directedness. In *Proceedings of the 3rd International Conference on Intelligent and Fuzzy Systems, INFUS 2021 Conference, Izmir, Turkey*, ed. C. Kahraman et al. *Intelligent and Fuzzy Techniques for Emerging Conditions and Digital Transformation*, vol. 1. LNNS, vol. 307, 173–181. Springer.

DUJ23 Dujmović, J., and V. Torra. 2023. Logic Aggregators and Their Implementation. In *MDAI 2023*, ed. V. Torra, and Y. Narukawa. LNAI, vol. 13890, 3–42. Springer.

FOD94 Fodor, J., and M. Roubens. 1994. *Fuzzy Preference Modelling and Multicriteria Decision Support*. Dordrecht, The Netherlands: Kluwer Academic Publishers.

GIN58 Gini, C., et al. 1958. *Le Medie, (Means) Unione Tipografico-Editorice Torinese*. Milano. (Russian Translation: Srednie Velichiny, Statistika, Moskva, 1970).

GRA09 Grabisch, M., J.-L. Marichal, R. Mesiar, and E. Pap. 2009. *Aggregation Functions*. Cambridge: Cambridge University Press.

KAC15 Kacprzyk, J., and W. Pedrycz (eds.). 2015. *Springer Handbook of Computational Intelligence*. Springer.

KLI95 Klir, G.J., and B. Yuan. 1995. Fuzzy Sets and Fuzzy Logic. Prentice-Hall.

KLE99 Klement, E.P., and M. Navara. 1999. Propositional Fuzzy Logics Based on Frank T-Norms: A Comparison. In: *Fuzzy Sets, Logics and Reasoning About Knowledge*, ed. D. Dubois, H. Prade, and E.P. Klement. Applied Logic Series, vol. 15. Dordrecht: Springer. https://doi.org/10.1007/978-94-017-1652-9_3.

KLE00 Klement, E.P., R. Mesiar, and E. Pap. 2000. *Triangular Norms*. Dordrecht: K;uwer Academic Publishers.

LAR03 Larsen, H.L. 2003. Efficient Andness-Directed Importance Weighted Averaging Operators. *International Journal of Uncertainty, Fuzziness and Knowledge-Based Systems* 12 (Suppl.): 67–82.

LAR09 Larsen, H.L. 2009. Multiplicative and Implicative Importance Weighted Averaging Aggregation Operators with Accurate Andness Direction. In Proceedings of the Joint 2009 International Fuzzy Systems Association World Congress and 2009 European Society of Fuzzy Logic and Technology Conference, Lisbon, Portugal, July 20–24, 2009, ed. J.P. Carvalho, D. Dubois, U. Kaymak, and J.M. da Costa Sousa, 402–407. ISBN: 978-989-95079-6-8.

LAR12 Larsen, H.L. 2012. Importance Weighting and Andness Control in De Morgan Dual Power Means and OWA Operators. *Fuzzy Sets and Systems* 196: 17–32.

MEN01 Mendel, J.M. 2001. *Uncertain Rule-Based Fuzzy Logic Systems: Introduction and New Directions*. Prentice-Hall.

MES15 Mesiar, R., A. Kolesarova, and M. Komornikova. 2015. Aggregation Functions on [0,1]. In *Springer Handbook of Computational Intelligence*, ed. J. Kacprzyk, and W. Pedrycz, 61–74. Springer.

MIL56 Miller, G.A. 1956. The Magical Number Seven, Plus or Minus Two: Some Limits on Our Capacity for Processing Information. *The Psychological Review* 63: 81–97.

MIT69 Mitrinović, D.S., and P.M. Vasić. 1969. Means (In Serbo-Croatian). *Mathematical Library*, vol. 40. Belgrade.

MIT77 Mitrinović, D.S., P.S. Bullen, and P.M. Vasić. 1977. *Means and Related Inequalities (In Serbo-Croatian)*. Belgrade: Publications of the University of Belgrade EE Dept., No. 600.

MIT89 Mitrinović, D.S., and J.E. Pečarić. 1989. *Mean Values in Mathematics*. Belgrade: Scientific Book Publishers (in Serbo-Croatian: Srednje vrednosti u matematici, Naučna knjiga, Beograd, 1989).

NGU06 Nguyen, H.T., and E.A. Walker. 2006. *A First Course in Fuzzy Logic*, 3rd ed. Chapman & Hall/CRC.

OHA87 O'Hagan, M. 1987. Fuzzy Decision Aids. In Proceedings of 21st Annual Asilomar Conference on Signals, Systems, and Computers, vol. 2, 624–628. IEEE and Maple Press (published in 1988).

RAD00 Radojević, D.G. 2000. [0,1]-Valued Logic: A Natural Generalization of Boolean Logic. *Yugoslav Journal of Operations Research* 10 (2): 185–216.

RAL97 Ralescu, A.L., and D.A. Ralescu. 1997. Extension of Fuzzy Aggregation. *Fuzzy Sets and Systems* 86: 321–330.

REI34 Reichenbach, H. 1934. Wahrscheinlichtkeitslogic. *Erkenntnis* 5 (1–3): 37–43.

ROS10 Ross, T.J. 2010. Fuzzy Logic With Engineering Applications, 3rd ed. Wiley.

SEA11 SEAS Co. 2011. AGOPcalc—Calculator for Analysis and Design of Aggregation Operators. *AGOPcalc User Manual*. SEAS.

SLA74 Slavić, D.V., and J. Dujmović. 1974. Numerical Computation of Weighted Power Means. *Journal of the University of Belgrade, EE Dept., Series Mathematics and Physics*, 485: 167–171.

TOR97 Torra, V. 1997. The Weighted OWA operator. *International Journal of Intelligent Systems* 12: 153–166.

TOR21 Torra, V. 2021. Andness Directedness for Operators of the OWA and WOWA Families. *Fuzzy Sets and Systems* 414: 28–37.

TOR22 Torra, V. 2022. Andness Directedness for t-Norms and t-Conorms. *Mathematics* 10: 1598. https://doi.org/10.3390/math10091598.

TOR24 Torra, V. 2024. Y-Values: Power Indices à La Orness for Nonadditive Measures. *IEEE Transactions on Fuzzy Systems* 32 (7): 4099–4108.

TOR07 Torra, V., and Y. Narukawa. 2007. *Modeling Decisions*. Springer.

TRO16 Troiano, L., and I. Díaz. 2016. An Analytical Solution to Dujmovic's Iterative OWA. *International Journal of Uncertainty, Fuzziness and Knowledge-Based Systems* 24 (Suppl. 2): 165–179.

WIK25 Wikipedia. Fuzzy Logic. https://en.wikipedia.org/wiki/Fuzzy_logic. Retrieved 18 Jan 2025.

WIK22 Wikipedia. 2022. *The t-Norm Fuzzy Logics*. https://en.wikipedia.org/wiki/T-norm_fuzzy_logics.

WIK24b Wikipedia. *T-Norm* (Retrieved 2 Sep 2024).

WIK24c Wikipedia. *Construction of t-Norms* (Retrieved 2 Sep 2024).

YAG88 Yager, R.R. 1988. On Ordered Weighted Averaging Aggregation Operators in Multi-Criteria Decision-making. *IEEE Transactions on SMC* 18: 183–190.

YAG98 Yager, R.R. 1998. Including Importances in OWA Aggregations Using Fuzzy Systems Modeling. *IEEE Transactions on Fuzzy Systems* 6 (2): 286–294.

YAG04b Yager, R.R. 2004. Generalized OWA Aggregation Operators. *Fuzzy Optimization and Decision Making* 3: 93–107.

YAG04 Yager, R.R. 2004. Weighted Triangular Norms Using Generating Functions. *International Journal of Intelligent Systems* 19: 217–231.

YAG94 Yager, R.R., and D.P. Filev. 1994. *Essentials of Fuzzy Modeling and Control*. Wiley.

YAG97 Yager, R.R., and J. Kacprzyk (eds.). 1997. *The Ordered Weighted Averaging Operators—Theory and Applications*. Kluwer Academic Publishers.

ZIM96 Zimmermann, H.-J. 1996. *Fuzzy Set Theory and Its Applications*. New York: Springer Science+Business Media.

ZIM87 Zimmermann, H.-J. 1987. *Fuzzy Sets, Decision-Making, and Expert Systems*. Kluwer Academic Publishers.

ZIM79 Zimmermann, H.-J., and P. Zysno. 1979. Decisions and Evaluations by hierarchical Aggregation of Information. *Fuzzy Sets and Systems* 10: 243–266.

ZIM80 Zimmermann, H.-J., and P. Zysno. 1980. Latent Connectives in Human Decision-Making. *Fuzzy Sets and Systems* 4: 37–51.

ZYS79 Zysno, P. 1979. *One Class of Operators for the Aggregation of Fuzzy Sets*. Amsterdam: EURO III Congress.

Chapter 16
Partial Absorption: Mandatory, Sufficient, and Optional Requirements

In previous chapters, we have discussed the hard and soft simultaneity and substitutability implemented as the GCD function. The hard simultaneity is a logic aggregator where *all input values must be to some extent satisfied*. So, their satisfaction is mandatory (otherwise the resulting aggregated value would be zero). The soft simultaneity deals with optional inputs: their simultaneous satisfaction is desirable, but not mandatory. A single positive input yields a positive output.

In a dual case, the hard substitutability GCD functions aggregate inputs that are significant to such an extent that *the full satisfaction of any of them is sufficient to satisfy the whole hard disjunctive criterion*. In the case of soft substitutability, the individual inputs are not capable of fully satisfying the disjunctive criterion, but any of them can offer a partial satisfaction. In all these cases all inputs are symmetrical, in the sense that all of them have the same logical impact: either all are mandatory, or all are sufficient, or all are optional.

In the real commonsense reasoning, it is easy to find logic problems where the simultaneity or the substitutability can be asymmetric. We frequently aggregate mandatory and optional inputs. For example, an unemployed worker who supports a family might simultaneously need a well-paid and attractive job. However, a necessary salary level is a mandatory requirement (insufficient salary unconditionally yields the rejection of a job offer). On the other hand, the attractiveness of job is desirable, but optional (an unattractive but well-paid job may be accepted).

Likewise, we can aggregate sufficient and optional inputs. As an example of asymmetric substitutability, a homebuyer in a dense urban area might need a safe parking space for a car, and a home garage is sufficient to solve that problem, while a good street parking is an optional but less desirable solution. The availability of a home garage can fully satisfy the homebuyer's needs regardless of the unavailability of street parking. In the absence of a home garage, the street parking can partially (but not fully) satisfy the homebuyer's needs.

The asymmetric conjunction, denoted $z = x \trianglerighteq y$, and called *Conjunctive Partial Absorption* (CPA), aggregates two inputs: a mandatory input x and an optional input y.

J. Dujmović, *Graded Logic*, Intelligent Systems Reference Library 273,
https://doi.org/10.1007/978-3-031-88558-7_16

Table 16.1 Basic properties of the partial absorption functions

Conjunctive Partial Absorption (CPA)		
Inputs		Output
Mandatory x	Optional y	$z = x \trianglerighteq y$
0	$0 \le y \le 1$	0
$0 < x \le 1$	0	$x - penalty$
$0 < x < 1$	1	$x + reward$

Disjunctive Partial Absorption (DPA)		
Inputs		Output
Sufficient x	Optional y	$z = x \,\overline{\trianglerighteq}\, y$
1	$0 \le y \le 1$	1
$0 < x < 1$	0	$x - penalty$
$0 \le x < 1$	1	$x + reward$

Similarly, the asymmetric disjunction, denoted $z = x \overline{\trianglerighteq} y$, and called *Disjunctive Partial Absorption* (DPA), aggregates two inputs: a sufficient input x and an optional input y. Both CPA and DPA are idempotent functions: $\forall a \in [0, 1] \Rightarrow a \trianglerighteq a = a,\ a \overline{\trianglerighteq} a = a$.

The partial absorption functions generalize the classical Boolean absorption theorems:

$$
\begin{aligned}
x \wedge (x \vee y) &= (x \wedge x) \vee (x \wedge y) \\
&= x \vee (x \wedge y) = x \wedge (1 \vee y) = x \wedge 1 = x,
\end{aligned}
$$

$$
\begin{aligned}
x \vee (x \wedge y) &= (x \vee x) \wedge (x \vee y) \\
&= x \wedge (x \vee y) = x,\ \ x \in \{0, 1\},\ \ y \in \{0, 1\}.
\end{aligned}
$$

In both Boolean cases, x is the primary input and y is the secondary input that is completely absorbed (it has no impact). As opposed to that, in the case of CPA and DPA, we have $x \in [0, 1], y \in [0, 1]$, and the secondary (optional) variable y is partially absorbed, i.e., its impact is reduced, adjustable, and characterized using the parameters called *penalty* and *reward*, shown in Table 16.1. This property is consistent with the concept of the fully continuum-valued logic.

The partial absorption function is very frequent in GL applications. It was introduced simultaneously with the first industrial applications in [DUJ74b] and [DUJ75a], studied in detail in [DUJ79], and refined and expanded in [DUJ14a, DUJ18]. Generally, the penalty and reward are adjustable and depend on the way the CPA and DPA functions are implemented. A standard way of designing the CPA and DPA functions is to use two GCD functions implemented as weighted power means and four adjustable parameters shown in Fig. 16.1 where $\overline{W}_1 = 1 - W_1$, $\overline{W}_2 = 1 - W_2$ (in computer programs, the weights are denoted W1 and W2). The adjustable parameters are q, r, W_1, W_2, and the analytic form is the following:

$$
\begin{aligned}
z(x, y; q, r, W_1, W_2) &= \left\{(1 - W_2)\left[W_1 x^q + (1 - W_1) y^q\right]^{r/q} + W_2 x^r\right\}^{1/r} \\
&= \begin{cases} x - P(x, y) & , y < x \\ x & , y = x \\ x + R(x, y) & , y > x \end{cases}
\end{aligned}
$$

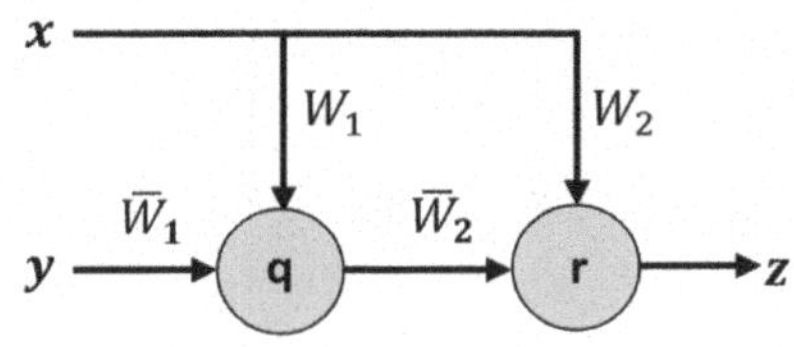

Fig. 16.1 A general form of the partial absorption function based on weighted power means

The mandatory (or sufficient) input is x and the optional input is y. The penalty function is $P(x, y)$, and the reward function is $R(x, y)$. Following are important special cases:

$$\begin{aligned}
z(x, 0; q, r, W_1, W_2) &= x\left(\overline{W}_2 W_1^{r/q} + W_2\right)^{1/r} \\
&= x - P_{\max}(x, 0; q, r, W_1, W_2) \\
z(x, 1; q, r, W_1, W_2) &= \left\{\overline{W}_2\left[W_1 x^q + \overline{W}_1\right]^{r/q} + W_2 x^r\right\}^{1/r} \\
&= x + R_{\max}(x, 0; q, r, W_1, W_2) \\
z(x, 0; 1, r, W_1, W_2) &= x\left(\overline{W}_2 W_1^r + W_2\right)^{1/r} \\
&= x - P_{\max}(x, 0; 1, r, W_1, W_2) \\
z(x, 1; 1, r, W_1, W_2) &= \left\{\overline{W}_2\left[W_1 x + \overline{W}_1\right]^r + W_2 x^r\right\}^{1/r} \\
&= x + R_{\max}(x, 1; 1, r, W_1, W_2)
\end{aligned}$$

The maximum penalty is proportional to the value of x and the maximum reward is a nonlinear function of x:

$$\begin{aligned}
P_{max}(x, 0; q, r, W_1, W_2) &= x\left[1 - \left(\overline{W}_2 W_1^{r/q} + W_2\right)^{1/r}\right] \\
R_{max}(x, 0; q, r, W_1, W_2) &= \left\{\overline{W}_2\left[W_1 x^q + \overline{W}_1\right]^{r/q} + W_2 x^r\right\}^{1/r} - x
\end{aligned}$$

The case $q = 1$ is simple (based on arithmetic mean) and frequent in applications.

The simplest way to demonstrate the properties of a partial absorption function is to analyze simple versions of the conjunctive partial absorption (CPA) based on the arithmetic/geometric means (AG), where $q = 1, r = 0$, or the arithmetic/harmonic means (AH), were $q = 1, r = -1$. They are shown in Fig. 16.2. In both AG and AH cases x is a mandatory input and y is an optional input. This is an asymmetric aggregator: if $x = 0$ then the output is also 0 because both geometric and harmonic means are hard aggregators. However, if $0 < x \leq 1$ then causes $0 < z < x$ (the decrement of output value, $P(x, y) = x - z(x, y)$ is called the *penalty*). Similarly, if $0 < x < 1$, then $y = 1$ causes $x < z < 1$ (the increment of output value $R(x, y) = z(x, y) - x$ is called the *reward*). Obviously, y is desirable but not mandatory. Thus, x is the main input and y is an auxiliary input.

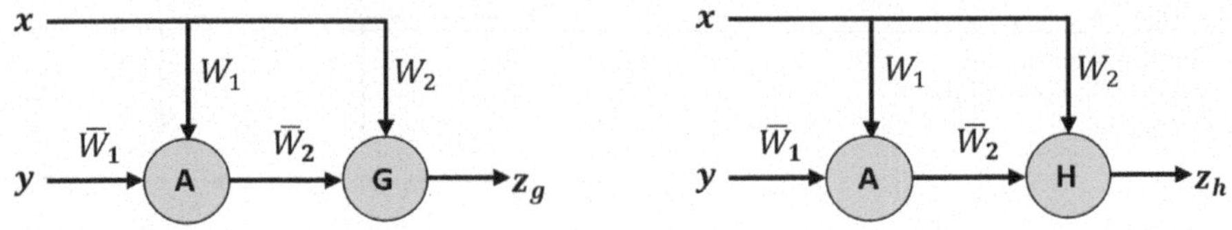

Fig. 16.2 AG and AH versions of the conjunctive partial absorption function

The values of weights, and the necessary andness of the conjunctive aggregator can be computed from desired values of penalty and reward. That can be done in the following ways: (1) using the mean penalty/reward tables introduced in [DUJ79], and also shown in [DUJ18], (2) using analytic expressions derived for simple special cases, primarily for the AH version [DUJ18], (3) using the neuron-training software tool AnSy (analysis and synthesis of preferential neurons [DUJ91, SEA10]), and using the mean penalty/reward calculator built in the software tool LSP.NT [SEA17]. We are now going to exemplify some of these techniques.

Simple Analytic Models

For simplicity, let us assume that the weights in Fig. 16.2 satisfy the conditions $0 < W_1 < 1$ and $W_2 = \overline{W}_2 = 0.5$. In such a case we have the following simple expressions, where P denotes the penalty, and R denotes the reward:

$$z_g(x, y; W_1) = \sqrt{(W_1 x + \overline{W}_1 y)x}, \quad \overline{W}_1 = 1 - W_1,$$

$$z_g(0, y; W_1) = 0$$

$$z_g(x, 0; W_1) = x\sqrt{W_1} < x; \quad P_g = x - z_g(x, 0; W_1) = \mathrm{x}\left(1 - \sqrt{W_1}\right)$$

$$z_g(x, 1; W_1) = \sqrt{(W_1 x + \overline{W}_1)x} \geq x; \quad R_g = z_g - (x, 1; W_1)\mathrm{x} = \sqrt{(W_1 x + \overline{W}_1)x} - \mathrm{x}$$

$$z_h(x, y; W_1) = \frac{2x\left(W_1 x + \overline{W}_1 y\right)}{x + W_1 x + \overline{W}_1 y}$$

$$z_h(0, y; W_1) = 0$$

$$z_h(x, 0; W_1) = \frac{2W_1 x}{1 + W_1} < x; \quad P_h = x - z_h(x, 0; W_1) = x\frac{1 - W_1}{1 + W_1}$$

$$z_h(x, 1; W_1) = \frac{2x\left(W_1 x + \overline{W}_1\right)}{x + W_1 x + \overline{W}_1} \geq x; \quad R_h = z_h(x, 1; W_1) - x = 2x\frac{W_1 x + \overline{W}_1 - x}{W_1 x + \overline{W}_1 + x}.$$

In a special case where x and y are binary values (0 or 1) the rewards are not possible ($R_g = R_h = 0$), and the penalties are obtained for $x = 1, y = 0$:

$$P_g = 1 - \sqrt{W_1}, \, W_1 = \left(1 - P_g\right)^2, \quad P_h = \frac{1 - W_1}{1 + W_1}, \quad W_1 = \frac{1 - P_h}{1 + P_h}.$$

For example, if we want a penalty of 50% (i.e. ½), then we can achieve that by using the AG aggregator with the weight $W_1 = 1/4$, or the AH aggregator with the weight $W_1 = 1/3$. In a general case, we can combine the arithmetic mean and a conjunctive GCD.

In the case of AG and AH implementations of CPA, if $W_2 = \overline{W}_2 = 0.5, y = 0$ then $z_g < x$ and $z_h < x$, and we can define the following relative penalty indicators (expressed as percentages):

$$\begin{aligned}
P_g^{rel}(x; W_1) &= 100\frac{x - z_g(x, 0; W_1)}{x} \\
&= 100\left(1 - \sqrt{W_1}\right) \\
P_h^{rel}(x; W_1) &= 100\frac{x - z_h(x, 0; W_1)}{x} \\
&= 100\left(1 - \frac{2W_1}{1 + W_1}\right) = 100\frac{1 - W_1}{1 + W_1}
\end{aligned}$$

Note that the relative penalty depends on weights, but it is not a function of x. So, if we want a desired penalty, we can realize it by selecting an appropriate value of weight.

If $y = 1$ then $z_g \geq x$ and $z_h \geq x$, and we can define the following relative reward indicators:

$$\begin{aligned}
R_g^{rel}(x; W_1) &= 100\frac{z_g(x, 1; W_1) - x}{x} \\
&= 100\frac{\sqrt{(W_1 x + \overline{W}_1)x} - x}{x} = 100\left(\sqrt{(W_1 + \overline{W}_1/x)} - 1\right) \\
R_h^{rel}(x; W_1) &= 100\frac{z_h(x, 1; W_1) - x}{x} \\
&= 100\left(\frac{2(W_1 x + \overline{W}_1)}{x + W_1 x + \overline{W}_1} - 1\right) = 100\frac{\overline{W}_1 \overline{x}}{x(W_1 + 1) + \overline{W}_1},
\end{aligned}$$

In the special case $x = W_1 = 1/2$ we have

$$\begin{aligned}
P_g^{rel}(0.5; 0.5) &= 29.3\%, \quad R_g^{rel}(0.5; 0.5) = 22.5\% \\
P_h^{rel}(0.5; 0.5) &= 33.3\%, \quad R_h^{rel}(0.5; 0.5) = 20\%.
\end{aligned}$$

If the penalty and reward depend on the values of weight W_1 and input x, then the resulting penalties and rewards are shown in Fig. 16.3. The penalty (for readability, it is plotted as a negative value) does not depend on x but the rewards are functions of x. The penalty can be increased and the reward can be decreased if we increase the andness of the output conjunctive aggregator of CPA.

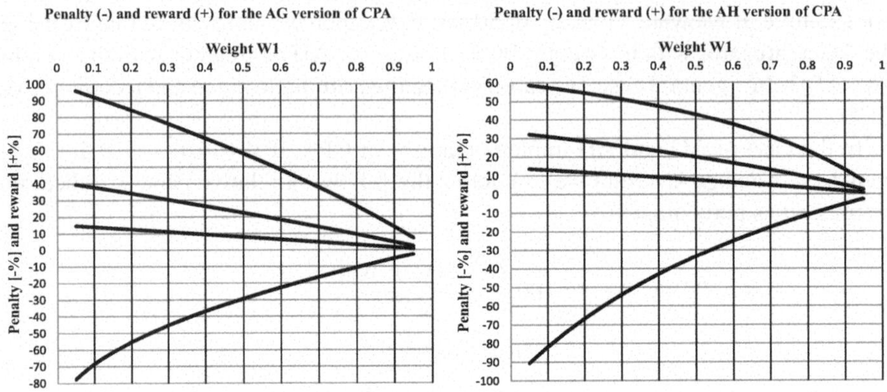

Fig. 16.3 Penalty (negative) and rewards (positive) for $x = 0.25, 0.5, 0.75$ (top-down)

The presented examples in Fig. 16.3 (for readability, the penalty is denoted as negative) show that the parameters of the CPA aggregator can be selected from the desired values of the relative penalty and reward. For example, if we want $P_g^{rel} = 40\%$ for $x = 0.5$, then we should use $W_1 = (1 - P_g^{rel}/100)^2 = (1 - 0.4)^2 = 0.36$.

The AH version of CPA ($q = 1, r = -1$) has a unique property that weights can be directly computed from the desired value of penalty. According to [DUJ14a, DUJ18], the computation of the *maximum relative penalty* and reward for CPA/AH can be done in the central point $x = {}^1\!/_2$ as follows:

$$
\begin{aligned}
p_{\max} &= \frac{P_{max}({}^1\!/_2, 0; 1, -1, W_1, W_2)}{{}^1\!/_2} \\
&= \frac{{}^1\!/_2 - z({}^1\!/_2, 0; 1, -1, W_1, W_2)}{{}^1\!/_2} \\
&= 1 - 2z({}^1\!/_2, 0) = \frac{\overline{W}_1 \overline{W}_2}{W_1 W_2 + \overline{W}_2}, \\
r_{\max} &= \frac{R_{max}({}^1\!/_2, 1; 1, -1, W_1, W_2)}{{}^1\!/_2} \\
&= \frac{z({}^1\!/_2, 1; 1, -1, W_1, W_2) - {}^1\!/_2}{{}^1\!/_2} \\
&= 2z({}^1\!/_2, 1) - 1 = \frac{\overline{W}_1 \overline{W}_2}{\overline{W}_1 W_2 + 1}.
\end{aligned}
$$

Now, we can compute the necessary weights as functions of desired values of p_{max} and r_{max}:

$$W_1 = \frac{2r_{\max}(1-p_{\max})}{r_{\max}+p_{\max}},$$
$$W_2 = \frac{p_{\max}-r_{\max}}{p_{\max}-r_{\max}+2r_{\max}p_{\max}}, \quad p_{\max} > r_{\max}$$

For example, suppose that we need $z(1/2, 0) = 0.3$ and $z(1/2, 1) = 0.6$. The maximum relative penalty in the central point $x = 1/2$ is $p_{max} = 1 - 0.6 = 0.4$(i.e. the penalty of 40%). The maximum relative reward in the central point is $r_{\max} = 1.2 - 1 = 0.2$ (i.e., the reward of 20%). Consequently $W_1 = \frac{2\times0.2\times0.6}{0.2+0.4} = 0.4$, $W_2 = \frac{0.4-0.2}{0.4-0.2+2\times0.2\times0.4} = 0.555556$.

Instead of harmonic mean in the AH aggregator which aggregates the mandatory input x and the output t of the arithmetic mean ($z = 2xt/(x+t)$) we can use its disjunctive De Morgan dual ($z = 1 - 2(1-x)(1-t)/(2-x-t)$) and the resulting aggregator is a DPA/AH. Using the similar calculation [DUJ18], the weights can be computed as functions of desired maximum relative penalty and reward as follows:

$$W_1 = \frac{2p_{\max}(1-r_{\max})}{r_{\max}+p_{\max}},$$
$$W_2 = \frac{r_{\max}-p_{\max}}{r_{\max}-p_{\max}+2r_{\max}p_{\max}}, \quad r_{\max} > p_{\max}.$$

AnSy and LSP.NT—Software Tools for Generating Partial Absorption Functions

In practice, it is not possible to intuitively select four parameters of various partial absorption aggregators. However, evaluators can easily select two independent aggregated parameters: a desired penalty and a desired reward that can be specified in a specific point (e.g. $x = 1/2$) or as a mean value inside $[0, 1]^2$. The role of penalty and reward is easily understandable, and the percepts of desired penalty and reward can be used to provide inputs for various partial absorption calculators. Therefore, we need methods and software tools to compute the parameters of the partial absorption aggregator from the desired values of the penalty and reward.

A standard way of the analysis and synthesis of GCD, CPA and DPA functions is to use software tools, and one of them is AnSy [DUJ91, SEA10]. The dialog with this tool is initialized as shown in Fig. 16.4, where the partial absorption is interpreted as a neuron which uses two weighted power means and four adjustable parameters: $q, r, W1, W2$. The mandatory (or sufficient) input is x and the optional input is y. If we provide a training set of desired x, y, z triplets, AnSy will adjust the parameters $q, r, W1, W2$ to obtain the function $z(x, y)$ closest to the desired mapping specified by the training set. Some parameters can be fixed, and remaining parameters are adjustable. A typical example of this process is shown in Fig. 16.4 where we decided to use the fixed parameters $q = 1, r = -1$ (i.e., the AH aggregator) and AnSy must find the optimum values of the remaining two parameters ($W1, W2$). Using the desired penalty and reward in the central point $x = 0.5$ AnSy verified the results $W1 = 0.4$ and W2 = 0.555557 obtained using the analytic model in the previous section.

```
THE SYNTHESIS OF PARTIAL ABSORPTION FUNCTION

  x ------>-------.-------->---------.
                  |                  |          PA Neuron
                  |W1                |W2
                  |                  |
        1-W1    .---.      1-W2    .---.
  y ---------->| q |------------->| r |------>  z
                `---'              `---'

 This function has the following parameters:

 Identifier:     1    2    3    4
--------------------------------
 Parameter :     W1   W2   q    r

 How many parameters do you want to determine? 2
 Enter the identifier(s) of selected parameter(s): 1 2

 Enter fixed values of remaining parameters:
 q  =  1
 r  = -1

 Enter a training set containing any number of desired x,y,z values
 (a value outside the [0,1] interval denotes the end of data)

  x   y   z
 -----------
  0   1   0
 0.5  1  0.6
 0.5  0  0.3
  9   9   9

 T r a i n i n g ...      Please wait !

THE RESULTING OPTIMUM VALUES OF PARAMETERS

 W1 =    0.400000
 W2 =    0.555557

 The minimum value of criterion function = 0.164E-06
 The number of function computations      =  73
 The achieved maximum error of result     = 0.000000

 =======================================================
  T r a i n i n g    s e t   |     Trained PA neuron
 ----------------------------+--------------------------
     x        y        z     |      z          Error
 ----------------------------+--------------------------
   0.000   1.000   0.0000    |    0.0000    0.0000000
   0.500   1.000   0.6000    |    0.6000   -0.0000003
   0.500   0.000   0.3000    |    0.3000    0.0000001
 =======================================================
```

Fig. 16.4 Using AnSy for training the partial absorption neuron and computing weights

Another typical example of this process is shown in Fig. 16.5 where we decided to use the fixed parameter $q = 1$ (i.e. the first aggregator is the arithmetic mean) and AnSy has to find the optimum values of the remaining three parameters ($W1$, $W2$, and r). The first row of the training set defines the mapping $z(0, 1) = 0$, which specifies that the desired function is a CPA. The second row, $z(0.5, 1) = 0.6$ shows the desired maximum reward of 20% for $x = 0.5$. The third row, $z(1, 0) = 0.5$ shows the desired maximum penalty of 50% for $x = 1$. Based on these specific requirements, AnSy provides the resulting weights $W1 = 0.222404$, $W2 = 0.645766$, and the exponent $r = -0.588087$. These parameters exactly provide the desired mapping specified in points $(x, y) = (0, 1)$, $(x, y) = (0.5, 1)$ and $(x, y) = (1, 0)$.

Another tool implementing general GCD-based CPA/DPA aggregators is LSP.NT [SEA17]. The penalty and reward functions are the following:

$$P(x, y) = x - z(x, y), \quad y < x$$
$$R(x, y) = z(x, y) - x, \quad y > x$$
$$P(x, y) = R(x, y) = 0, \quad x = y$$

These functions can be averaged and desired average values $\overline{P}$ and $\overline{R}$ can be used as inputs for the LSP.NT tool for computing the parameters of the CPA or DPA functions as exemplified in Tables 16.2 and 16.3. These examples show that LSP.NT CPA and DPA calculators provide all parameters of the general CPA and DPA functions based on GCD (not only AH versions). The resulting mean penalty and reward are regularly very close to the desired mean penalty and reward (differences are typically less than 1%). The cases $P > R$ and $P < R$ are supported for both CPA and DPA.

The properties of CPA and DPA functions can be visualized using the sensitivity curves that show the aggregated overall suitability as a function of one selected input. Examples of such curves for a CPA function and parameters $P/R = 30\%/20\%$ are $CPA(x, 75\%)$ and $CPA(75\%, y)$, shown in Fig. 16.6. CPA is an idempotent function and consequently $CPA(75\%, 75\%) = 75\%$. The properties of a DPA function are exemplified in Fig. 16.7 using $DPA(x, 50\%)$ and $DPA(50\%, y)$ for penalty and reward $P/R = 20\%/30\%$. DPA is also an idempotent function and consequently $DPA(50\%, 50\%) = 50\%$.

$CPA(mandatory, optional)$ and $DPA(sufficient, optional)$ are logically asymmetric functions because their inputs are logically different. CPA is a conjunctive function and consequently more sensitive to decrements of input values than to increments of input values. This is the reason why it is natural to use penalty which is greater than reward ($P > R$). Similarly, DPA is a disjunctive function and consequently more sensitive to increments of input values than to decrements of input values. Therefore, in the case of DPA we expect that the desired reward is greater than the desired penalty ($R > P$).

Specialized evaluation tools based on the partial absorption function include the suitability map system LSPmap [PIT10], a simplified LSP mandatory/optional evaluator LSPeval [KAP17, DUJ22b], and a hotel recommender tool proposed in [SOL24].

```
THE SYNTHESIS OF PARTIAL ABSORPTION FUNCTION

  x ------>-------.-------->---------.
                  |                  |            PA Neuron
                  |W1                |W2
                  |                  |
        1-W1     .---.      1-W2    .---.
  y ----------->| q |------------->| r |------>  z
                 `---'              `---'

 This function has the following parameters:

 Identifier:    1    2    3    4
--------------------------------
 Parameter :    W1   W2   q    r

 How many parameters do you want to determine? 3
 Enter the identifier(s) of selected parameter(s): 1 2 4

 Enter fixed values of remaining parameters:
 q  = 1

 Enter a training set containing any number of desired x,y,z values
 (a value outside the [0,1] interval denotes the end of data)

 x    y    z
 -----------
 0    1    0
 0.5  1    0.6
 1    0    0.5
 9    9    9

 T r a i n i n g ...      Please wait !

THE RESULTING OPTIMUM VALUES OF PARAMETERS

 W1 =    0.222404
 W2 =    0.645766
 r  =   -0.588087

 The minimum value of criterion function = 0.139E-07
 The number of function computations     = 165
 The achieved maximum error of result    = 0.000000

 ======================================================
  T r a i n i n g    s e t    |     Trained PA neuron
 -----------------------------+------------------------
     x       y         z      |     z          Error
 -----------------------------+------------------------
   0.000   1.000    0.0000    |   0.0000    0.0000000
   0.500   1.000    0.6000    |   0.6000    0.0000000
   1.000   0.000    0.5000    |   0.5000    0.0000000
 ======================================================
```

Fig. 16.5 Computation of parameters of a CPA function (penalty 50% and reward 20%)

Table 16.2 Computation of CPA parameters for desired mean values of penalty (P) and reward (R)

CPA: Desired values	Resulting values: mean penalty, reward, and parameters					
$P[\%]$ / R [%]	Penalty [%]	Reward [%]	$W1$	$W2$	q	r
40 / 30	39.9384	29.8611	55.5	16	1	-0.72
40 / 20	39.8813	20.0031	45.5	42	1	-0.72
40 / 10	40.1173	10.0887	27	71.5	1	-0.72
30 / 20	29.9174	20.0311	62	29	1	-0.72
30 / 10	30.1127	9.97026	40	68.5	1	-0.72
20 / 10	19.9652	10.0814	61.5	58.5	1	-0.72
10/5	10.0188	4.97991	7.5	87	3.929	-0.72

Table 16.3 Computation of DPA parameters for desired mean values of penalty (P) and reward (R)

DPA: Desired values	Resulting values: mean penalty, reward, and parameters					
$P[\%]$ / R [%]	Penalty [%]	Reward [%]	W1	W2	q	r
30 / 40	30.0912	39.9444	25.5	24.5	-0.72	3.929
20 / 40	20.0831	40.0884	49.5	37.5	1	3.929
10 / 40	9.86261	40.0308	10.5	66.5	-0.72	3.929
20 / 30	20.1564	29.9854	64	29	1	3.929
10 / 30	9.90974	29.9514	50	64	1	3.929
10 / 20	10.0356	20.0303	33	66	-0.72	3.929
5 / 10	5.04959	10.0013	79	69.5	1	3.929

All these tools are based on the aggregation structure shown in Fig. 16.8. The aggregation includes $k \geq 1$ mandatory inputs aggregated using a weighted hard conjunctive GCD and $m \geq 1$ optional inputs aggregated using the weighted arithmetic mean. These values are then aggregated using the CPA function with default penalty/reward combinations denoted as *relaxed* (low penalty), *standard* (medium penalty) and *strict* (high penalty), as follows:

- Relaxed partial absorption model: penalty = 15%, reward = 30%
- Standard partial absorption model: penalty = 30%, reward = 25%
- Strict partial absorption model: penalty = 45%, reward = 20%.

The users of canonical decision model shown in Fig. 16.8 need a minimum effort to specify k mandatory inputs and their importance weights, and m optional inputs and their weights. The resulting suitability is computed for relaxed, standard, and strict CPA aggregation and can be used for comparison of any number of competitive objects/alternatives. Of course, this is an approximate evaluation model. For more precise evaluation models we use the general LSP method presented in the Chap. 17.

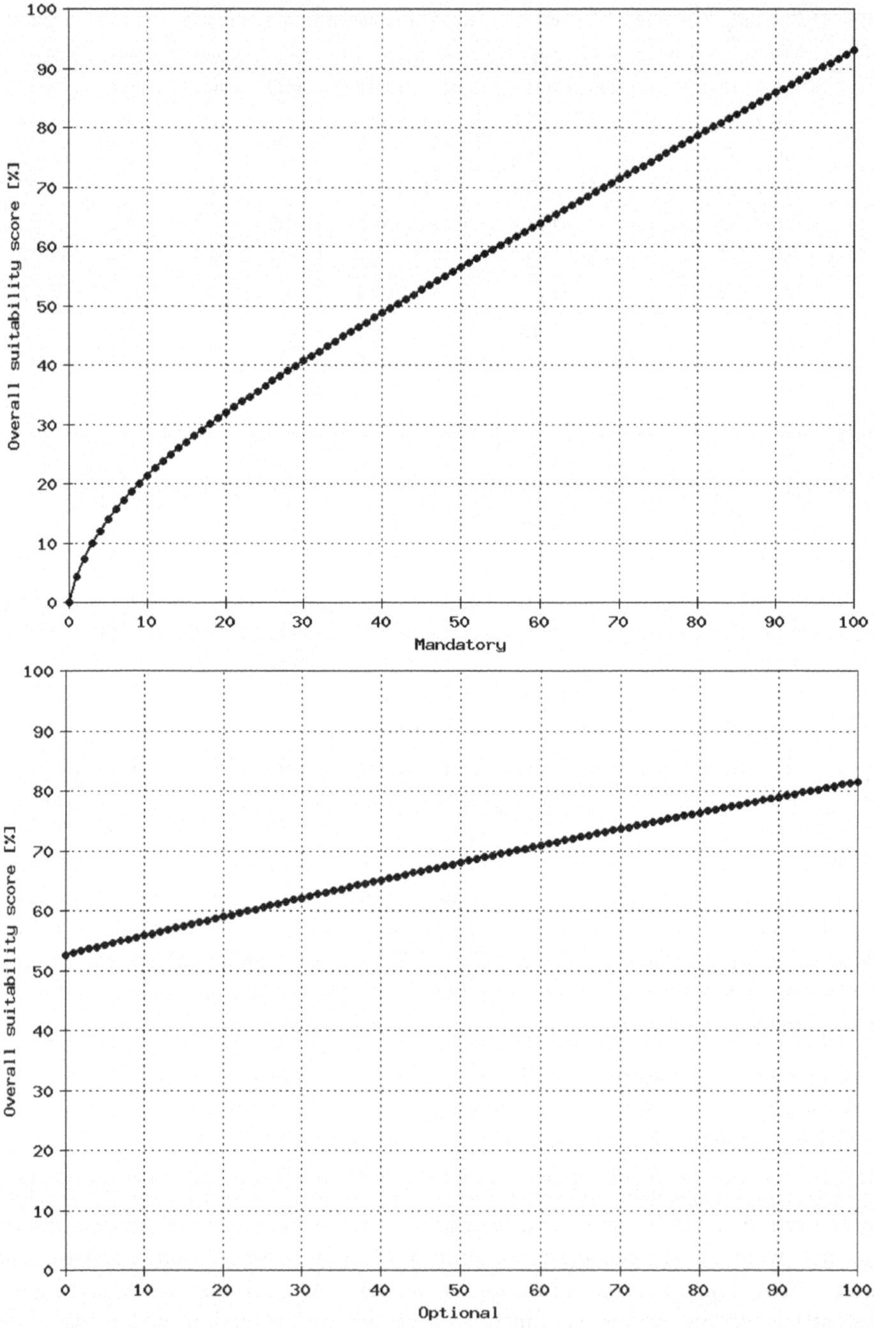

Fig. 16.6 Typical sensitivity curves for the CPA function with penalty of 30% and reward of 20% (input values are set to 75%)

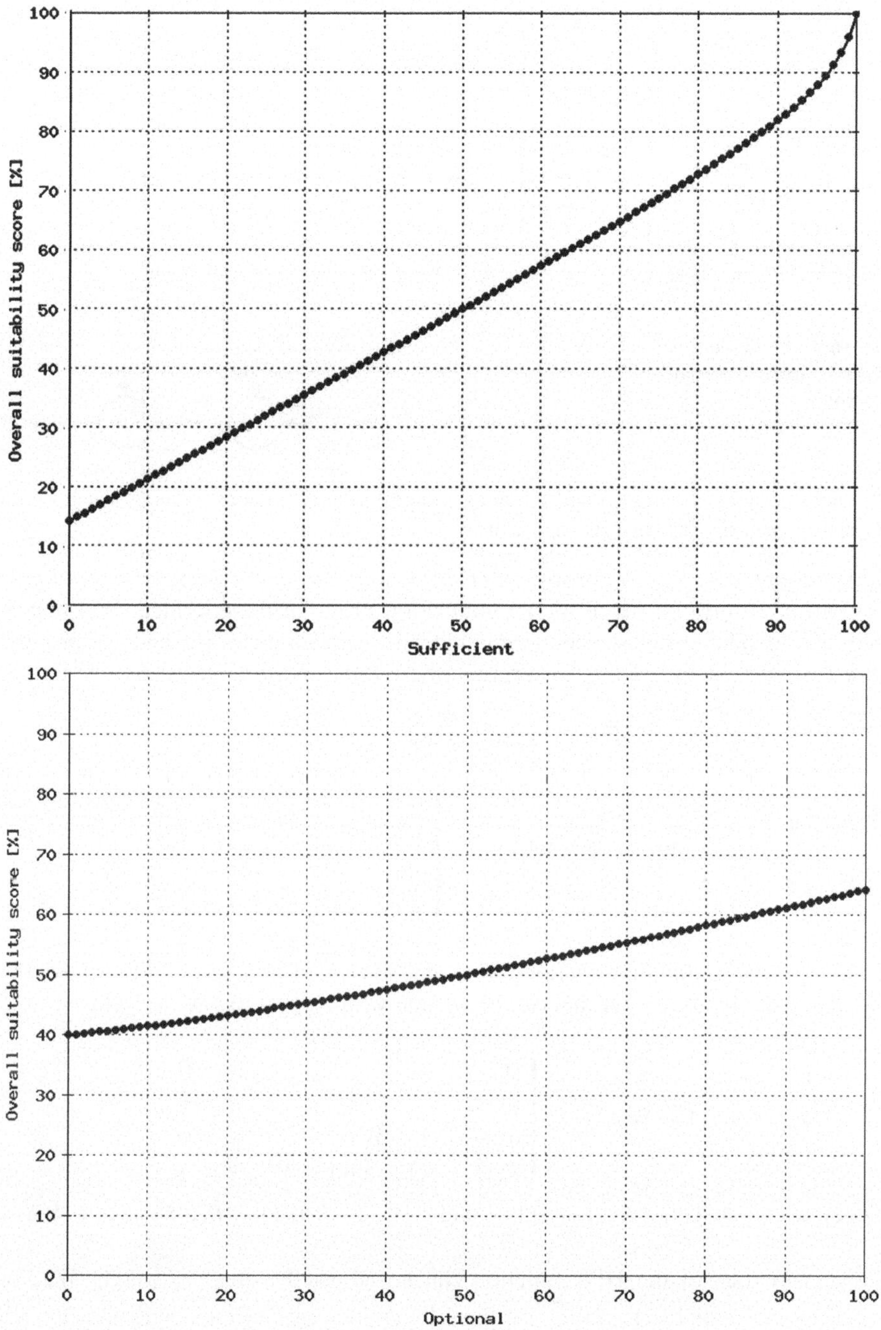

Fig. 16.7 Typical sensitivity curves for DPA with penalty of 20% and reward of 30% (input values are set to 50%)

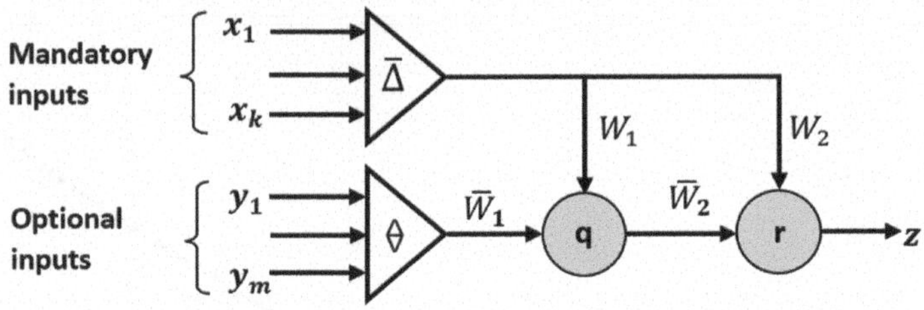

Fig. 16.8 Simplified aggregation structure for k mandatory and m optional inputs

Fig. 16.9 A sample hyperconjunctive partial absorption

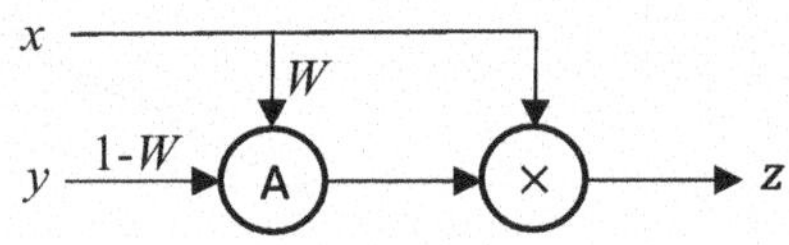

Hyperconjunctive Partial Absorption

Hyperconjunctive partial absorption ($HPA(x, y), x = mandatory, y = optional$) is a partial absorption based on a combination of the arithmetic mean and hyperconjunction. The simplest version combines the arithmetic mean and the product t-norm shown in Fig. 16.9. Unsurprisingly, HPA is not an idempotent function: $HPA(x, x) = x^2$.

The analytic form of HPA is the following:

$$z(x, y) = (Wx + (1 - W)y)x = \begin{cases} 0, & x = 0 \\ x^2, & x = y \\ Wx^2, & y = 0 \\ x^2 + x(1 - x)(1 - W), & y = 1 \end{cases}$$

If $x = y$, then the diagonal section function is $\delta(x) = z(x, x) = x^2$. So, we have the penalty P and the reward R that can be written with respect to $\delta(x)$ as follows:

$$z(x, y) = (Wx + (1 - W)y)x = \begin{cases} 0, & x = 0 \\ \delta(x), & x = y \\ \delta(x) - (1 - W)x^2, & y = 0 \\ \delta(x) + x(1 - x)(1 - W), & y = 1 \end{cases}$$

$$P(x) = (1 - W)x^2, \quad R(x) = x(1 - x)(1 - W), \quad P(0.5) = R(0.5) = (1 - W)/4.$$

Basic properties of the HPA function can be adjusted using the weight W. For example, in the point $x = 0.5$ we have the same values of penalty and reward. The value of penalty can be adjusted using the weight $W = 1 - 4P$, and if we want $P = R = 0.2$ (i.e., 20%), then we must use $W = 0.2$. While the concept and analytic forms of the HPA function are theoretically consistent with the concepts and analytic forms of CPA and DPA, it is important to note that experimental verification of the role of HPA function in human commonsense reasoning is currently a part of the future research.

References

DUJ14a Dujmović, J. 2014. An analysis of penalty and reward for partial absorption aggregators. In *Proceeding of the 2014 World Conference on Soft Computing*, 126–133, Berkeley, CA, May 25–27, 2014.

DUJ18 Dujmović, J. 2018. *Soft computing evaluation logic*. J. Wiley

DUJ22b Dujmović, J., and K. Kapre. 2022. LSPeval—An educational tool for studying soft computing logic aggregation structures. In *Proceedings of the World Conference on Soft Computing 2021*, Feb 2–6, 2022.

DUJ74b Dujmović, J. 1974. New results in the development of the "Mixed Averaging by Levels" method for system evaluation, (in Serbo-Croatian). In *Proceedings of the Informatica Conference 1974*, paper 4.36, Bled, Yugoslavia.

DUJ75a Dujmović, J. 1975. Extended continuous logic and the theory of complex criteria. *Journal of the University of Belgrade,* 537: 197–216 (EE Department, Series Mathematics and Physics).

DUJ79 Dujmović, J. 1979. Partial absorption function. *Journal of the University of Belgrade* 659: 156–163 (EE Department, Series Mathematics and Physics).

DUJ91 Dujmović, J. 1991. Preferential Neural Networks. *Chapter 7 in Neural networks—Concepts, applications, and implementations*, vol. II. ed. P. Antognetti, and V. Milutinović, 155–206. Prentice-Hall Advanced Reference Series, Prentice-Hall

KAP17 Kapre, K. 2017. *LSP evaluator—A suitability evaluation tool based on the LSP method.* Computer Science Department, San Francisco State University, Report SFSU-CS-CE-17.18

PIT10 Pittas, M.A. 2010. *Location suitability maps based on the LSP method.* Department of Computer Science, San Francisco State University, Report SFSU-CS-CE-10.08.

SEA10 SEAS Co. 2010. *Analysis and synthesis of aggregation operators.* AnSy User Manual V.3.2. SEAS

SEA17 SEAS Co. 2017. *LSP.NT—LSP method for evaluation over the Internet.* LSP.NT User Manual. SEAS

SOL24 Solano-Barliza, A., A. Valls, A. Moreno, M. Acosta-Coll, J. Escorcia-Gutierrez, J. Dujmović,. and E. De-La-Hoz-Franco, Personalized Hotel Recommender System Based on Graded Logic with Asymmetric Criteria, KES2024 conference,, Procedia Computer Science, Elsevier, 2024.

Chapter 17
The LSP Decision Method

The Logic Scoring of Preference (LSP) method is a graded logic decision method for evaluation, optimization, and comparison of complex objects/alternatives from the standpoint of requirements of a specific stakeholder/decision-maker. The LSP method is human-centric, in the sense that it strictly implements the basic model of human decision making discussed in Chap. 6 and summarized in Fig. 6.1. In its initial form, the LSP method was introduced in 1973 [DUJ73c]. The most detailed presentation of the current status and use of this method can be found in [DUJ18]. In this chapter we present the LSP method in a condensed form, emphasizing its human-centric roots and relationships with Graded Logic.

A Summary of the LSP Method

Five basic steps of the LSP method are summarized in Fig. 17.1. In the first step we identify the stakeholder/decision-maker, stakeholder's goals, and $m \geq 1$ alternatives for achieving these goals. These alternatives are objects of evaluation, and the decision problem is to find the most suitable alternative that is at the end of evaluation process proposed for realization. In all cases, suitability is defined as the degree of truth of assertion that an evaluated object (or one of its components) completely satisfies stakeholder's requirements. The overall suitability $X \in [0, 1]$ (or $X[\%] \in [0, 100]$) is the degree of truth of the assertion that the evaluated object/alternative completely satisfies all stakeholder's requirements.

The number of alternatives can be 1 and that is the main determinant that decides the properties of the LSP method. If $m = 1$, then the LSP method must be able to compute the overall suitability of a single object/alternative (e.g., a company that has a job opening can receive the application of only one candidate). Since the number of alternatives can be one or more, the evaluation problem is to determine the absolute extent to which each alternative satisfies stakeholder's requirements, and not to determine only the relative ranking of alternatives (which assumes two or more alternatives), because all compared alternatives can be below the stakeholder's

J. Dujmović, *Graded Logic*, Intelligent Systems Reference Library 273,
https://doi.org/10.1007/978-3-031-88558-7_17

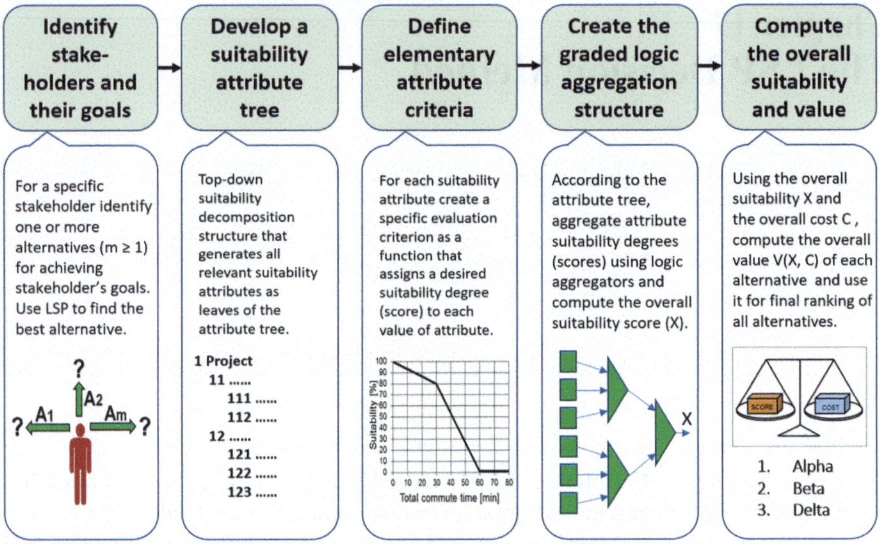

Fig. 17.1 Five basic steps of the LSP method

acceptability threshold and in such a case they should be justifiably rejected. This is the way we define the evaluation decision problem.

In the second step of the LSP method, our goal is to develop an array of suitability attributes $a_1, \ldots, a_n$ $(a_i \in \mathbb{R},\ \ i = 1, \ldots, n)$ containing all attributes that affect the suitability of evaluated object for the given use. The suitability attributes are the leaves of the suitability attribute tree. Initially, the root of the tree is the overall suitability which is then decomposed into two or more components. Next, each component is decomposed into simpler components until we reach simple components that can be individually evaluated. These are the suitability attributes.

In the third step of the LSP method, for each suitability attribute, we create an attribute criterion $g_i : \mathbb{R} \to [0, 1],\quad i = 1, \ldots, n$ that determines the degree of suitability $x_i = g_i(a_i) \in [0, 1]$ that corresponds to the value a_i of the ith attribute. The value $x_i = 1$ (or 100%) denotes the highest suitability and the value $x_i = 0$ denotes the complete unsuitability. The degree of suitability has logic interpretation as the degree of truth of a value statement that asserts the complete satisfaction of stakeholder's requirements for the analyzed attribute.

In the fourth step of the LSP method, we create a compound graded logic function $L : [0, 1]^n \to [0, 1]$ and use it to aggregate all suitability degrees $x_1, \ldots, x_n$ and compute the overall suitability $X = L(x_1, \ldots, x_n), X \in [0, 1]$. The function L. is a compound graded propositional calculus formula based on superposition of the GCD aggregator and the standard negation $not(x) = \overline{x} = 1 - x$.

In the fifth step of the LSP method, there are two possibilities: either the evaluated object/alternative does not have cost,[1] or there is a compound cost C, obtained as the result of cost analysis that considers all cost components.[2] If there is no cost, then competitive objects can be compared using only the overall suitability scores. If an evaluated object has a total cost C, then its overall value (based on the desired simultaneity of high suitability and low cost) is a function $V(X, C)$, $X \geq X_{\min}$, $C \leq C_{\max}$. Here $X_{\min}$ denotes the acceptability threshold, i.e., the minimum acceptable satisfaction of stakeholder's requirements (if $X < X_{\min}$ the evaluated object is rejected). Similarly, $C_{\max}$ denotes the maximum approved compound cost of the evaluated object. Therefore, if $C > C_{\max}$, then the evaluated object is rejected. Therefore, all evaluated objects must satisfy the acceptability condition $(X_i, C_i) \in [X_{\min}, 1] \times [0, C_{\max}]$, $i = 1, \ldots, m$ (objects that do not satisfy this condition must be rejected). In the context of Graded Logic, the value function $V(X_i, C_i)$ can be defined as a logic condition that requires simultaneously high suitability and high affordability, as follows:

$$V(X_i, C_i) = \left(\frac{X_i}{\max(X_1, \ldots, X_m)}\right)\overline{\Delta}\left(\frac{\min(C_1, \ldots, C_m)}{C_i}\right) \in [0, 1], \quad i = 1, \ldots m.$$

The hard GCD aggregator $\overline{\Delta}$ is usually weighted (e.g. a weighted geometric mean) to express different importance of suitability and affordability. In cases where $m > 1$ we can perform a relative comparison of value indicators by dividing $V(X_i, C_i)$ of each object by the same value of the best competitor:

$$V_{rel}(X_i, C_i) = \frac{V(X_i, C_i)}{\max(V(X_1, C_1), \ldots, V(X_m, C_m))} \in [0, 1], \quad i = 1, \ldots, m.$$

So, the relative value $V_{rel} = 1$ (or 100%) denotes the highest value achieved in a group of m competitors. The final ranking of competitors is based on decreasing relative values. In the case $m = 1$ the acceptability criterion is simply $X \geq X_{\min}$, $C \leq C_{\max}$ and the value function is $V{=}X\overline{\Delta}g(C)$, where $g(C)$ denotes an attribute criterion for the suitability of cost. E.g., if C_{min} is a fully acceptable cost, then $g(C){=}\max(0, \min(1,(C_{max}{-}C)/C_{max}{-}C_{min})))$.

In some cases, after determining the most suitable alternative, it is necessary to validate the proposed decision by identifying and explaining all reasons for the obtained result. Explainability (Chap. 18) justifies the results of evaluation and helps the stakeholder to confidently accept and defend the selected most suitable alternative. Explainability also justifies the rejection of inferior alternatives.

Based on this summary, we can now present details of the LSP method using a simple example where the stakeholder/decision-maker is an individual who is looking

[1] For example, a student can evaluate the web sites of various universities for information contents, readability, and completeness. Web sites have different suitability but not cost.

[2] For example, a car buyer can compute the total cost by combining the cost of car, financing, registration, insurance, fuel, maintenan, and an estimated resale value.

- 1 JOB
  - **11 Monetary compensations**
    - 111 Offered salary and benefits
      - 1111 Salary [%]
      - 1112 Benefits [%]
    - 112 Anticipated future salary [%]
  - **12 Main characteristics of the job**
    - 121 Attractiveness of job [score 0-6]
    - 122 Total time
      - 1221 Average work week [h]
      - 1222 Daily commute time [min]

<table>
<tr><th colspan="4">Suitability aggregation tree for job offer evaluation</th></tr>
<tr><td>1111 Salary [%]</td><td rowspan="2">111 Offered salary and benefits</td><td rowspan="3">11 Monetary compensations</td><td rowspan="6">1 JOB</td></tr>
<tr><td>1112 Benefits [%]</td></tr>
<tr><td colspan="2">112 Anticipated future salary [%]</td></tr>
<tr><td colspan="2">121 Attractiveness of job [score 0-6]</td><td rowspan="3">12 Main characteristics of the job</td></tr>
<tr><td>1221 Average work week [h]</td><td rowspan="2">122 Total time</td></tr>
<tr><td>1222 Daily commute time [min]</td></tr>
</table>

Fig. 17.2 Two notations of the suitability attribute tree (decomposition and aggregation)

for a new job and compares job offers. Of course, this is the case where the number of obtained job offers m can be one or more.

Selection of Suitability Attributes

To streamline the presentation and to provide semantic identity of all variables, let us use a simplified job selection problem based on only six attributes (a very detailed study of such a problem can be found in [MIL70]; a similar example can also be found in [DUJ18]). So, our stakeholder is an employee who wants to replace his/her current job with a job that is better paid and more attractive. The stakeholder has job offers that should be evaluated and compared, and for each offer, the final result should be the overall suitability that shows the degree of satisfaction of the stakeholder's requirements.

The initial step consists of decomposition of the overall suitability into simpler suitability categories, as exemplified in Fig. 17.2. The decomposition starts from the root (denoted as node 1 Job) where we identified two components denoted in the tree structure as nodes 11 (*monetary compensations*) and 12 (*job characteristics*). Both the monetary compensations and the job characteristics are compound attributes that have their own components. Thus, each compound attribute can be further decomposed into simpler suitability attributes. The decomposition process terminates when we reach components that cannot be further decomposed. These components are leaves of the attribute tree, and we use them as suitability attributes that can be individually evaluated. In Fig. 17.2 we have six suitability attributes $(a_1, a_2, a_3, a_4, a_5, a_6)$ denoted in the tree structure as leaf nodes 1111, 1112, 112, 121, 1221, and 1222.

The attribute decomposition tree can also be written in the opposite direction as the suitability aggregation tree, going from leaves towards the root. In Fig. 17.2, the aggregation tree is shown as a table. The aggregation structure will be used later, when we compute the suitability of attributes, and then insert aggregators, node by node, going towards the root.

Each monetary compensation attribute is expressed as a percentage of the current job value (e.g. the salary of 120% means that the offered salary is 20% above the current salary). The attractiveness of an offered job is expressed using a scale of scores from the lowest (0) to the highest (6), and the remaining attributes are basic components of the time spent at work.

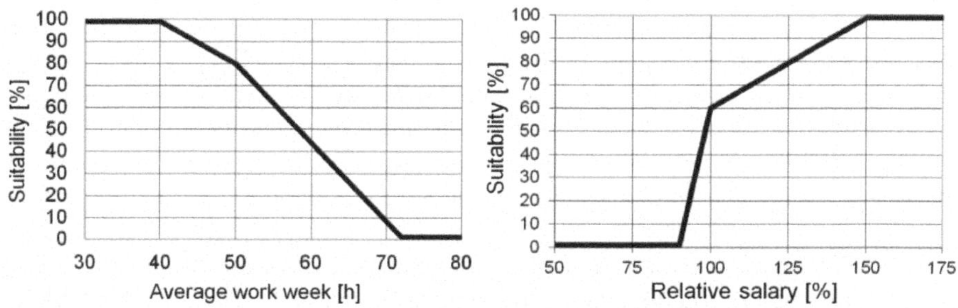

Fig. 17.3 Piecewise linear forms of attribute criteria for the work week and salary

Development of Attribute Criteria

In the next step of the LSP method, for each suitability attribute a_i, $i = 1, ..., n$ the stakeholder/decision-maker defines the attribute evaluation criterion $x_i = g_i(a_i)$ as a function that maps the value of an attribute ($a_i \in \mathbb{R}$) to the corresponding suitability degree ($x_i \in [0, 1]$). The attribute suitability degree x_i specifies the degree of satisfaction of stakeholder's requirements. This function (usually called the "elementary attribute criterion") can be expressed in four ways: (a) graphically, (b) using vertex notation, (c) using documentation tables, and (d) analytically.

The graphic notation is exemplified in Fig. 17.3 using a piecewise linear interpolation between several vertex points. Expectedly, the suitability of the average work week is a decreasing function of the number of hours spent weekly at work. The stakeholder is completely satisfied (100%) with the standard work week of 40 h (or less), and would be 80% satisfied with 50 h, but 72 h or more are completely unacceptable. This function can be defined in a very compact form as a sequence of coordinates (<hours> , <suitability>) of given vertex points. Such a sequence, ordered according to the increasing values of hours, is called the *vertex notation*. So, according to Fig. 17.3, the criterion for an average work week time T, in the vertex notation, is the following:

$$Crit(T[h]) = \{(40, 100), (50, 80), (72, 0)\}; (ordering : 40 < 50 < 72).$$

The vertex notation assumes that if the value of attribute is less than the value in the first point in the sequence ($T < 40$), then the suitability (always expressed as %) is the same as for the first point (i.e., 100%). Similarly, for the last point, if the value of the attribute is greater than the value in the last point ($T > 72$), then the suitability is the same as the value in the last point (i.e., 0%). In this way the compact vertex notation can be used for any value of the attribute T.

In the case of offered salary, the new salary S_{new} should be compared with the current salary S_{old} and the relative salary is defined as $S = 100S_{new}/S_{old}$ [%]. According to Fig. 17.3, the stakeholder obviously expects the salary increase up to 50% and the corresponding salary criterion is $Crit(S[\%]) = \{(90, 0), (100, 60), (150, 100)\}$. In other words, the unchanged salary satisfies only

60% of stakeholder's requirements, but it is not unacceptable if the new job offers other spectacular improvements (a lower work week, an excellent location, a high attractiveness, etc.). However, the reduction of salary to 90% of the current salary (or less than 90%) is considered unacceptable. The linear interpolation used in Fig. 17.3 gives the suitability for any value of the relative salary. For example, if the increase of salary is 25% (the relative salary is 125%), then the stakeholder's satisfaction would be 80%.

All LSP evaluation projects must have documentation. In the case of job selection, the attribute criteria documentation table is shown in Table 17.1. This table is produced using the LSP.NT tool [SEA17]. For each attribute, LSP.NT provides a small table that contains the attribute numerical identifier, the attribute name, a small vertex values table (the increasing attribute values and the corresponding suitability values, and a description field that contains the precise definition of the attribute evaluation process, units of measurement, and other necessary information.

Attribute criteria illustrated in Table 17.1 show the basic goals of stakeholder. Primarily, the monetary compensation is expected to significantly increase because if they stay at the current level, the stakeholder's satisfaction will be around 60%, and that is not sufficient for an enthusiastic acceptance of such a job offer. The stakeholder also wants to get a job where s/he can grow: the desired and expected promotions in the period of the next 3–5 years should produce an expected increase in income of 25%.

The attractiveness of the new job (A) can be evaluated in two ways. If the offered job is precisely defined, then it is possible to identify its components and make a small point-additive criterion for its evaluation: for example, the stakeholder can assign 20 points to training s/he will obtain at the new job, 15 points for a management job title and responsibility, and 10 points to the use of each of three software tools that the candidate is familiar from the current employment. The total sum of such points is $20 + 15 + 10 + 10 + 10 = 65$ and the corresponding criterion can be $Crit(A) = \{(0, 0), (65, 100)\}$. If the job is not precisely defined, then the evaluation can be based on the 7 grades rating scale $Crit(A) = \{(0, 0), (6, 100)\}$, as shown in Table 17.1.

Employment-related time away from home is divided into work time and commuting time. The possibility to work online from home is evaluated as the commute time 0, and in the criterion 1222 that is the most desirable option. Long commuting times are not acceptable, as well as the exaggerated work week. Obviously, our stakeholder is not looking for employment in a risky startup.

Attribute criteria can have a variety of forms and a detailed presentation of 12 characteristic forms can be found in [DUJ18]. The most frequent forms are the "absolute criteria" that can be applied to a single evaluated object as we did in Table 17.1. Such attribute criterion functions map the attribute value into the suitability degree and can be easily specified analytically, either by using some specific functions, or by interpolating between $k_i \geq 2$ vertex points $(a_{i,1}, x_{i,1}), \ldots, (a_{i,k_i}, x_{i,k_i})$. Here $a_{i,j}$ denotes a selected value of the ith attribute and $x_{i,j} \in [0, 1]$ (or $x_{i,j} \in [0, 100\%]$) denotes the degree of suitability that corresponds to the attribute value $a_{i,j}$. The simplest form is the vertex notation:

Table 17.1 Documentation table for attribute suitability criteria

1111		**Salary [%]**
Value	%	Evaluated as the following relative salary: Srel = 100*Snew/Sold where: Snew = offered new salary Sold = salary at the current job
90	0	
100	60	
150	100	

1112		**Benefits [%]**
Value	%	Evaluated as the following relative benefits: Brel = 100*Bnew/Bold where: Bnew = offered health+retirement+vacation benefits Bold = current health+retirement+vacation benefits
70	0	
100	60	
150	100	

112		**Anticipated future salary [%]**
Value	%	Evaluated as the following relative anticipated salary: RAS = 100*A/Snew where: A = 0.5*(anticipated 3-year salary) + 0.5*(anticipated 5-year salary) Snew = the offered starting salary
100	0	
125	100	

121		**Attractiveness of job [0 .. 6]**
Value	%	Evaluated using the following rating scale: 0 = lowest, 1 = very low, 2 = low, 3 = average, 4 = high, 5 = very high, 6 = highest
0	0	
6	100	

1221		**Average work week [h]**
Value	%	Average work week time, measured in hours: Time <= 40 h : standard (or reduced) workload Time = 50 h : high load (5 days * 10 hours) Time = 72 h : startup mode (6 days * 12 hours)
40	100	
50	80	
72	0	

1222		**Daily commute time [min]**
Value	%	Total average daily commute time measured in minutes: 0 = working online from home 30 min = 2 * 15 min (working close to home) 90 min = 2 * 45 min (not acceptable)
0	100	
30	80	
90	0	

$$Crit(a_i) = \{(a_{i,1}, x_{i,1}), (a_{i,2}, x_{i,2}), \ldots, (a_{i,k_i}, x_{i,k_i})\}, \quad a_{i,1} < a_{i,2} < \cdots < a_{i,k_i}, \quad k_i \geq 2.$$

This notation is very convenient in cases where interpolation is not necessary/possible, like in cases where we have discrete attributes: $a_i \in \{a_{i,1}, \ldots, a_{i,k_i}\}$. The simplest case is the binary criterion $Crit(a_i) = \{(0, 0), (1, 100)\}$ where a_i denotes the availability of specific service (e.g. a hospital can have or not have the emergency services). Similarly, a homebuyer who needs 3 or 4 bedrooms can evaluate the number of bedrooms using the criterion $Crit(a_i) = \{(2, 0), (3, 75), (4, 100)\}$. In all these examples, the interpolation between the vertex points is not meaningful. If

Table 17.2 Four versions of the attribute criterion for job attractiveness

Job attractiveness scale		Suitability degree for different interest in job attractiveness			
Linguistic label (interpretation)	Numeric label	Very low int. ($p = 0.5$)	Moderate int. ($p = 1$)	High interest ($p = 1.5$)	Very high int. ($p = 2$)
Lowest	0	0	0	0	0
Very low	1	40.8	16.7	6.8	2.8
Low	2	57.7	33.3	19.2	11.1
Average	3	70.7	50	35.4	25
High	4	81.6	66.7	54.4	44.4
Very high	5	91.3	83.3	76.1	69.4
Highest	6	100	100	100	100

the attribute a_i is a continuous variable, then it is useful to use the vertex points and interpolations between them, as follows:

$$x_i = g_i(a_i) = \begin{cases} x_{i,1}, & a_i \le a_{i,1} \\ x_{i,1} + \left(x_{i,2} - x_{i,1}\right)\frac{a_i - a_{i,1}}{a_{i,2} - a_{i,1}}, & a_{i,1} \le a_i \le a_{i,2} \\ \cdots & \cdots \\ x_{i,k_i-1} + \left(x_{i,k_i} - x_{i,k_i-1}\right)\frac{a_i - a_{i,k_i-1}}{a_{i,k_i} - a_{i,k_i-1}}, & a_{i,k_i-1} \le a_i \le a_{i,k_i} \\ x_{i,k_i}, & a_{i,k_i} \le a_i, \end{cases}$$

$$a_{i,1} < a_{i,2} < \cdots < a_{i,k_i}, \quad k_i \ge 2, \quad i = 1, \ldots, n.$$

The number of vertex points $k_i \ge 2$. is selected according to the desired precision of the attribute criterion. For simplicity, the values of attribute suitability $x_{i,j}$ are usually denoted for as percentages of satisfied requirements.

Attribute criteria can sometimes use nonlinear rating scales. For example, the attractiveness of job can range from the lowest to the highest, but the corresponding suitability can also express the stakeholder's interest in the job attractiveness, as shown in Table 17.2. If A denotes the numeric label of the job attractiveness in a 7-degree scale ($0 \le A \le 6 = A_{\max}$), then the corresponding suitability can be expressed as $S = 100(A/A_{\max})^p$ where p is a parameter used to express the stakeholder's interest in job attractiveness.

Some people expect from their job both the monetary compensation and the satisfaction derived from the job attractiveness to the extent that the high job attractiveness can be the primary motivation for the job search; that can be expressed as $p > 1$, where only very attractive jobs are highly suitable. Other people sometimes hate their job, but must accept it primarily for monety compensation, and that can be expressed using $p < 1$, where almost any job is acceptable (as long as it brings salary), as shown in Table 17.2. In most cases, people have moderate interest in job attractiveness, balanced with the interest in monetary compensation; in such cases $p = 1$ and $Crit(A[0..6]) = \{(0, 0), (6, 100)\}$. If $p \neq 1$, then $Crit(A[0..6])$. must clude 7 vertex points.

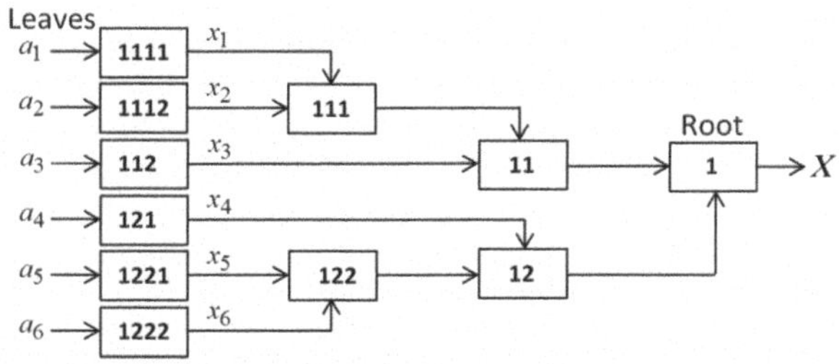

Fig. 17.4 The structure of suitability aggregation tree

Logic Aggregation of Suitability

The next step of the LSP method is the logic aggregation of attribute suitability scores, performed to compute the overall suitability $X = L(x_1, ..., x_n)$. This step is the most delicate, and it is performed using the Graded Logic functions. In Fig. 17.2 we decomposed the overall suitability to reach leaves of the attribute tree, and now we perform aggregation going in the opposite direction, from leaves to the root, as shown in the suitability aggregation tree table in Fig. 17.2, and in Fig. 17.4. The blocks 1111, 1112, 112, 121, 1221, and 1222 are the attribute criteria that provide attribute suitability $x_1, \ldots, x_6$. The blocks 111, 11, 122, 12, and 1 are the logic aggregators that are used to compute the overall suitability $X = L(x_1, \ldots, x_6)$. Generally, each logic aggregator combines two or more input suitability scores and implements specific graded logic properties. In most cases the aggregators are selected from the medium precision list presented in Fig. 17.5. The users who do not have access to the LSP.NT tool, can use the analytic forms of these aggregators based on Chap. 15, All versions of partial absorption presented in Chap. 16, and numerical analysis from [DUJ18], resulting in analytic forms shown in Table 17.4. Characteristic forms of these aggregators are shown in Fig. 15.4 and sample numerical tables of typical idempotent functions of two variables are presented in Tables 17.5 and 17.6

The first step in selecting aggregators 111, 11, 122, 12, and 1 can be to go through the list of input attributes and ask the question whether each input can be classified as mandatory, or optional, or sufficient. If the mandatory input is $x_i = 0$, then the overall result must be $X = 0$ regardless the values of other attributes. If the sufficient attribute is $x_i = 1$, then the output should be $X = 1$. Optional inputs affect the output but cannot cause strong effects reserved for mandatory or sufficient inputs. A sample classification/explanation is shown in Tables 17.3 and 18.2.

The classification in Table 17.3 facilitates the selection of aggregators. The aggregators are selected from the list of 21 aggregators shown in Fig. 17.5 according to the following Q&A dialogue (alternative answers are separated by symbol "|"):

GRADED LOGIC AGGREGATORS (version with 21 basic aggregator types)

GRADED CONJUNCTION/DISJUNCTION (GCD)

Simultaneity (conjunction: graded *and* function)
- Drastic conjunction .... **CC** (andness > 1)
- Product (t-norm) .... **CP** (andness > 1)
- Full Conjunction .... **C** (andness = 1)
- Partial Conjunction
  - Hard Partial Conjunction [mandatory]
    - High .... **HC+**
    - Medium .... **HC**
    - Low .... **HC-**
  - Soft Partial Conjunction [optional]
    - High .... **SC+**
    - Medium .... **SC**
    - Low .... **SC-**

↑ Increasing andness (degree of simultaneity)

Logic neutrality (arithmetic mean) .... **A** (andness=orness=½)

Substitutability (disjunction: graded *or* function)
- Partial Disjunction
  - Soft Partial Disjunction [optional]
    - Low .... **SD-**
    - Medium .... **SD**
    - High .... **SD+**
  - Hard Partial Disjunction [sufficient]
    - Low .... **HD-**
    - Medium .... **HD**
    - High .... **HD+**
- Full Disjunction .... **D** (orness = 1)
- Dual product (t-conorm) .... **DP** (orness > 1)
- Drastic disjunction.............................. **DD** (orness > 1)

↓ Increasing orness (degree of substitut-ability)

PARTIAL ABSORPTION (mandatory/optional and sufficient/optional)
- **Conjunctive Partial Absorption** (M/O) .... **CPA** (penalty > reward)
- **Disjunctive Partial Absorption** (S/O) .... **DPA** (reward > penalty)

Mandatory = "must have"; Optional = "nice to have"; Sufficient = "enough to have"

Fig. 17.5 Classification and symbols of basic logic aggregators (medium precision)

Table 17.3 Classification of input attributes

ID	Suitability attribute	Type
1111	Salary	Mandatory (must be satisfied)
1112	Benefits	Optional (desirable but not necessary)
112	Anticipated future salary	Optional (desirable but not necessary)
121	Attractiveness of job	Mandatory (must be satisfied)
1221	Average work week	Mandatory (must be satisfied)
1222	Daily commute time	Mandatory (must be satisfied)

Table 17.4 Interpolative implementation of 21 medium precision LSP logic aggregators

Aggre-gator	Type	Implementation $[\boldsymbol{x} = (x_1,\dots,x_n),\ \boldsymbol{1}-\boldsymbol{x} = (1-x_1,\dots,1-x_n),\ \boldsymbol{W} = (W_1,\dots,W_n)]$
CC	C/H/NI	$\boldsymbol{CC}(\boldsymbol{x}) = \lfloor \Pi_{i=1}^{n} x_i \rfloor$
CP	C/H/NI	$\boldsymbol{CP}(\boldsymbol{x}) = \Pi_{i=1}^{n} x_i$
C	C/H/ID	$\boldsymbol{C}(\boldsymbol{x}) = min(x_1,\dots,x_n)$
HC+	C/H/ID	$\boldsymbol{HCP}(\boldsymbol{x};\boldsymbol{W}) = \left(\Sigma_{i=1}^{n} W_i x_i^{r(n)}\right)^{1/r(n)}$ $r(2) = -7.675,\quad r(3) = -6.53,\quad r(4) = -5.75,\quad r(5) = -5.19$
HC	C/H/ID	$\boldsymbol{HC}(\boldsymbol{x};\boldsymbol{W}) = \left(\Sigma_{i=1}^{n} W_i x_i^{r(n)}\right)^{1/r(n)}$ $r(2) = -2.813,\quad r(3) = -2.539,\quad r(4) = -2.327,\quad r(5) = -2.165$
HC-	C/H/ID	$\boldsymbol{HCM}(\boldsymbol{x};\boldsymbol{W}) = \left(\Sigma_{i=1}^{n} W_i x_i^{r(n)}\right)^{1/r(n)}$ $r(2) = -1.188,\quad r(3) = -1.151,\quad r(4) = -1.101,\quad r(5) = -1.058$
SC+	C/S/ID	$\boldsymbol{SCP}(\boldsymbol{x};\boldsymbol{W}) = \frac{1}{7}(\Sigma_{i=1}^{n} W_i x_i) + \frac{6}{7}\left(\Sigma_{i=1}^{n} W_i x_i^{R(n)}\right)^{1/R(n)}$ $R(2) = -0.721, R(3) = -0.737,\quad R(4) = -0.727,\quad R(5) = -0.712$
SC	C/S/ID	$\boldsymbol{SC}(\boldsymbol{x};\boldsymbol{W}) = \frac{3}{7}(\Sigma_{i=1}^{n} W_i x_i) + \frac{4}{7}\left(\Sigma_{i=1}^{n} W_i x_i^{R(n)}\right)^{1/R(n)}$
SC-	C/S/ID	$\boldsymbol{SCM}(\boldsymbol{x};\boldsymbol{W}) = \frac{5}{7}(\Sigma_{i=1}^{n} W_i x_i) + \frac{2}{7}\left(\Sigma_{i=1}^{n} W_i x_i^{R(n)}\right)^{1/R(n)}$
A	N/S/ID	$\boldsymbol{A}(\boldsymbol{x};\boldsymbol{W}) = W_1 x_1 + \cdots + W_n x_n$
SD-	D/S/ID	$\boldsymbol{SDM}(\boldsymbol{x};\boldsymbol{W}) = 1 - \boldsymbol{SCM}(\boldsymbol{1}-\boldsymbol{x};\boldsymbol{W})$
SD	D/S/ID	$\boldsymbol{SD}(\boldsymbol{x};\boldsymbol{W}) = 1 - \boldsymbol{SC}(\boldsymbol{1}-\boldsymbol{x};\boldsymbol{W})$
SD+	D/S/ID	$\boldsymbol{SDP}(\boldsymbol{x};\boldsymbol{W}) = 1 - \boldsymbol{SCP}(\boldsymbol{1}-\boldsymbol{x};\boldsymbol{W})$
HD-	D/H/ID	$\boldsymbol{HDM}(\boldsymbol{x};\boldsymbol{W}) = 1 - \boldsymbol{HCM}(\boldsymbol{1}-\boldsymbol{x};\boldsymbol{W})$
HD	D/H/ID	$\boldsymbol{HD}(\boldsymbol{x};\boldsymbol{W}) = 1 - \boldsymbol{HC}(\boldsymbol{1}-\boldsymbol{x};\boldsymbol{W})$
HD+	D/H/ID	$\boldsymbol{HDP}(\boldsymbol{x};\boldsymbol{W}) = 1 - \boldsymbol{HCP}(\boldsymbol{1}-\boldsymbol{x};\boldsymbol{W})$
D	D/H/ID	$\boldsymbol{D}(\boldsymbol{x}) = max(x_1,\dots,x_n)$
DP	D/H/NI	$\boldsymbol{DP}(\boldsymbol{x}) = 1 - \Pi_{i=1}^{n}(1-x_i)$
DD	D/H/NI	$\boldsymbol{DD}(\boldsymbol{x}) = 1 - \lfloor \Pi_{i=1}^{n}(1-x_i) \rfloor$
CPA [AH]	C/A/ID	$\boldsymbol{CPA}(x,y;P,R) = \frac{x[W_1 x + (1-W_1)y]}{W_1 W_2 x + (1-W_1)W_2 y + (1-W_2)x}$, $W_1 = \frac{2R(1-P)}{P+R},\ W_2 = \frac{P-R}{P-R+2PR},\quad P > R$; $P = \frac{½ - \boldsymbol{CPA}(½,0)}{½} = 1 - 2\boldsymbol{CPA}(½,0),\quad R = \frac{\boldsymbol{CPA}(½,1) - ½}{½} = 2\boldsymbol{CPA}(½,1) - 1$
DPA [AH]	D/A/ID	$\boldsymbol{DPA}(x,y;P,R) = 1 - \frac{(1-x)[1-(W_1 x + (1-W_1)y)]}{(1-W_2)(1-x) + [1-(W_1 x + (1-W_1)y)]W_2}$, $W_1 = \frac{2P(1-R)}{P+R},\ W_2 = \frac{R-P}{R-P+2PR},\ R > P$; $P = \frac{½ - \boldsymbol{DPA}(½,0)}{½} = 1 - 2\boldsymbol{DPA}(½,0);\ R = \frac{\boldsymbol{DPA}(½,1) - ½}{½} = 2\boldsymbol{DPA}(½,1) - 1$
Type codes: C = conjunctive, D = disjunctive, N = neutral, H = hard, S = soft, A = asymmetric, ID = idempotent, NI = nonidempotent, P = maximum relative penalty in the central point $x = 0.5$, R = maximum relative reward in the central point $x = 0.5$.		

Type codes: C = conjunctive, D = disjunctive, N = neutral, H = hard, S = soft, A = asymmetric, ID = idempotent, NI = nonidempotent, P = maximum relative penalty in the central point $x = 0.5$, R = maximum relative reward in the central point $x = 0.5$

Table 17.5 SC, HC, HD, and SD aggregators of two equally important inputs (all values are [%])

SC: y \ x	0	10	20	30	40	50	60	70	80	90	100
0	0	2.1	4.3	6.4	8.6	10.7	12.9	15	17.1	19.3	21.4
10	2.1	10	14.2	17.5	20.4	23.1	25.7	28.2	30.6	33	35.3
20	4.3	14.2	20	24.5	28.3	31.8	34.9	37.9	40.8	43.5	46.2
30	6.4	17.5	24.5	30	34.6	38.8	42.5	46	49.3	52.4	55.4
40	8.6	20.4	28.3	34.6	40	44.7	49	53	56.7	60.2	63.6
50	10.7	23.1	31.8	38.8	44.7	50	54.8	59.2	63.3	67.2	70.9
60	12.9	25.7	34.9	42.5	49	54.8	60	64.8	69.3	73.5	77.5
70	15	28.2	37.9	46	53	59.2	64.8	70	74.8	79.4	83.7
80	17.1	30.6	40.8	49.3	56.7	63.3	69.3	74.8	80	84.9	89.5
90	19.3	33	43.5	52.4	60.2	67.2	73.5	79.4	84.9	90	94.9
100	21.4	35.3	46.2	55.4	63.6	70.9	77.5	83.7	89.5	94.9	100

HC: y \ x	0	10	20	30	40	50	60	70	80	90	100
0	0	0	0	0	0	0	0	0	0	0	0
10	0	10	12.2	12.6	12.7	12.7	12.8	12.8	12.8	12.8	12.8
20	0	12.2	20	23.2	24.4	24.9	25.2	25.3	25.4	25.5	25.5
30	0	12.6	23.2	30	33.7	35.6	36.6	37.2	37.6	37.8	37.9
40	0	12.7	24.4	33.7	40	44	46.4	47.9	48.8	49.4	49.9
50	0	12.7	24.9	35.6	44	50	54.1	56.9	58.8	60.1	61
60	0	12.8	25.2	36.6	46.4	54.1	60	64.3	67.3	69.6	71.2
70	0	12.8	25.3	37.2	47.9	56.9	64.3	70	74.4	77.7	80.1
80	0	12.8	25.4	37.6	48.8	58.8	67.3	74.4	80	84.4	87.9
90	0	12.8	25.5	37.8	49.4	60.1	69.6	77.7	84.4	90	94.5
100	0	12.8	25.5	37.9	49.9	61	71.2	80.1	87.9	94.5	100

HD: y \ x	0	10	20	30	40	50	60	70	80	90	100
0	0	5.5	12.1	19.9	28.8	39	50.1	62.1	74.5	87.2	100
10	5.5	10	15.6	22.3	30.4	39.9	50.6	62.2	74.5	87.2	100
20	12.1	15.6	20	25.6	32.7	41.2	51.2	62.4	74.6	87.2	100
30	19.9	22.3	25.6	30	35.7	43.1	52.1	62.8	74.7	87.2	100
40	28.8	30.4	32.7	35.7	40	45.9	53.6	63.4	74.8	87.2	100
50	39	39.9	41.2	43.1	45.9	50	56	64.4	75.1	87.3	100
60	50.1	50.6	51.2	52.1	53.6	56	60	66.3	75.6	87.3	100
70	62.1	62.2	62.4	62.8	63.4	64.4	66.3	70	76.8	87.4	100
80	74.5	74.5	74.6	74.7	74.8	75.1	75.6	76.8	80	87.8	100
90	87.2	87.2	87.2	87.2	87.2	87.3	87.3	87.4	87.8	90	100
100	100	100	100	100	100	100	100	100	100	100	100

SD: y \ x	0	10	20	30	40	50	60	70	80	90	100
0	0	5.1	10.5	16.3	22.5	29.1	36.4	44.6	53.8	64.7	78.6
10	5.1	10	15.1	20.6	26.5	32.8	39.8	47.6	56.5	67	80.7
20	10.5	15.1	20	25.2	30.7	36.7	43.3	50.7	59.2	69.4	82.9
30	16.3	20.6	25.2	30	35.2	40.8	47	54	62.1	71.8	85
40	22.5	26.5	30.7	35.2	40	45.2	51	57.5	65.1	74.3	87.1
50	29.1	32.8	36.7	40.8	45.2	50	55.3	61.2	68.2	76.9	89.3
60	36.4	39.8	43.3	47	51	55.3	60	65.4	71.7	79.6	91.4
70	44.6	47.6	50.7	54	57.5	61.2	65.4	70	75.5	82.5	93.6
80	53.8	56.5	59.2	62.1	65.1	68.2	71.7	75.5	80	85.8	95.7
90	64.7	67	69.4	71.8	74.3	76.9	79.6	82.5	85.8	90	97.9
100	78.6	80.7	82.9	85	87.1	89.3	91.4	93.6	95.7	97.9	100

Table 17.6 Sample CPA/AH, and DPA/AH aggregators with maximum relative penalty P and maximum relative reward R defined in the central point $x = 0.5$ (all values are [%])

CPA P=30% R=10%: y (optional) \ x (mandatory)	0	10	20	30	40	50	60	70	80	90	100
0	0	7	14	21	28	**35**	42	49	56	63	70
10	0	10	18	25.5	32.8	40	47.1	54.3	61.3	68.4	75.5
20	0	11	20	28.2	36	43.6	51	58.3	65.6	72.8	80
30	0	11.5	21.2	30	38.3	46.2	54	61.6	69.1	76.5	83.8
40	0	11.8	22	31.3	40	48.3	56.4	64.3	72	79.6	87.1
50	0	12	22.6	32.2	41.3	**50**	58.4	66.5	74.5	82.3	90
60	0	12.1	23	33	42.4	51.4	60	68.4	76.6	84.6	92.5
70	0	12.3	23.3	33.6	43.3	52.5	61.4	70	78.4	86.6	94.7
80	0	12.3	23.6	34.1	44	53.5	62.6	71.4	80	88.4	96.7
90	0	12.4	23.8	34.5	44.6	54.3	63.6	72.6	81.4	90	98.4
100	0	12.5	24	34.8	45.1	**55**	64.5	73.7	82.7	91.4	100

DPA P=15% R=30%: y (optional) \ x (sufficient)	0	10	20	30	40	50	60	70	80	90	100
0	0	8.1	16.3	24.7	33.5	**42.5**	52	62.1	73.1	85.5	100
10	2.1	10	18.1	26.3	34.9	43.7	52.9	62.8	73.5	85.6	100
20	4.3	12.1	20	28.1	36.4	45	54	63.6	74	85.8	100
30	6.7	14.3	22.1	30	38.1	46.5	55.2	64.5	74.5	86	100
40	9.2	16.8	24.4	32.1	40	48.1	56.6	65.5	75.2	86.3	100
50	12	19.4	26.9	34.4	42.1	**50**	58.2	66.7	76	86.6	100
60	15	22.3	29.6	37	44.5	52.1	60	68.2	77	87	100
70	18.3	25.4	32.6	39.9	47.2	54.6	62.2	70	78.3	87.6	100
80	21.8	28.9	36	43.1	50.3	57.5	64.8	72.3	80	88.5	100
90	25.7	32.7	39.8	46.8	53.8	60.9	68	75.1	82.4	90	100
100	30	37	44	51	58	**65**	72	79	86	93	100

CPA P=40% R=30%: y (optional) \ x (mandatory)	0	10	20	30	40	50	60	70	80	90	100
0	0	6	12	18	24	**30**	36	42	48	54	60
10	0	10	16	22	28	35	41	47	53	59	65
20	0	13	20	26	33	39	45	51	57	63	69
30	0	15	23	30	36	43	49	55	61	67	73
40	0	17	26	33	40	46	53	59	65	71	78
50	0	19	28	36	43	**50**	56	63	69	75	82
60	0	20	31	39	46	53	60	67	73	79	85
70	0	21	33	42	49	56	63	70	77	83	89
80	0	22	34	44	52	59	67	73	80	87	93
90	0	23	36	46	55	62	70	77	83	90	97
100	0	24	37	48	57	**65**	73	80	87	93	100

DPA P=30% R=40%: y (optional) \ x (sufficient)	0	10	20	30	40	50	60	70	80	90	100
0	0	7	13	20	27	**35**	43	52	63	77	100
10	3	10	17	23	30	38	45	54	64	77	100
20	7	13	20	27	33	41	48	56	66	78	100
30	11	17	23	30	37	44	51	58	67	79	100
40	15	21	27	34	40	47	54	61	69	80	100
50	18	25	31	37	44	**50**	57	64	72	81	100
60	22	29	35	41	47	54	60	67	74	83	100
70	27	33	39	45	51	57	64	70	77	85	100
80	31	37	43	49	55	61	67	74	80	87	100
90	35	41	47	53	59	65	72	78	84	90	100
100	40	46	52	58	64	**70**	76	82	88	94	100

Q1: Is the aggregator symmetric or asymmetric?
A1: Asymmetric (partial absorption PA, number of inputs = 2)
Q2: Is the PA conjunctive or disjunctive?
A2: Conjunctive (mandatory/optional) | Disjunctive (sufficient/optional)
Q3: Select the desired penalty (P) and reward (R)
A3: Selected P and R, $P > R$ | Selected P and R, $P < R$
End of dialogue

A1: Symmetric (GCD)
Q2: Is the GCD conjunctive or disjunctive?
A2: Conjunctive | disjunctive
Q3: Is the aggregator hard or soft?
A3: Hard | soft
Q4: Select the desired andness
A4: Selected andness $\alpha \in \left[-\frac{1}{k-1}, \frac{k}{k-1}\right]$, k = number of inputs
Q5: Select the desired importance weights
A5: Selected weights $W_1, \ldots, W_k$, $k \geq 2$
End of dialogue

In the case of GCD there are 5 Q&A steps, and in the case of PA there are 3 Q&A steps. Thus, assuming the basic understanding of symmetric and asymmetric aggregators presented in Chaps. 15 and 16, the selection of logic aggregators and development of the logic aggregation structure is rather simple. Based on selected basic types of input attributes (mandatory/optional/sufficient), the stakeholder/decision-maker can create the aggregation structure for the job offer evaluation criterion exemplified in Fig. 17.6. All numeric values in the aggregation structure (as well as all attribute criteria shown in Table 17.1) reflect specific requirements of the stakeholder. The aggregation structure is built aggregator by aggregator, considering the role and meaning of those attributes that are used as inputs in each aggregator. For example, the output of aggregator 11 represents the aggregated suitability of attributes 1111, 1112, and 112, i.e., the complete monetary compensations. Similarly, the aggregator 12 is the representative of attributes 121, 1221, and 1222, i.e., it represents all main characteristics of the job. Then the stakeholder can compare the importance of these two inputs and select the most appropriate parameters of the root aggregator 1. Of course, the root aggregator must be hard, and the weights 55, 45% reflect the stakeholder who is interested in monetary compensation slightly more than in job attractiveness.

The CPA/AH is selected for aggregator 111 because a good salary is a mandatory requirement and benefits are significant but not mandatory: in the role of stakeholder, we want to have benefits without giving them the power to significantly compensate

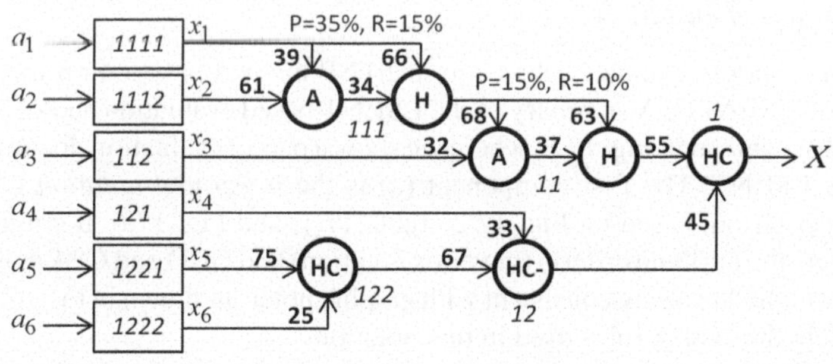

Fig. 17.6 A complete LSP criterion: the attribute criteria and the suitability aggregation structure (a GL propositional calculus formula) for the job selection problem

an insufficient salary. So, let us penalize poor benefits 35% and reward excellent benefits only 15%. Using $P = 0.35$ and $R = 0.15$, and formulas from Table 17.4 we find $W_1 = 0.3 \times 0.65/0.5 = 0.39$ and $W_2 = 0.2/(0.2 + 0.7 \times 0.15) \approx 0.66$. Similarly, we use CPA/AH for aggregator 11 because we assume that the primary motivation for job change is a mandatory improvement at the start of the new job and much less a risky promise of a bright future, which is desirable, but not mandatory. Reflecting that perspective, we will use modest values $P = 0.15$ and $R = 0.10$, yielding $W_1 = 0.68$ and $W_2 \approx 0.63$.

Regarding the characteristics of the new job, all three attributes (121, 1221, 1222) should be mandatory and they need hard aggregators. It would be unacceptable to take a fully unattractive job, or to either work or commute each day beyond reasonable limits [SAA16]. That yields the low-level hard partial conjunction HC- as the aggregator for both 122 and 12. The weights reflect the standpoint that work time is three times more important than the commute time, and the certain and measurable total time at work is two times more important than the estimated (but possibly uncertain) attractiveness of the new job. Finally, the root node 1 must be a stronger conjunction compared to preceding aggregators, yielding the hard partial conjunction HC; the minor differences in weights reflect the situation where the offered monetary compensation is considered slightly more important than the other job characteristics.

There are three important observations that can be derived from this example. First, the criterion shown in Fig. 17.6 directly reflects the above narrative, which is a verbalized description of stakeholder's requirements. Of course, a different narrative would yield a different criterion, showing that each LSP criterion must precisely reflect stakeholder's goals and requirements. The second observation is that even in the toy problem with only $n = 6$ inputs there is a need for aggregation based on Graded Logic and aggregation structures composed by combining appropriate GL functions selected from the spectrum of LSP aggregators shown in Fig. 17.5). The third observation is that the LSP method, in the case of n input attributes, is essentially a problem that needs a linear effort $O(n)$, as shown below. The total effort for developing LSP criteria and using the LSP method is linear because the LSP criterion is systematically created attribute by attribute and aggregator by aggregator.

Evaluation Based on LSP.NT

The simplest way to evaluate jobs using the LSP method is based on the LSP.NT software tool [SEA17]. A summary of the LSP.NT-based evaluation and comparison of three jobs is shown in Fig. 17.8, where we show a part of evaluation documentation created by LSP.NT. The first component (a) is the *integrated notation* of the job evaluation criterion, based on Fig. 17.2, Table 17.1, and Fig. 17.6. It combines the attribute tree, the attribute criteria (in vertex notation) and GCD and CPA aggregators (type and weights) in a way consistent with graphic notation shown in Fig. 17.6. First, let us present the syntax rules used in this notation.

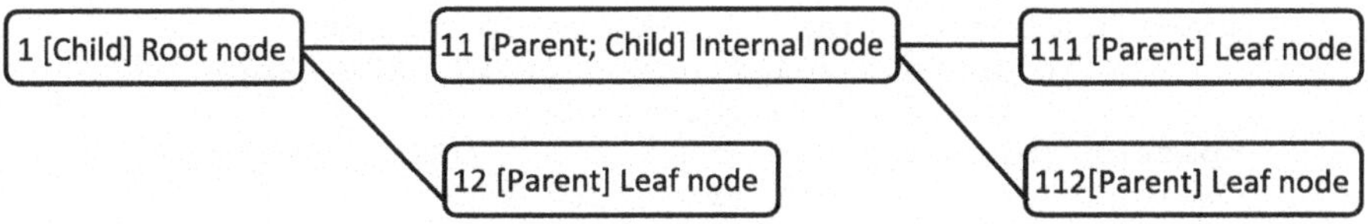

Fig. 17.7 The parent/child notation of aggregation tree nodes (root, internal nodes and leaves)

The integrated notation of an LSP criterion is based on the aggregation tree structure (exemplified in Fig. 17.4). Each node has the following node specifier, defined using the BNF notation[3]:

```
<node specifier>::=<numeric ID>[<relationships with
neighbors>] <node name>
```

The suitability aggregation tree structure consists of the following three types of nodes shown in Fig. 17.7:

- *Root node*: a single node without parent that has two or more children nodes
- *Leaf node*: a terminal node without children that has a single parent node
- *Internal node*: a node with a single parent and two or more children nodes.

This classification shows that the root node needs only the specification of relationship with the children nodes; that relationship means the specification of the operator used for aggregating the suitability of child nodes. For idempotent aggregators based on granularity 15 the proposed BNF syntax of the root node (identified by the fixed ID number 1) is the following:

```
<root node specifier> ::= 1[<child relation>]<node name>
<child relation > ::= C | HC+ | HC | HC- | SC+ | SC |
        SC- | A | SD- | SD | SD+ | HD- | HD | HD+ | D |
        CPA: P=<percent>, R=<percent> |
        DPA: P=<percent>, R=<percent>
<percent> ::= <integer from 1 to 99>%
<node name> ::= <simple character string>
```

According to this syntax, an example of root node is 1 [HC] Job.

The leaf nodes need only the specification of relationship with the parent node; that relationship means the specification of either the weight (for symmetric conjunctive or disjunctive aggregators) or the status (mandatory, sufficient, or optional) in the case of partial absorption:

[3] Backus-Naur Form (BNF) is the most frequent syntax notation based on three components: (1) < x > denotes an object called x, (2) vertical line | denotes exclusive or (e.g. < digit > | <letter > denotes any digit or letter), and (3) the assignment (the definition of an object) is denoted:: = (e.g. < binary digit > :: = 0 | 1).

```
<leaf node specifier> ::= <node ID number>
        [ <parent relation> ] <leaf node name>
        [ <unit of measurement> ]
<parent relation> ::= MAN | SUF | OPT | <weight>
<weight> ::= <integer from 1 to 99>%
```

In the proposed integrated notation the leaf nodes include also attribute criterion function.

```
<leaf node> ::= <leaf node specifier>:
                <criterion function>
<criterion function> ::= { <list of vertices> }
<list of vertices> ::= (<attribute>, <score>),
        ( <attribute>, <score>) |
        <list of vertices> , (<attribute> , <score>)
```

We assume that attributes are increasing real numbers, and suitability scores are values between 0 and 100%.

Internal nodes are all nodes located between the root and leaves; each of them has both a parent and children, yielding the following specifier:

```
<internal node specifier> ::= <node ID number>
        [<parent relation>; <child relation>]
        <node name>
```

The parent and child relations are same as those defined in the leaf and the root node respectively. The integrated notation of the criterion shown in Fig. 17.6 with attribute criteria introduced in Table 17.1 is presented in Fig. 17.8a.

The example in Fig. 17.8 shows the comparison of three jobs. We assume that the stakeholder has a current job (donated "OldJob") that is very attractive, needs 40 h of work per week, and it is based on work from home. Unfortunately, the financial compensation is insufficient and anticipated future salary is very modest. This is the reason for looking for a new job and the stakeholder received two new job offers. The new jobs require relocation to places where the cost of living is higher than the current location. This is shown in the row denoted as "Cost." The NewJob1 is in the area where the cost of relocation/living is 5% above the current location, and the NewJob2 is in the area where the cost of relocation/living is 10% above the current location. However, the NewJob2 offers a significant improvement in current income both initially and in the future. The NewJob1 offers the best benefits and the possibility of working from home but the increase of income is moderate. That is visible in the Fig. 17.8b.

The suitability evaluation results (Fig. 17.8c) show that the most convenient is the NewJob2 that satisfies almost 78% of the stakeholder's requirements, compared to 66.9% requirements satisfied by the current job (OldJob). The NewJob1 is also better than the OldJob (the overall suitability is 72.59%).

If we want to consider the cost of relocation/living (C_i), the LSP.NT provides the cost/preference analysis that is presented in the two last tables shown in Fig. 17.8d.

1 [HC] JOB
  11 [55%; CPA: P=15%, R=10%] Monetary compensation
    111 [MAN; CPA: P=35%, R=15%] Offered salary and benefits
      1111 [MAN] Salary [%]: {(90,0), (100,60), (150,100)}
      1112 [OPT] Benefits [%]: {(70,0), (100,60), (150,100)}
    112 [OPT] Anticipated future salary [%]: {(100,0), (125,100)}
  12 [45%; HC-] Main characteristics of the job
    121 [33%] Attractiveness of job [0..6]: {(0,0), (6,100)}
    122 [67%; HC-] Total time
      1221 [75%] Average work week [h]: {(40,100), (50,80), (72,0)}
      1222 [25%] Daily commute time [min]: {(0,100), (30,80), (90,0)}

(a)
Job evaluation criterion function (integrated notation)

Id	Attribute	NewJob1	NewJob2	OldJob
	Cost	1.0500	1.1000	1.0000
1111	Salary [%]	110	130	100
1112	Benefits [%]	125	120	100
112	Anticipated future salary [%]	115	125	110
121	Attractiveness of job [0..6]	4	3	6
1221	Average work week [h]	50	40	40
1222	Daily commute time [min]	0	30	0

(b)
Suitability attributes of available jobs (old job and two new jobs)

Id	Attribute	NewJob1	NewJob2	OldJob
1	JOB	72.59	77.97	66.90
11	Monetary compensation	69.42	84.32	57.19
12	Main characteristics of the job	77.35	72.19	100.00
111	Offered salary and benefits	70.93	81.91	60.00
122	Total time	84.14	94.03	100.00
1222	Daily commute time [min]	100.00	80.00	100.00
1221	Average work week [h]	80.00	100.00	100.00
121	Attractiveness of job [0..6]	66.67	50.00	100.00
112	Anticipated future salary [%]	60.00	100.00	40.00
1112	Benefits [%]	80.00	76.00	60.00
1111	Salary [%]	68.00	84.00	60.00

(c)
Suitability evaluation results [%]

Overall value: 100 * $(\text{Score}_k/\text{Score}_{max})^W$ $(\text{Cost}_{min}/\text{Cost}_k)^{1-W}$ [%]

System	Cost	Relative importance of high score (w)											Overall score [%]
		0%	10%	20%	30%	40%	50%	60%	70%	80%	90%	100%	
NewJob1	1.0500	95.24	95.02	94.81	94.59	94.38	94.17	93.95	93.74	93.53	93.32	93.11	72.59
NewJob2	1.1000	90.91	91.78	92.66	93.55	94.44	95.35	96.26	97.18	98.11	99.05	100.0	77.97
OldJob	1.0000	100.0	98.48	96.98	95.51	94.06	92.63	91.22	89.84	88.47	87.13	85.80	66.90

Normalized value

System	Cost	Relative importance of high score (w)											Overall score [%]
		0%	10%	20%	30%	40%	50%	60%	70%	80%	90%	100%	
NewJob1	1.0500	95.24	96.49	97.76	99.04	99.94	98.76	97.61	96.46	95.33	94.21	93.11	72.59
NewJob2	1.1000	90.91	93.20	95.54	97.94	100.0	100.0	100.0	100.0	100.0	100.0	100.0	77.97
OldJob	1.0000	100.0	100.0	100.0	100.0	99.59	97.15	94.77	92.44	90.17	87.96	85.80	66.90

(d)
Cost/preference analysis: total value and normalized value

Fig. 17.8 Results of job selection obtained using the LSP.NT

The total value (V) is defined using the weighted geometric mean as a hard conjunctive aggregator of two inputs: the relative suitability score and the relative relocation/living cost. The total value of each job offer is computed as follows:

$$V(X_i, C_i) = \left(\frac{X_i}{\max(X_1, \ldots, X_m)}\right)^w \left(\frac{\min(C_1, \ldots, C_m)}{C_i}\right)^{1-w} \in [0, 1], \quad i = 1, \ldots, m.$$

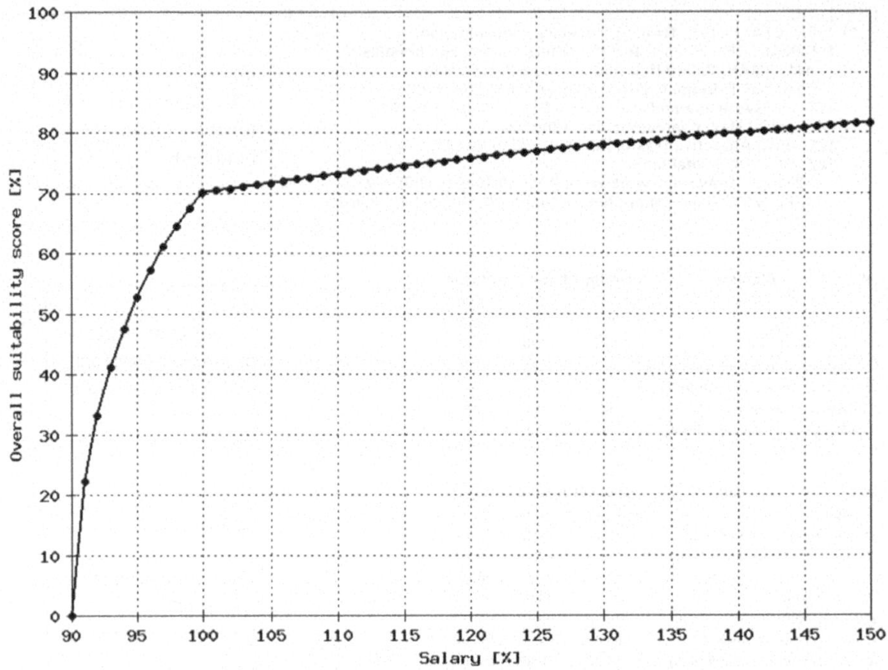

Fig. 17.9 Overall suitability score of NewJob2 as a function of the offered salary

The total value is computed for $m = 3$, and all weights ($w \in [0, 1]$) of the overall suitability score, from 0 to 100%, with the step of 10%. In the last table the relative value is normalized so that the highest value for each weight is always 100%. That shows the following final decision of the job selection problem: if the importance of the high value is in the range $40\% \leq w \leq 100\%$ the winner is NewJob2 and the stakeholder should quit the OldJob. However, if $0\% \leq w \leq 30\%$, then there are no sufficient reasons for taking the NewJob2. In such a case, the stakeholder can continue the job search and/or negotiate the monetary compensation with the NewJob2 employer to get better offer that would more convincingly outperform the other competitors. In such an analysis it is useful that the decision-maker considers the sensitivity curve shown in Fig. 17.9. For example, for any salary above 140% the overall satisfaction with the NewJob2 would be above 80%.[4] More results of the sensitivity analysis can be found in Chap. 18.

[4] It is useful to note that LSP.NT does not use for partial absorption the simplistic AH aggregator (Fig. 16.2). Instead of the arithmetic and harmonic means LSP.NT uses two adjustable weighted power means (Fig. 16.1) which offer wider range of desired penalty and reward combinations. Consequently, numerical results obtained by LSP.NT can be slightly (insignificantly) different from results obtained using the simple AH version of the partial absorption aggregator used in Fig. 17.6.

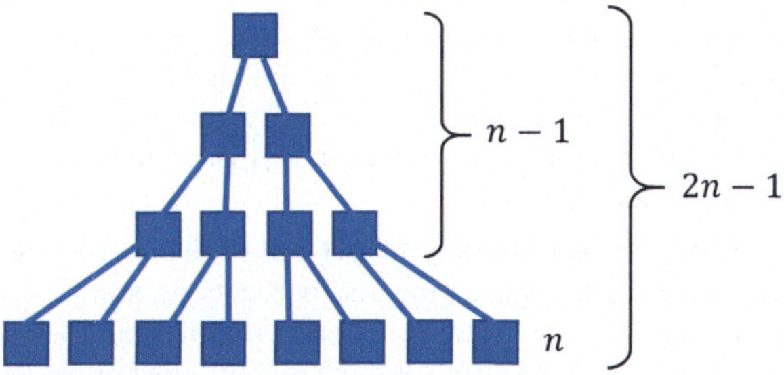

Fig. 17.10 A sample binary structure of an LSP criterion with n input suitability attributes

Effort Needed for Evaluation Based on the LSP Method

The highest LSP development effort is obtained if the LSP tree structure has the form of a binary tree, as shown in Fig. 17.10 where the number of input suitability attribute criteria is $N_{ac} = n$. The number of logic aggregators is $N_{la} = 1+2+4+\cdots+\frac{n}{2} = n-1$. In this case the root node is decomposed in two components and this process continues creating the number of suitability attributes $N_{sa} = 1+2+4+\cdots+n = 2n-1$. Suppose that the individual efforts (measured in time units) for creating an attribute criterion, creating a logic aggregator, and computing a suitability degree are respectively $E_{ac}, E_{la,}, E_{sa}$. So, the total effort of the LSP method is $E_{LSP} = nE_{ac} + (n-1)E_{la,} + (2n-1)E_{sa} = n(E_{ac} + E_{la} + 2E_{sa}) - E_{la} - E_{sa}$. The total effort is proportional to n, and the time complexity of the LSP method is $O(n)$.

The presented binary case is more complex than the real life LSP structures that are regularly not binary. E.g., the expression $y = \left(x_1^{1/3} \times x_2^{2/3}\right)^{0.3} \times \left(x_3^{3/7} \times x_4^{4/7}\right)^{0.7}$ includes three binary aggregators, and it can be considered a more complex structure than the equivalent single aggregator $y = x_1^{0.1}x_2^{0.2}x_3^{0.3}x_4^{0.4}$. So, in all cases, the LSP time complexity is never above $O(n)$. The linear time complexity is very important because it enables simple and systematic creation of real life LSP evaluation criteria where typically $20 < n < 600$, and the large numbers of attributes regularly reflect complex professional evaluation problems.

LSP Method and Intuitive Human Reasoning

It is important to note that all steps of the LSP method are fully consistent with observable properties of intuitive human reasoning. That includes the use of parameters taken from adjustable coordinates of human commonsense logical reasoning shown in Fig. 2.1, and the use of the standard model of human intuitive decision making shown in Fig. 6.1.

According to the decision-making model shown in Fig. 6.1, when humans compare alternatives (e.g., evaluate and compare alternative cars they consider buying), the first step is to define relevant attributes that must be taken into consideration (e.g., the power of engine, the maximum number of passengers, fuel consumption, color, the volume of trunk, etc.). There is no doubt that each attribute is individually evaluated according to a justifiable stakeholder's criterion (e.g., each car buyer knows the desired number of seats in a car, necessary for buyer's family), and for each alternative the stakeholder creates the percept of attribute suitability, which is a degree of satisfaction of the stakeholder's justifiable requirements. At the end of individual evaluation of suitability attributes, the intuitive evaluator has a set of suitability degree percepts of all attributes. Some of them can be high and some of them can below.

There is no doubt that the percept of the overall suitability of an evaluated object is obtained by aggregating the percepts of suitability of individual attributes and it is easy to see that the process of intuitive aggregation of suitability degrees is a logic process. First, it is obvious that all attribute suitability degrees are not equally important for a specific stakeholder: some of them can be very important (e.g., the power of engine) and some of them are less important (e.g., the power adjustment of rear mirrors). Next, human stakeholders most frequently require a high simultaneity of satisfaction of important attribute requirements. The degree of simultaneity is adjustable in the range from high (e.g. the car must simultaneously have the necessary number of seats, and the necessary power of engine, and all cars that fail one of these requirements will be rejected) to low (it is nice to have good speakers for front seats and good speakers for rear seats, but an attractive car will not be rejected because of the inferior sound in the rearmost row).

In many cases human suitability aggregation is based on substitutability (e.g., a home garage can be replaced by a shared garage or a good street parking). Mandatory and optional requirements are frequently encountered in human reasoning (e.g., for a professional cello player, it is mandatory that the volume of trunk is adequate to accommodate the case of the instrument, and good light and other trunk features may be optional). Sufficient requirements are also present in human intuitive reasoning (e.g., the all-wheel drive is sufficient to satisfy the buyer who prefers the front wheel drive). All presented intuitive logic aggregators are regularly interpreted in a verbalized form that is summarized in Table 17.7. The presented logic aggregators correspond to Fig. 17.5 and Table 17.4.

During the evaluation process, humans create the percept of overall suitability of the evaluated object or alternative. They then relate this percept to their percept of the object's or alternative's affordability. All human decision-makers prefer simultaneously good quality and low cost. There is no doubt that human decision-makers use conjunctive criteria for aggregation of suitability and affordability of each alternative, and there is no doubt that this aggregator must belong to soft computing with adjustable degree of simultaneity and the adjustable degrees of importance of input arguments.

All components of the LSP method are strictly derived from observable properties of human decision making, which is based on evaluation and comparison of alternatives. Since decision making is human mental activity, the credibility of any decision method must be based on provable consistency with observable properties of human reasoning. The credibility of the LSP method results is based on this consistency.

Table 17.7 Verbalized interpretation of seven basic types of logic aggregators

Verbalized interpretation of logic aggregation	Number of inputs	Logic aggregators
We must have ***all input requirements highly satisfied.*** If a single input is not satisfied (0), the result of aggregation is 0. The output value has high negative sensitivity to low input values. In the extreme case of CC, all inputs must be completely (perfectly) satisfied (1).	2 or more	Hard graded conjunctions: C, CP, CC, HC+, HC, HC-
We want to have ***most input requirements highly satisfied.*** A single 0 input can notably reduce the output value, but not to cause the result 0. The sensitivity to low input values is less than in the case of HC-, but greater than in the case of A. The negative impact of small input values is greater than the positive impact of large input values.	2 or more	Soft graded conjunctions: SC-, SC, SC+
We want to satisfy input requirements in a ***balanced*** way, so that simultaneity and substitutability requirements are ***equally present.*** The result of aggregation is 0 only if all inputs are 0, and 1 only if all inputs are 1.The negative impact of small input values is equal to the positive impact of large input values.	2 or more	Logic neutrality: A
We want to have at least ***some inputs highly satisfied.*** A single fully satisfied input (1) can notably increase the output value, but the output can be 1 only if all inputs are 1, and 0 only if all inputs are 0. The negative impact of small input values is less than the positive impact of large input values.	2 or more	Soft graded disjunctions: SD-, SD, SD+
It is enough to have ***any input highly satisfied.*** A single fully satisfied input (1) is ***sufficient*** to fully satisfy given requirements. The output value has high positive sensitivity to high input values. In the extreme case of DD the result of aggregation is 0 only if all inputs are 0 (in all other cases the output is 1).	2 or more	Hard graded disjunctions: HD-, HD, HD+ D, DP, DD
We have two inputs: *primary* and *secondary*. The primary input is ***necessary***: it ***must be at least partially satisfied.*** The satisfaction of the secondary input is ***optional*** (not necessary). The negative impact (penalty) caused by the optional value 0 is regularly greater than the positive impact (reward) caused by the highest optional value 1.	2	Conjunctive partial absorption: CPA
We have two inputs: *primary* and *secondary*. The primary input is ***sufficient*** to completely satisfy the given requirements. The satisfaction of the secondary input is ***optional*** (desired but not sufficient): it can only partially satisfy the given requirements. The negative impact (penalty) caused by the optional value 0 is regularly less than the positive impact (reward) caused by the maximum optional value 1.	2	Disjunctive partial absorption: DPA

Summary of the LSP Method Properties

The LSP method presented in this chapter is sufficiently simple and can be successfully used after a modest preparation of a decision-maker. Main benefits of the LSP method can be summarized as follows:

- The LSP method can be used for the suitability evaluation of a single object/candidate.

- The process of evaluation and comparison of $m > 1$ objects/alternatives is reduced to m evaluations of a single object (symbolically, $evaluate(m) = m \times evaluate(1)$).
- Evaluation is based on suitability attributes that are developed using a systematic suitability decomposition process.
- The number of input suitability attributes n can be large (e.g., the largest LSP evaluation projects so far had $n \approx 600$)
- The necessary effort for developing LSP criteria with n input attributes and using the LSP method is $O(n)$. LSP criteria are systematically developed in small and simple steps, and the corresponding effort is linearly proportional to n.
- Attribute criteria offer high precision in defining input suitability attribute requirements.
- Logic aggregation of suitability based on andness-directed graded logic functions offers ultimate flexibility in modeling logic requirements that can precisely reflect specific stakeholder's goals and requirements.
- LSP method provides a complete decision process for evaluation and selection of competitive objects, including both functional characteristics of evaluated objects, and financial aspects of decision making.
- LSP method is provably consistent with basic observable properties of human commonsense evaluation reasoning.
- The tree structure of suitability aggregation offers semantic identity of each aggregation node, providing interpretability and explainability of evaluation results.
- LSP method is supported by necessary software tools and successfully verified in a spectrum of industrial applications.

These properties show that the LSP method is particularly suitable for complex industrial applications. Using the basic LSP functionality presented in this chapter, and the presented mathematical models of aggregators, as well as the available LSP software support, it is possible to solve practical evaluation decision problems in variety of areas (see Chapter 19). In addition, it is possible to advance to more complex LSP techniques [DUJ18], such as optimization, sensitivity analysis, reliability analysis, missingness-tolerant aggregation, explainability, modeling of human cognitive processes, and refining of mathematical aspects of logic aggregation.

References

DUJ73c Dujmović, J. 1973. Mixed averaging by levels (MAL)—A system and computer evaluation method (in Serbo-Croatian). In *Proceedings of the Informatica Conference*, paper d28, Bled, Yugoslavia.

DUJ18 Dujmović, J. 2018. *Soft computing evaluation logic*. Wiley.

MIL70 Miller, J.R., III. 1970. *Professional decision-making*. New York: Praeger.

SEA17 SEAS Co. 2017. *LSP.NT—LSP method for evaluation over the Internet*. LSP.NT User Manual. SEAS.

SAA16 Saad, L. 2016. The "40-hour" workweek is actually longer - by seven hours. *Gallup*. http://www.gallup.com/poll/175286/hour-workweek-actually-longer-seven-hours.aspx

Chapter 18
Explainability and Confidence

The process of evaluation of one or more objects creates results that stakeholders must carefully investigate to verify their credibility and acceptability. Generally, during the evaluation process, the stakeholders need to verify that the evaluation criterion and the obtained results properly support the attainment of desired goals that the stakeholders identified at the beginning of the evaluation process. These goals were used to specify requirements that are built in the LSP criterion function and used to evaluate and compare the set of alternatives. In this chapter we expand the LSP method with techniques for explanation and justification of LSP criterion and evaluation results. Such an expanded method, based on [DUJ20, DUJ22a], is called "Explainable LSP" and in this chapter denoted XLSP.

In AI, explainability is a necessary component that supports automatically generated decisions and predictions [LES24]. In the case of using black box models like deep neural networks, it is difficult to understand and interpret the operation of complex AI models, and stakeholders cannot always accept decisions and other results with confidence. Fortunately, in the case of XLSP decision models, the explainability is an inherent property: all XLSP models are visible and understandable. Based on human-centric approach, decision models are consistent with observable human decision making, and consequently, explainability must be consistent with observable actions that humans use to explain their decisions.

In the area of XLSP decision making, it is useful to identify and separate the following two linked explainability problems:

(1) *Explainability of a specific XLSP criterion* that reflects the goals and requirements of a specific stakeholders.
(2) *Explainability of evaluation and comparison results*, obtained for a specific set of competitive objects/alternatives.

The first problem characterizes the initial phase of an evaluation project. The goal is to correctly (completely and accurately) specify the needs of a stakeholder/decision-maker. In decision engineering, the resulting verified evaluation criterion is given

J. Dujmović, *Graded Logic*, Intelligent Systems Reference Library 273,
https://doi.org/10.1007/978-3-031-88558-7_18

to potential competitors, to get the best offers, closest to the stakeholder's needs. The goal of the second problem is to provide verbal explanation of reasons for the obtained results of an evaluation project.

A typical process of explanation and justification of an XLSP criterion and its results consists of providing answers to various verification questions which for most evaluation projects include the following:

- What is the ranking of suitability attributes according to their overall importance? What requirements are mandatory, and what requirements are optional?
- What is the ranking of suitability attributes according to their contribution to the overall suitability of an object/alternative?
- What are the strengths and weaknesses of each object, and what are the range and ranking of suitability contributors?
- What is the potential for the improvement of individual objects?
- What are the principal discriminators that affect the comparison of two objects?
- What are the degrees of appropriateness of the costs of leading objects?

Explainability can be defined as the capability of a decision method to provide answers to the verification questions. The corresponding narrative can be automatically generated by a software tool called the *evaluation explanation report generator* or briefly the *verbalizer* of the XLSP method [DUJ22a].

High Contributors, Low Contributors, and the Concept of Concordance

The fundamental property of each attribute suitability degree is its relationship with other attribute suitability degreess: superiority, inferiority, or equilibrium. This relationship is like the situation where a new player joins an existing sports team. The existing team has an overall quality[1] denoted $X = x^* \in [0, 1]$. For example, consider a team that has n players. The quality of players is $x_1, \ldots, x_n, x_i \in [0, 1], i = 1, \ldots, n$. So, in a simplest case, the overall team quality could be $X = \frac{x_1+\cdots+x_n}{n} = x^*$. Let the quality of the new player be $x_{n+1} \in [0, 1]$. Now, we have a new overall team quality $X = \frac{x_1+\cdots+x_{n+1}}{n+1}$. If $x_{n+1} = x^*$, then the team quality remains unchanged:

$$X = \frac{x_1 + \cdots + x_n + x_{n+1}}{n+1} = \frac{x_1 + \cdots + x_n + (x_1 + \cdots + x_n)/n}{n+1}$$
$$= (x_1 + \cdots + x_n)\frac{1 + 1/n}{n+1} = x^*.$$

In this case, the new player is in *perfect concordance* with existing players and the quality of the team remains unchanged (x^*). Looking in the opposite direction, if such an average player leaves the team, then again, the team value remains unchanged. So, x^* is called the *concordance value*. If $x_{n+1} > x^*$. , then the new player can be considered a *high contributor*, causing an increase of the overall team quality. Similarly, if $x_{n+1} < x^*$, then the new player can be considered a *low contributor*,,

[1] X is the percept of quality, i.e., the degree of truth of the statement "the team has the highest overall value."

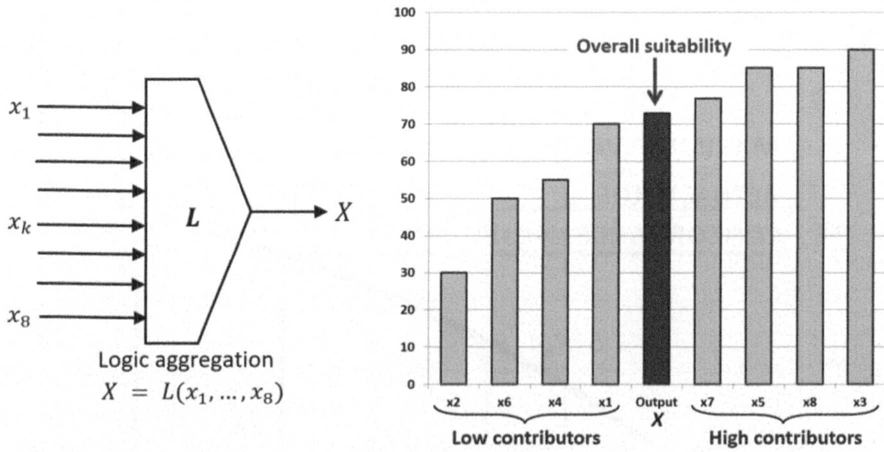

Fig. 18.1 Logic aggregation, low contributors, and high contributors

causing a decrease of the overall team quality. These relationships are exemplified in Fig. 18.1, where the overall quality/suitability X is obtained using a logic aggregation of input quality/suitability scores $x_1, \ldots, x_8$. Of course, some of inputs are below the overall value X and some of inputs are above the overall value.

For idempotent aggregators, let us now examine the properties of the kth input attribute shown in Fig. 18.2. The individual impact of attribute x_k can be analyzed using the *attribute impact function* $X = S(x_k) = L(x_1, \ldots, x_k, \ldots, x_n), k \in \{1, \ldots, n\}$. The function $S(x_k)$ provides the resulting aggregated suitability X as a function of the selected argument x_k, assuming constant values of all other arguments. Four characteristic shapes of the attribute impact functions are shown in Fig. 18.3. Each attribute has its individual concordance (or neutral) value x_k^*. The concordance value is the result of evaluation based on remaining $n - 1$ arguments (the value of team without the player k). According to Fig. 18.2, for each argument x_k there are four characteristic values visible on the $S(x_k)$ curve:

- The minimum value $X_k^{\min} = S(0)$.
- The actual value $X = S(x_k)$.
- The concordance value $S\left(x_k^*\right) = x_k^*$.
- The maximum value $X_k^{\max} = S(1)$.

If $S(0) = 0$, as in Fig. 18.2, then the annihilator 0 is supported and the corresponding attribute is a hard input in the analyzed aggregation structure. If $x_k = x_k^*$, then the aggregated suitability $S(x_k)$ has the same value x_k^*. That is the intersection of line $X = x_k$ and the attribute impact curve $X = S(x_k)$ shown in Figs. 18.2 and 18.3. Consequently, for each input argument x_k we can find the corresponding concordance value x_k^* as the fixed point of aggregation function, i.e., as the solution of the following equation:

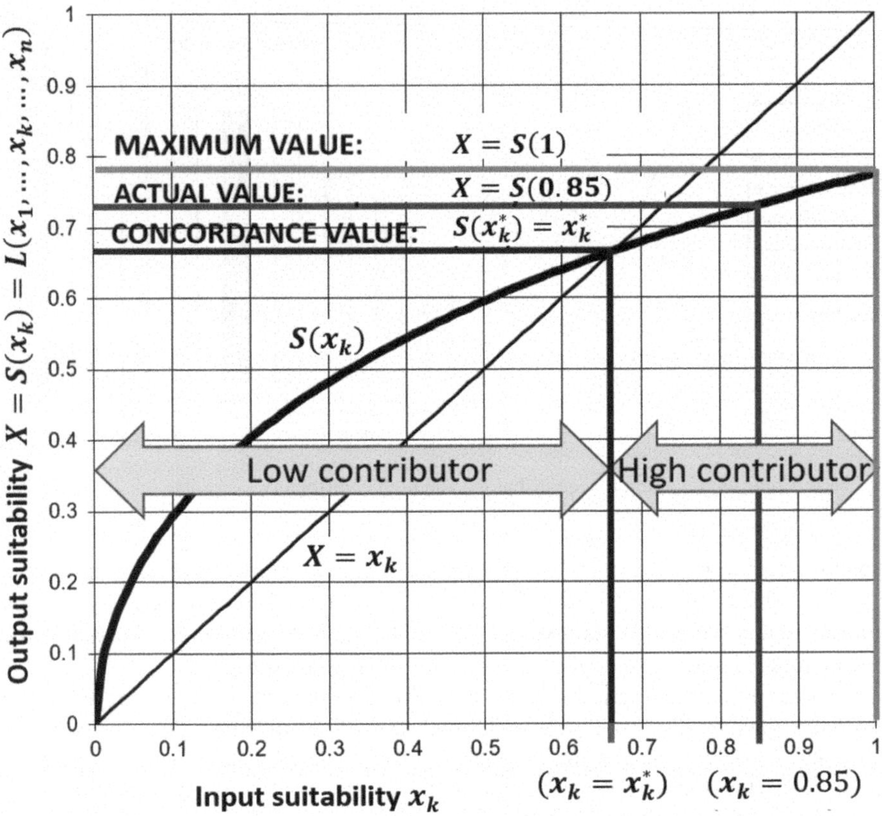

Fig. 18.2 The concepts of concordance value, actual value, minimum and maximum values, low contributors, high contributors, and a typical impact function of a hard attribute in the case of idempotent aggregators

$$x_k^* = L\big(x_1, \dots, x_k^*, \dots, x_n\big), k \in \{1, \dots, n\}.$$

According to the fixed-point iteration concept [POL07], the solution of equation $x = f(x)$ can be obtained using the convergent iteration $x_j = f\big(x_{j-1}\big), j = 1, 2, \dots$ and the initial value $0 < x_0 < 1$. The result of this iteration is the concordance value $\lim_{j \to +\infty} x_j = x^* = f(x^*)$. Here f denotes a nondecreasing function, and x_0 is an arbitrary initial value (e.g., $x_0 = 0.5$). Since L is a nondecreasing function of x_k^*, the concordance value can be computed as shown in Fig. 18.4.

Sensitivity curves (e.g., those shown in Figs. 17.9, 18.5, 19.8, 19.9, and 19.10) are functions of input attributes and their shapes depend on attribute criteria (such as those shown in Figs. 6.4 and 17.3); consequently, they can have arbitrary shapes. As opposed to that, each idempotent aggregation structure L: $[0, 1]^n \to [0, 1]$ is a GL propositional calculus formula, where arguments are degrees of truth, and

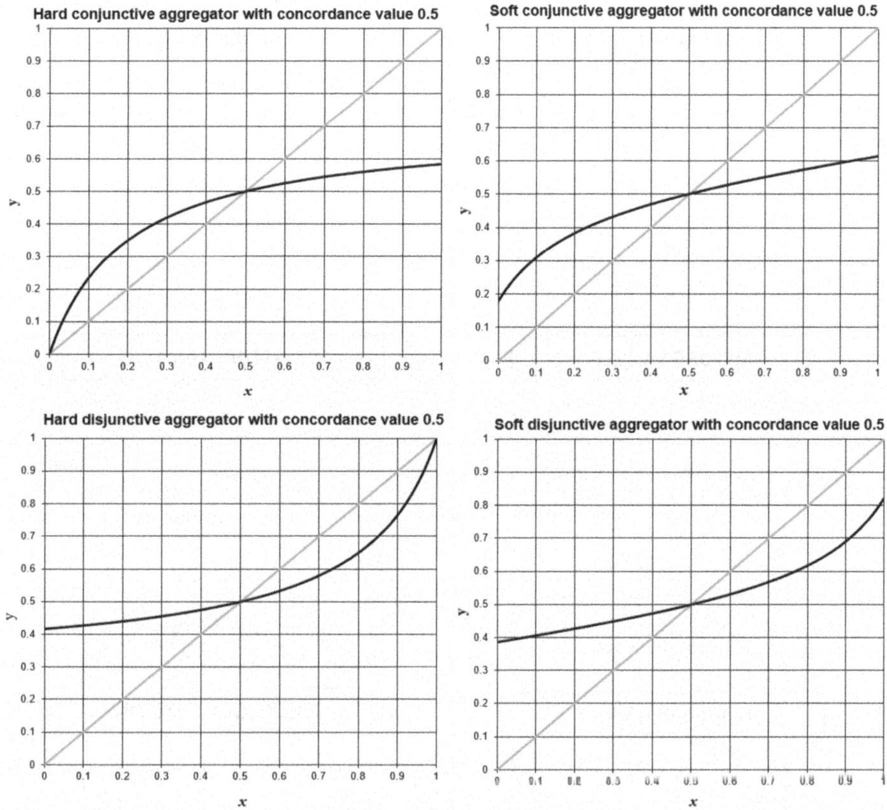

Fig. 18.3 Typical forms of idempotent hard and soft conjunctive and disjunctive aggregators

Fig. 18.4 A simple C+ + function for computing the fixed point of a nondecreasing function f

```
double FixedPoint(double (* f)(double))
{    double x, eps = 0.0000001;
     x = 0.5;
     do x = f(x);
     while (fabs(x - f(x)) > eps);
     return x;
}
```

L is nondecreasing in each argument.[2] As a general case, let us consider aggregation based on the quasi-arithmetic mean $y = F^{-1}\left(\sum_{i=1}^{n} W_i F(x_i)\right), 0 < W_i < 1, \sum_{i=1}^{n} W_i = 1, n > 1, k \in \{1, \ldots, n\}$. As shown in [DUJ12a, DUJ18], the solution of equation $x_k = F^{-1}\left(\sum_{i=1}^{n} W_i F(x_i)\right)$ is $x_k^* = F^{-1}\left((1 - W_k)^{-1} \sum_{i=1, i \neq k}^{n} W_i F(x_i)\right)$.

---

[2] The overall suitability of an object cannot decrease when the suitability of an input argument increases.

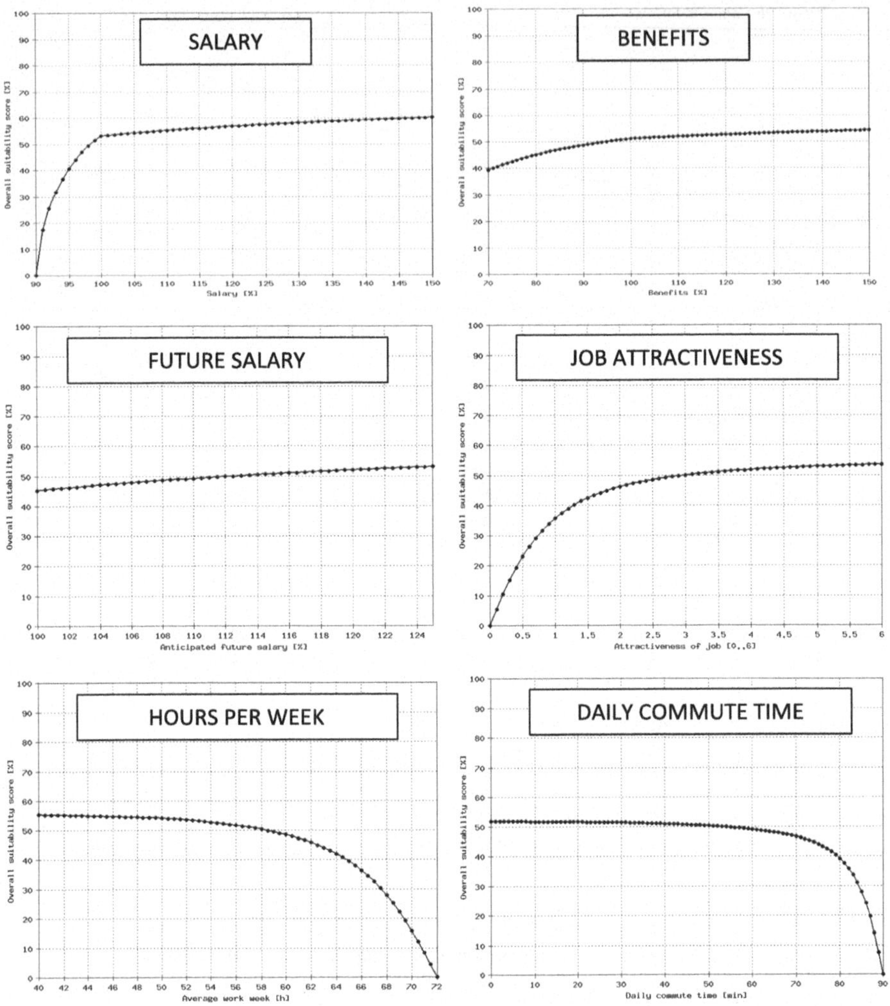

Fig. 18.5 Sensitivity analysis of job selection criterion for the middle point ($c = 50\%$)

The resulting value x_k^* is an aggregated indicator of the collective quality of all inputs other than x_k. Thus, x_k^* is the *neutral concordance value* that is in perfect balance with the aggregated value of the remaining $n - 1$ inputs: if $x_k = x_k^*$, then $F^{-1}\left(\sum_{i=1}^{n} W_iF(x_i)\right) = x_k^*$. If $x_k > x_k^*$, then $F^{-1}\left(\sum_{i=1}^{n'} W_iF(x_i)\right) \geq x_k^*$ and if $x_k < x_k^*$, then $F^{-1}\left(\sum_{i=1}^{n} W_iF(x_i)\right) \leq x_k^*$.

Using the analogy with sports teams, if the quality of a new player equals the previous average quality of a team, then the new average team quality remains unchanged. Otherwise, the new player can increase or decrease the overall team quality, as shown in Fig. 18.2. If $x_k < x_k^*$, then the attribute k is called a *low* *contributor*, and if $x_k > x_k^*$ (as the sample value $x_k = 0.85$ shown in Fig. 18.2), then the attribute k is called a *high contributor*. High contributors and low contributors (Fig. 18.1) explain the strengths and weaknesses of each evaluated object/alternative and indicate areas where improvements would be the most beneficial. This classification of attributes contributes to the explainability of results.

According to the $X = S(x_k)$ curve in Fig. 18.2, it is easy to see that if $x_k > x_k^*$ then $x_k^* < X < x_k$. Likewise, if $x_k < x_k^*$ then $x_k < X < x_k^*$. In other words, the equivalent conditions for high contributor are either $x_k > x_k^*$ (visible from the $S(x_k)$ curve), or $x_k > X$ (visible from the histogram in Fig. 18.1). The equivalent conditions for low contributor are either $x_k < x_k^*$, or $x_k < X$. This holds for all idempotent aggregation structures. Such aggregation structures are organized as a superposition of means; consequently, they are idempotent [BUL03].

The concordance value x_k^* is a fundamental indicator that explains the relationships of the kth attribute and $n - 1$ other attributes of an analyzed object. . As shown in Table 18.1, the concordance value is necessary for definitions of various other explainability indicators. Of course, the concordance values reflect the properties of specific evaluated object, and they are used for explainability of evaluation and comparison results.

Explainability Indicators

The *evaluation explanation report* (EER) is based on explainability indicators presented in Table 18.1. These indicators have clear and intuitively acceptable interpretation and hold for all idempotent aggregation structures. In the case of using nonidempotent aggregators, such as CP and DP, in unlikely extreme cases, it is possible that all contributors are high ($X \le x_{(1)} \le \cdots \le x_{(n)}$), or low ($x_{(1)} \le \cdots \le x_{(n)} \le X$). In all cases, the graphs of attribute sensitivity functions are useful to understand and explain the overall importance of aggregators. Sensitivity functions for the job selection problem introduced in Chap. 17 (Fig. 17.6) and constant input suitability $c = 0.5$ (obtained for selected values in the value column) are shown in Fig. 18.5.

The first part of EER is independent of evaluated objects and devoted to an explanation of properties of the XLSP evaluation criterion. It contains three components: (1) the presentation of the suitability attribute tree (exemplified in Fig. 17.2), (2) the classification and explanation of the type of attributes shown in Table 18.2, and (3) the ranking of overall importance of suitability attributes, exemplified in Table 18.3, which is based on suitability curves from Fig. 18.5. The column $r_k = X_k^{\max} - X_k^{\min}, k = 1, \ldots, 6$ denotes the overall importance of individual attributes. The attributes where $X_k^{\min} = 0$ denote mandatory requirements. For optional attributes we have $X_k^{min} > 0$. The overall importance can be normalized: $R_k[\%] = 100 r_k / \max(r_1, \ldots, r_n)$. The resulting values $R_1, \ldots, R_n$ denote the relative importance and can be sorted $100 = R_{(1)} \ge \cdots \ge R_{(n)}$, as shown in Table 18.3.

The sensitivity curves in Fig. 18.5 and the relative importance ranking of attributes in Table 18.3 should be investigated by the stakeholder and decision-making team before the practical use of the XLSP criterion. The curves must have acceptable shapes, and the ranking of importance should be consistent with the stakeholder's

Table 18.1 A survey of explainability indicators for idempotent aggregation structures

Explainability indicator definition	Explainability indicator interpretation and use
$x_k \in [0,1], k \in \{1,\dots,n\}, x_i = c, i \neq k$ $X_k^{max}(c) = L(x_1,\dots,x_{k-1},1,x_{k+1},\dots,x_n)$ $X_k^{min}(c) = L(x_1,\dots,x_{k-1},0,x_{k+1},\dots,x_n)$ $r_k(c) = X_k^{max}(c) - X_k^{min}(c)$ $R_k(c) = 100r_k/\max(r_1,\dots,r_n)$, Ranking: $R_{(1)} \ge \dots \ge R_{(n)}$ $R_{(k)} \in \{R_1(c),\dots,R_n(c)\}$, $k = 1,\dots,n$	Overall importance of a suitability attribute. This indicator, $R_{(k)}$, reflects the general capability of attribute k to contribute to the overall suitability of an object. If all attributes have the constant value c (typically, $c = 0.5$), the overall suitability is $X = c$. If x_k=1, then $X = X_k^{max} > c$. Similarly, if $x_k = 0$ then $X = X_k^{min} < c$. The range $r_k = X_k^{max} - X_k^{min}$ reflects the overall importance of x_k. For easy comparison of attributes, r_k is normalized and sorted. This indicator reflects general properties of a criterion function.
$x_{(1)} \le x_{(2)} \le \dots \le x_{(m)} \le X \le$ $\le x_{(m+1)} \le \dots \le x_{(n-1)} \le x_{(n)}$ $x_{(k)} \in \{x_1,\dots,x_n\}, \quad k = 1,\dots,n$	Ranking of suitability attributes identifies *m low contributors* and *n-m high contributors*. In the evaluation explanation report, low contributors are indicated as inputs that need improvement.
$x_k^{\bullet} = L(x_1,\dots,x_k^{\bullet},\dots,x_n)$, $k = 1,\dots,n$ Ranking: $x_{(1)}^{\bullet} \le \dots \le x_{(n)}^{\bullet}$	Concordance value indicates the balanced neutral quality of an attribute. The value $x_k \gtrless x_k^{\bullet}$ makes that input a *high or low contributor*. The ranking of neutral values indicates the need to make more efforts to satisfy high neutral inputs.
$A_k = x_k - x_k^{\bullet}, \quad k = 1,\dots,n$ Ranking: $A_{(1)} \ge \dots \ge A_{(n)}$ $A_{(1)} > 0, \; A_{(n)} < 0$	Accomplishment: the difference between delivered suitability x_k and neutral (concordance) suitability $x_k^{\bullet}$. The improvement priority should be given to attributes with low (i.e., negative) accomplishment and high potential for improvement.
$C_k = X - x_k^{\bullet}, \quad k = 1,\dots,n$ Ranking: $C_{(1)} \ge \dots \ge C_{(n)}$ $C_{(1)} > 0, \; C_{(n)} < 0$	Contribution: a part of the overall suitability X that is separately contributed by x_k. The attribute that corresponds to $C_{(1)}$ is the *highest contributor*. Negative values of C_k indicate attributes that are low contributors and need to be improved. The *lowest contributor* is the attribute that corresponds to $C_{(n)}$.
$D_k^+ = X_k^{max} - X, \quad k = 1,\dots,n$ $X_k^{max} = L(x_1,\dots,x_{k-1},1,x_{k+1},\dots,x_n)$ $X = L(x_1,\dots,x_k,\dots,x_n)$	Total potential for improvement (or the output suitability increment) caused by increasing the input suitability of selected attribute from the current value x_k to the maximum value 1.
$D_k^- = X - X_k^{min}, \quad k = 1,\dots,n$ $X_k^{min} = L(x_1,\dots,x_{k-1},0,x_{k+1},\dots,x_n)$ $X = L(x_1,\dots,x_k,\dots,x_n)$	Output suitability decrement caused by decreasing input suitability from the current value to the minimum value 0. If $X_k^{min} = 0$, then the satisfaction of x_k is an important *mandatory* requirement. If $X_k^{min} > 0$, then x_k is an *optional* attribute.
$D_k = D_k^+ + D_k^- = X_k^{max} - X_k^{min}$, $k = 1,\dots,n$ Ranking: $D_{(1)} \ge \dots \ge D_{(n)}$	Total impact (or output suitability range) is an indicator of the output suitability change caused by changing the value of selected attribute of an object in the whole range from 0 to 1.
$P_k = \frac{X_k^{max} - X}{1 - x_k} = \frac{D_k^+}{1 - x_k}, \quad x_k < 1$ $P_k = 0, \; x_k = 1, \; k = 1,\dots,n$ Ranking: $P_{(1)} \le \dots \le P_{(n)}$	Power (or relative improvement potential) indicates the capability of an attribute to have *positive impact* and improve the overall suitability of an analyzed object. Note: highly accomplished attributes can have low values of P_k.
$B_k = x_k/x_k^{\bullet}, \quad k = 1,\dots,n$ Ranking: $B_{(1)} \le \dots \le B_{(n)}$	Balance of selected attribute: $B_k > 1$ indicates the quality above expectations; $B_k < 1$ indicates the quality below expectations; $B_k \approx 1$ denotes an average well-balanced (neutral) attribute.
$V = 100\sigma_B/\overline{B}$; ($\sigma$ denotes the standard deviation)	Imbalance of attributes: the coefficient of variation of all attribute balance values computed for an analyzed object.

expectations. In such a case the criterion can be accepted and practically used. Among unacceptable properties are attributes that have negligible relative importance, and in such cases, it is necessary to identify the reasons for unacceptable properties and correct the structure and parameters of the XLSP criterion.

The second part of EER depends on the properties of competitive objects. In the job selection example let us consider two available jobs denoted A (the old job from Fig. 17.9), and B (the new job 2 from Fig. 17.9). The next section of EER starts with

Table 18.2 Explanation of the type of input arguments

ID	Type	Explanation
1111	Mandatory	An increase of salary is the primary motive for the job change
1112	Optional	Benefits are less significant than the salary
112	Optional	The future can be risky and unpredictable
121	Mandatory	An attractive job can be equally desirable as a high salary
1221	Mandatory	An excessive work week is not acceptable
1222	Mandatory	Excessive commute time significantly reduces the positive effects of a new job

Table 18.3 Overall importance r_k, and the importance ranking of attributes for $c = 50\%$

ID #	Attribute	Value	X_k^{min} [%]	X_k^{max} [%]	r_k [%]	R_k [%]	Rank
1111	Salary	98.3333	0	60.13	60.13	100	1
1221	Hours per week	58.25	0	55.24	55.24	91.88	2
121	Job attractiveness	3	0	53.49	53.49	88.96	3
1222	Daily commute time	52.5	0	51.78	51.78	86.11	4
1112	Benefits	95	39.22	54.16	14.94	24.85	5
112	Future salary	112.5	45.30	53.08	7.78	12.94	6

Table 18.4 Properties of the old job computed by the EER program

Attribute	x	Contrib	x*	x-x*	X-x*	Xmin	Xmax	Impact	Dplus	Dminus	Balance	Power
1111: Offered salary	60.00	Low	78.72	-18.72	-11.59	0.00	89.39	89.39	22.25	67.13	0.76	0.556
1112: Offered benefits	60.00	Low	68.60	-8.60	-1.46	47.42	72.68	25.27	5.55	19.72	0.87	0.139
112 : Expected future salary	40.00	Low	70.88	-30.88	-3.75	60.53	73.69	13.16	6.56	6.61	0.56	0.109
121 : Job attractiveness	100.00	High	65.16	34.84	1.98	0.00	67.13	67.13	0.00	67.13	1.53	0.000
1221: Work week [h/week]	100.00	High	63.79	36.21	3.34	0.00	67.13	67.13	0.00	67.13	1.57	0.000
1222: Commute time [min/day]	100.00	High	66.22	33.78	0.92	0.00	67.13	67.13	0.00	67.13	1.51	0.000

ranking of suitability attributes and pointing at the strong and weak components of each object/alternative. The properties of the old job are summarized in Table 18.4 and compared with the new job summarized in Table 18.5. There are similarities between the analysis of XLSP criterion in Tables 18.3, 18.4 and 18.5. However, Table 18.3 is created for artificial constant suitability $c = 50\%$ for all inputs, and if $x_k = c, k = 1, \dots, n$, then also $x_k^* = c$. So, the concordance value has no applicability in that case. In Tables 18.4 and 18.5, all values correspond to the actual inputs of competitive objects. The columns in Tables 18.4 and 18.5 show the explanation indicators introduced in Table 18.1. The old job has the overall suitability $X \approx 67\%$ and the new job offers the improvement of 11%, yielding $X \approx 78\%$.

In the case of the old job the low contributors are the monetary compensation indicators: the salary, benefits, and the expected future salary. The column $x - x^*$ is called the *accomplishment*: it shows to what extent a specific attribute outperforms the average quality of other attributes. Negative accomplishment values of the old

Table 18.5 Properties of the new job computed by the EER program

Attribute	x	Contrib	x*	x-x*	X-x*	Xmin	Xmax	Impact	Dplus	Dminus	Balance	Power
1111: Offered salary	84.00	High	75.26	8.74	2.70	0.00	81.81	81.81	3.85	77.96	1.12	0.241
1112: Offered benefits	76.00	Low	78.13	-2.13	-0.17	63.93	79.63	15.70	1.67	14.03	0.97	0.070
112 : Expected future salary	100.00	High	76.81	23.19	1.14	70.91	77.96	7.05	0.00	7.05	1.30	0.000
121 : Job attractiveness	50.00	Low	87.34	-37.34	-9.38	0.00	88.83	88.83	10.87	77.96	0.57	0.217
1221: Work week [h/week]	100.00	High	72.30	27.70	5.65	0.00	77.96	77.96	0.00	77.96	1.38	0.000
1222: Commute time [min/day]	80.00	High	77.77	2.23	0.19	0.00	79.23	79.23	1.27	77.96	1.03	0.064

job show attributes that have insufficient performance and need improvement. The concordance value x^* denotes the collective overall value of all attributes different from the specific attribute that has the concordance value x^*. If $X = x^*$, then the specific attribute has the value that is the same as the collective value of all other attributes. If we now increase the value of that attribute above its concordance value, $x > x^*$, then the positive accomplishment $x - x^*$ will cause an increase of the overall suitability $X > x^*$. Consequently, the difference $X - x^*$ is the individual *contribution* of that specific attribute. The negative contribution $-$ 11.59% of the salary of the old job shows that the low salary is the primary problem of the old job, responsible for decrementing the collective suitability of other attributes by 11.59%. Negative contributions quantitatively explain the effects caused by the low contributors. After increasing the salary by the new job, the new salary becomes the positive contributor, and the primary problem of the new job is now the low job attractiveness.

The columns denoted *Xmin, Xmax* and *Impact* in Tables 18.4 and 18.5 are equivalent to columns $X_k^{\min}$, $X_k^{\max}$, and $r_k = X_k^{\max} - X_k^{\min}$ in Table 18.3 except that the importance r_k corresponds to constant input suitability values while the *impact* (defined as the difference $X_{\max} - X_{\min}$) corresponds to the actual suitability values of the analyzed object and represents the overall suitability range of a specific attribute.

The column denoted Dplus is rather important because it shows the *potential for improvement* of specific attribute, i.e., the increment of the overall suitability X that can be achieved by increasing the suitability of specific attribute to its maximum value 1. In the case of the old job, the highest potential for improvement has the salary, which also has the highest impact. After the significant salary increase provided by the new job, the highest potential for improvement has the job attractiveness. The potential for improvement is sometimes related to increasing the cost of the new object (e.g., the increased cost of living in the location of the new job). In such a case, it is reasonable to invest in improving those attributes that provide the highest benefit per invested monetary unit.

The column of *balance* $B = x/x^*$ is an auxiliary indicator that shows those components that are better than others ($B > 1$) and those that are worse than others ($B < 1$). Generally, it is not convenient that evaluated objects are imbalanced in the sense that they include both very good and very poor components. The coefficient of variation (V) of the balance values is an *indicator of imbalance* and for the old job the imbalance is $V = 36.31\%$ and for the new job the imbalance is reduced to $V = 24.67\%$.

In most evaluation projects, decision-makers are interested in improving the evaluated alternatives/objects. Of course, all suitability attributes don't have the same power to improve the overall suitability of the evaluated object. Therefore, the potential to improve the overall suitability is like the first derivative of the overall suitability with respect to the suitability of a specific attribute. If the increase of an attribute from x to 1 causes the increase of the overall suitability from X to X^{max}, then the *potential for improvement* (the *power of attribute*) can be defined as $P = (X^{max} - X)/(1 - x)$. Expectedly, in the case of the analyzed old job, the highest potential for improvement has the salary. In the case of the new job, where the salary is improved, the highest potential for improvement has again the salary and very close is the job attractiveness.

The presented explainability indicators and their presentation in the context of EER provide information that complements and explains the results of an evaluation-based decision project. In addition, the explainability indicators are useful in situations where it is possible to negotiate the changes/improvements if available objects/ alternatives. More details and examples can be found in [DUJ20, DUJ22a].

Evaluation of Confidence

At the end of the decision-making process based on the evaluation of alternatives, the stakeholder organization examines the overall suitability of each alternative, along with explanations that justify the decision outcomes. In the case of a single object, a typical question the stakeholder organization might ask is, "How reliable is this result, and what degree of confidence can we have if we use it?" In the case of two or more objects, where the result indicates that object A outperforms object B (denoted A $\succ$ B), a typical stakeholder question is, "What is the degree of confidence in this ordering, and is it possible that B $\succ$ A?" These questions focus on confidence, a graded human percept necessary in both natural commonsense reasoning and complex professional decision-making.

In most professional XLSP decision problems, stakeholders, domain experts, and decision engineers form a cohesive, synchronized team, building confidence step-by step throughout all phases of the evaluation project. As a result, the issue of insufficient confidence does not arise, and final decisions are convincingly explained and justified. Naturally, this does not preclude research interest in the less common cases where high confidence is not automatically achieved. In this section, we survey selected confidence evaluation methods available in the literature.

There are three main sources of limited confidence:

- Limited accuracy of the structure and parameters of evaluation criteria.
- The problem of missing attribute data.
- The problem of data veracity.

The limited accuracy of evaluation criteria parameters is due to polygonal approximations of attribute criteria, the discretization of andness, and imprecision in selecting importance weights for logic aggregators. The reliability of evaluation results can be investigated by simulating criteria with random parameters, as proposed in [FAN04, DUJ04a, DUJ18].

The problem of missing data arises when one or more input attributes are unavailable (either unknown for acceptable reasons or intentionally hidden), while other attributes have favorable values, making it unjustified to disqualify objects with incomplete data. The penalty-controlled missingness-tolerant aggregation method is studied in [DUJ12a, DUJ18] and implemented in the LSP.NT tool [SEA17]. The basic idea behind the missingness-tolerant aggregation method is to use concordance values for missing attributes. If an attribute is justifiably missing, its unknown value can be replaced with the concordance value. In this way, the missing attribute is excluded from evaluation without diminishing the impact of other attributes. When this approach is not justified, missing data can be penalized, with an adjustable maximum penalty that, in extreme cases, can reduce the suitability of the missing attribute to zero.

The problem of data veracity is focused on the logic aggregation of data that have limited accuracy. It is studied in [DUJ20, DET21, DET22, DET23]. The basic idea of these papers is that each suitability degree should be accompanied by its sibling value, the confidence degree. These pairs can be interpreted as Zadeh's Z-numbers[3] [ZAD11] and the logic aggregation can be interpreted as a propositional logic calculus with Z-numbers [DET22].

The values of suitability attributes are sometimes not precisely known and fully reliable. For example, the job selection criterion includes imprecise attributes such as the anticipated future salary, and the attractiveness of the offered job. In both cases, the actual experience can differ from expectations. So, for most evaluations and decision results the stakeholders are interested to know the degree of confidence in their accuracy and applicability.

Similarly to suitability, the confidence is also an adjustable human percept. So, in practical evaluation projects, the result can be a pair (s, c), where $s \in [0, 1]$ denotes the overall degree of suitability of an evaluated object/alternative, and $c \in [0, 1]$ denotes the degree of confidence in accuracy and applicability of the value of s. So, we are interested in methods for computing the confidence degree c.

A simple approach to the evaluation of confidence, proposed in [DUJ20], is based on aggregating attribute confidence degrees using the same criterion used for aggregating suitability attributes. The idea behind this approach is that the precision of evaluation results is more affected by the precision of high importance/impact inputs than by the low importance/impact inputs. Likewise, if we can estimate the confidence degrees of input attributes, then the output confidence degree can be estimated by aggregating input confidence degrees using the same aggregation structure as the one used for aggregation of the suitability attributes. This is exemplified in Table 18.6 for the job evaluation problem. The offered salary has a fixed value, and we can have complete confidence that it will not change. The same holds for the commute time to known location and for fixed offered benefits. However, the expected future salary can change, and the estimated confidence is not higher than 75%. Job attractiveness will be known only after taking the job and experiencing the job conditions; so, it is considered not completely predictable, expressed as the confidence of 85%. There is also a minor insecurity about the work week (estimated confidence of 95%). The

[3] In the context of Graded Logic, Zadeh's concept of a Z-number Z is defined as an ordered pair $Z = (s, c)$, where s denotes a suitability degree, and c is the confidence, certainty, trust, or the strength of belief in the correctness of the given values.

Table 18.6 Estimated confidence degree of the job selection attributes

Attribute	Confidence degree
1111: Offered salary	1.00
1112: Offered benefits	1.00
112: Expected future salary	0.75
121: Job attractiveness	0.85
1221: Work week (h/week)	0.95
1222: Commute time (min/day)	1.00

result of aggregation of confidence degrees using the aggregation structure from Fig. 17.6 is the overall confidence 94.63%. This result shows that we can be rather confident that the overall suitability is a reliable result, but the positive or negative variations of approximately 5% should be tolerated.

Conjunctive Fusion of Suitability and Confidence: The Preference Degree

At the end of evaluation projects with Z-grades we know the overall suitability s and the corresponding confidence grade c. Suppose that a Z-grade for object O_i is $Z_i = (s_i, c_i)$, and a Z-grade for object O_j is $Z_j = \left(s_j, c_j\right)$. The expected next problem is the comparison of O_i and O_j. We may prefer O_i (denoted $O_i \succ O_j$) or to prefer O_j (denoted $O_i \prec O_j$). In rare special cases we may consider that O_i and O_j are equivalent (denoted $O_i \approx O_j$).

All stakeholders/decision-makers want simultaneously high suitability and high confidence. The preference $O_i \succ O_j$. can be easily assigned in the following cases of full dominance:

$$s_i > s_j, c_i \geq c_j \Rightarrow O_i \succ O_j$$
$$s_i = s_j, c_i > c_j \Rightarrow O_i \succ O_j.$$

However, the case $s_i > s_j, c_i < c_j$ remains problematic. If the confidence c_i is rather low, then the condition $s_i > s_j$ might be insufficient to claim $O_i \succ O_j$. In other words, decision-makers need enough confidence to accept the conclusion that $s_i > s_j$ is sufficient to claim $O_i \succ O_j$. One simple way would be to introduce the minimum threshold value $c_{\min}$ so that

$$s_i > s_j, \quad 0 < c_{\min} \leq c_i < c_j \Rightarrow O_i \succ O_j$$

Unfortunately, selecting $c_{\min}$ is not easy because the selected value depends on the difference $s_i - s_j$. Indeed, if the difference is very small ($s_i - s_j \ll 1$), then the threshold value $c_{\min}$ must be greater than in the case of large differences. In addition, in comparison problems, we are not only interested in conclusion $O_i \succ O_j$, but we also need to know how strong that preference is, considering both s and c values.

To solve the problem of quantitative comparison of competitive objects, we could use a legitimate requirement for simultaneous presence of high suitability and high confidence. In other words, we might conjunctively aggregate the suitability and confidence, creating the preference degree $p_i = \lambda(s_i, c_i)$, so that $p_i > p_j \Rightarrow O_i \succ O_j$. The preference degree can be defined as a "reduced (or effective) suitability," based on the following concepts:

- The preference degree cannot be greater than the computed suitability degree.
- For confidence close to 1, the preference equals the suitability degree.
- For lower confidence, the preference should be below the suitability.

These concepts have a convenient property that the preference has the same semantic identity (role and meaning for decision-maker) as the suitability based on the LSP model. The hard partial conjunction is an obvious candidate for the λ function. If the selected andness satisfies $\alpha < 1$, then the resulting aggregator will be a mean where $\min(s_i, c_i) < p_i$ yielding $s_i < p_i$ in all cases where $s_i < c_i$. This is not a "reduced suitability," that is our initial goal. Therefore, we must use hyper-conjunction where $\alpha > 1$. The simplest aggregator in that domain is the product t-norm:

$$p_i = s_i c_i, \quad \alpha = 1.25.$$

The resulting reduction is not adjustable, and it can be considered too strict or insufficiently strict. Thus, it might be more convenient to make the reduction adjustable by introducing the weight of confidence w, yielding the following *weighted product form*:

$$p_i = s_i c_i^w, \quad w > 0.$$

If $w > 1$, then the reduction of suitability for insufficient confidence is stronger than for the product t-norm. Similarly, for $0 < w < 1$, the reduction of suitability will be weaker than for the product t-norm. If for $s_i = \overline{s_i}$, and $c_i = \overline{c_i}$ we want to have $p_i = \overline{p_i}$, then the weight should be adjusted as $w = (\log \overline{p_i} - \log \overline{s_i}) / \log \overline{c_i}$.

The ideal (maximum) confidence is 1, but human estimates are never perfect, and we might select the threshold confidence $\underline{c} \leq 1$ so that for $c \geq \underline{c}$ the preference is equal to suitability (e.g., based on ideas introduced in [MIL56] the threshold confidence could be $\underline{c} > 6/7$). Thus, the confidence can be replaced by the following reduction function:

$$r(c) = \begin{cases} c/\underline{c}, & c < \underline{c} \\ 1, & c \geq \underline{c} \end{cases}$$

Now, we can generalize the weighted product form as follows:

$$\begin{aligned} p_i &= s_i r(c_i)^w \\ p_j &= s_j r(c_j)^w \\ p_i > p_j &\Rightarrow O_i \succ O_j \end{aligned}$$

The reduction factor $r(c)$ is the penalty if the confidence is below the threshold confidence c^*. According to this model, the preference and suitability have the same identity (*the degree of truth of a value statement*). If $\underline{c} = 1$ then the generalized model reduces to the weighted product form. The percentage of dominance of the superior object is

$$D = 100\big(\max(p_i/p_j,\ p_j/p_i) - 1\big)[\%].$$

This indicator of dominance is based on both suitability and confidence. Consequently, it is sufficiently reliable for making decisions about the status of the best alternative. This method is simpler than the similar approach based on weighted product $p_i = s_i^{2-w} c_i^w$, where the weight w is used to adjust the desired relationship of importance between suitability and confidence.

Experience in practical evaluation projects shows that suitability plays the primary role in the decision-making process. In most cases, the analysis of confidence is not necessary because all principal participants are satisfied with suitability results and believe that the ranking of competitors is realistic and not affected by insufficient confidence of some attributes. The calculation of confidence is justified only in cases where the evaluation process is performed under noticeable data veracity concerns, and the criterion function does not include attributes having so low confidence that it is more appropriate to exclude such insufficiently known attributes from the evaluation criterion.

For theoretical reasons, we can briefly note situations where preference can be defined as a figure-of-merit indicator, i.e., as an anonymous numeric score, and not as the reduced suitability. In such cases, the preference can be based on hard partial conjunction or on conjunctive partial absorption [DUJ18]. In the case of uniform hard partial conjunction, we have an andness in the range $0.75 < \alpha < 1$. Assuming that the suitability and confidence are equally important, the aggregator can be implemented as the harmonic mean $p_i = 2s_i c_i/(s_i + c_i)$, where $\alpha = 0.77$. In addition, some analysts might decide to assign different weights to suitability and confidence, yielding $p_i = 1/(w_s/s_i + w_c/c_i)$, $w_s + w_c = 1$. Such idempotent λ functions yield $p_i = s_i$ if $s_i = c_i$ and $p_i > s_i$ if $c_i > s_i$. These models yield a questionable result: if the suitability is low and the confidence is very high, that can significantly improve the status of evaluated object. E.g., if $c_i = 1$, then $p_i > s_i$, and p_i becomes undeserved "rewarded suitability." This obviously creates a contradiction: the high confidence means that s_i is a correct value, and the correct value should not be modified.

A similar figure-of-merit approach could be based on the conjunctive partial absorption function (Chap. 16). In this case, the assumption is that suitability s is a mandatory requirement ($s > 0$) and the confidence is an optional desirable input. Of course, in all evaluation problems we assume that the confidence is positive (either known or unknown). This form of the λ function is parameterized with two parameters: penalty and reward. The maximum penalty $P_{\max}$ is obtained from the condition $\lambda(s, 0) = s - P_{\max}$. Similarly, the maximum reward R_{max} is obtained from the condition $\lambda(s, 1) = s + R_{\max}$. Of course, both $P_{\max}$ and $R_{\max}$ are theoretical extremes that depend on s, and not the consequence of expectation to have the confidence 0 or 1. There are various forms of penalty and reward [DUJ14]. After selecting desired penalty and reward the priority degree can be computed as $p = C(A(s, c), s)$ where A denotes a weighted arithmetic mean and C denotes a weighted hard partial conjunction; all weights are computed from the desired penalty and reward. The value of the penalty should be moderate, and the value of the reward should be rather small.

In many cases, the confidence of an input can be unknown. In such cases we do not suggest unconditional abandoning of the computation of the overall confidence and cancelling the subsequent preference evaluation, because we can compute the overall confidence using the penalty-controlled missingness-tolerant aggregation method [DUJ12a]. According to this method, the confidence aggregation function replaces the missing confidence inputs with neutral concordance values that yield the aggregated confidence affected only by the existing input confidence degrees. If the missingness of an input is intentional (e.g., interpreted as an attempt to hide inconvenient properties), then it might be penalized. The penalization is possible and yields a reduction of the resulting overall confidence degree. Such approach contributes to the credibility of results, enabling the preference analysis in all situations.

References

BUL03 Bullen, P.S. 2003 and 2010. *Handbook of means and their inequalities*, Kluwer.

DET21 De Tré, G., and J. Dujmović. 2021. Dealing with data veracity in multiple criteria handling: An LSP-based sibling approach. In *Proceedings of the International Conference on Flexible Query Answering Systems, FQAS 2021*, Bratislava, Slovakia 82–96. ed. T. Andreasen et al. *Flexible query answering systems*,. Springer LNAI 12871.

DET22 De Tré, G., M. Peelman, and J. Dujmović. Logic operators and sibling aggregators for Z-grades. In *Proceedings of the IPMU 2022*. Ciucci Davide , Inés Couso, Jesús Medina, Dominik Ślęzak, Davide Petturiti, Bernadette Bouchon-Meunier and Ronald R. Yager. *Information Processing and Management of Uncertainty in Knowledge-Based Systems. 19th International Conference, IPMU 2022 Milan, Italy, July 11–15, 2022 Proceedings, Part II* 572–583. Springer, Communications in Computer and Information Science 1602.

DET23 De Tré, G., M. Peelman, and J. Dujmović. 2023. Logic reasoning under data veracity concerns. *International Journal of Approximate Reasoning*, 161: 108977https://doi.org/10.1016/j.ijar.2023.108977.

DUJ04a Dujmović, J., and W.Y. Fang. 2004. An empirical analysis of assessment errors for weights and andness in LSP criteria. In *Modeling Decisions for Artificial Intelligence* 139–150. ed. V. Torra, and Y. Narukawa. Springer LNAI 3131.

DUJ12a Dujmović, J. 2012. The problem of missing data in LSP aggregation. In *Proceeding of the 14th International Conference on Information Processing and Management of Uncertainty in Knowledge-Based Systems, IPMU, 2012*. ed. S. Greco et al. *Advances in Computational Intelligence, IPMU 2012, Part III, CCIS 299*, 336–346, Springer

DUJ14 Dujmović, J. 2014. Interpolative GCD aggregators. In *Proceedings of the 2014 IEEE World Congress on Computational Intelligence, Beijing, China July 6–11* 1778–1785, https://doi.org/10.1109/FUZZ-IEEE.2014.6891841

DUJ18 Dujmović, J. 2018. *Soft computing evaluation logic*. Wiley.

DUJ20 Dujmović, J. 2020. Interpretability and explainability of LSP evaluation criteria. In: Proceedings of the 2020 IEEE world congress on computational intelligence, 978-1-7281-6932-3/20, paper F-22042, July 2020.

DUJ22a Dujmović J., and W.L. Allen III 2022. . Explainable decision-making for water quality protection. *Information,* 13 (12): 551. https://doi.org/10.3390/info13120551

FAN04 Fang, W-Y. 2004. *Analysis of reliability of LSP models.* San Francisco State University, Computer Science Dept., Report SFSU-CS-TR-04.06

LES24 Leslie, D., C. Rincón, M. Briggs, A. Perini, S. Jayadeva, A. Borda, S.J. Bennett, C. Burr, M. Aitken, S. Mahomed, J. Wong, M. Waller, and C. Fischer. 2024. AI *Explainability in practice.* The Alan Turing Institute

MIL56 Miller, G.A. 1956. The magical number seven, plus or minus two: some limits on our capacity for processing information. *The Psychological Review,* 63: 81–97 (1956).

POL07 Polyanin, A.D., and A.V. Manzhirov. 2007. *Handbook of mathematics for engineers and scientists.* Chapman & Hall/CRC

SEA17 SEAS Co. 2017. *LSP.NT—LSP method for evaluation over the Internet.* LSP.NT User Manual. SEAS.

ZAD11 Zadeh, L.A. 2011. A note on Z-numbers. *Information Sciences,* 8 (3): 2923–2932

Chapter 19
Graded Logic and LSP in Action: Liver Transplantation Priority Evaluation

The primary applicability of Graded Logic is in the context of the LSP method, where GL is used to create logic aggregation structures. In this role, GL was used for evaluation and optimization of computer systems [DUJ73c, DUJ80, DUJ03a], evaluation and selection of software systems [SU87, DUJ03, DUJ06a], protection of endangered species [DRA21, DRA23], suitability maps for decision making in agriculture [MON16a, MON16b], optimum urban expansion analysis [DRA17], ecology [ALL11, MES17], urban densification [SHE21a, SHE21b] water quality protection [DUJ21f, DUJ22a], pollution analysis [REB14, REB16], urban suitability maps [PIT10], real estate [DUJ13], medical disability evaluation [DUJ08a], vaccination priority evaluation [DUJ21a], and other areas.

Evaluation and decision problems are very frequent in healthcare and medicine. Clinical decision making is the area of highest interest [SCH86, STR89, GRO00, SOX13, HUN14]. However, the delicate decision problems of organ transplantation prioritization exist only in research literature and surveys [CHO12, ABA21, SHI20, KIA12, EUR20]. Proposed methods for prioritization include AHP [COO90, SAA80] and fuzzy inference systems [TAK22, YAG94, KLI95], but most prioritization systems used in practice are based on simple additive scoring.

The liver transplantation priority evaluation is a problem in the category of professional decision making. Generally, the professional decision-making is based on a decision engineering process that specifies stakeholder's goals and requirements, creates or selects a set of alternatives that can satisfy the requirements and realize the goals, and then evaluates and compares the alternatives. Before accepting and implementing the most suitable alternative, it is necessary to clearly explain reasons that justify the proposed decision.

Each professional decision-making project requires the simultaneous presence and collaboration of three participants: the *stakeholder* interested in selecting the most suitable object/alternative, a *domain expert* familiar with functioning of evaluated objects, and a *decision engineer* familiar with all aspects of the LSP decision method and supporting software. In the area of organ transplantation, the stakeholder

J. Dujmović, *Graded Logic*, Intelligent Systems Reference Library 273,
https://doi.org/10.1007/978-3-031-88558-7_19

is regularly a *government* interested in the development of legal support for public health protection, the domain experts are *medical doctors* specialized in the organ transplantation area, and evaluated alternatives are *patients* waiting for a specific organ transplantation. The decision problem is the selection of the most appropriate patient so that the allocation of available organ(s) provides the best possible healthcare effects.

In this chapter we want to expand the areas of GL/LSP applicability by showing the use of GL and LSP method in the decision problems of prioritizing patients for liver transplantation. We selected this problem as a typical professional-grade complex and delicate decision problem that cannot be reduced to intuitive fast thinking of an individual decision maker, or to problems that can be solved without sufficient domain expertise.[1] Obviously, to determine the priority of patients who are in critical situation and need an organ transplantation, we need a sophisticated decision methodology that can justifiably combine many relevant decision components. Such decision problems are primary candidates for using GL and the LSP method.

Basic Concepts of Organ Transplantation Decision-Making

Diseases of vital human organs sometimes reach the end-stage where the only life-saving solution is the organ transplantation from a donor to a recipient. Successful transplantations of individual organs from deceased and/or living donors include kidney, liver, pancreas, kidney/pancreas, heart, lung, heart/lung, intestine, thymus, and uterus [WIK24, RIC21]. Modern transplantations also include functional units (VCA, Vascularized Composite Allograft), like a hand or face, where the transplantation includes a combination of tissues from a donor to a recipient. Various tissues that can be donated and transplanted include bone marrow, bone, skin, tendons, heart valves, cornea, cartilage, blood vessels and nerves. Finally, xenotransplantation is a very promising experimental treatment where organs can be taken from genetically engineered animals (pigs). In this chapter, we are interested only in decision aspects of prioritizing patients for organ transplantation, and our analysis will be focused on the decision problem of liver transplantation.

The scarcity of viable transplantation organs is a critical issue in organ transplantation: the demand far exceeds the available supply worldwide. Consequently, the decision problem of organ transplantation priority evaluation for patients who are waiting for organs is both delicate and complex. We can identify six aspects of this problem: (1) medical, (2) social, (3) ethical, (4) economic, (5) political, and (6) decision engineering. Below, we shortly explain these aspects.

1) *Medical factors* of organ transplantation are naturally the most important[2] and must be carefully investigated and included in each transplantation priority evaluation problem. Medical aspects are generally different for each organ.

[1] Domain expertise for liver transplantation was provided by Neil Mehta, MD, Division of Gastroenterology, University of California San Francisco School of Medicine. This chapter expands the work presented in [DUJ25].

[2] Of course, they are the most important, but they are not the only factors that affect the transplantation priority decisions.

The common priority factors for all organ transplantations are the medical urgency of transplantation, the donor/recipient compatibility, and the expected medical benefits of transplantation. Unsurprisingly, these criteria should be simultaneously satisfied and require aggregation using an appropriate graded conjunction.

(2) *Social factors* affect all human activities, and there is no proof that organ transplantation decision-making should be an exception. Each organ recipient interacts with society through his/her social role and responsibility, family circumstances and responsibilities, and contacts and relationships with the nearest residential and other environment. In most cases, social factors should not be neglected but considered a significant contributor to the patient's priority for organ transplantation. Unfortunately, neglecting social factors in the process of organ allocation is frequent, and primarily motivated by political goals of hiding the natural impact of wealth, privilege, and political power.

(3) *Ethical factors* should not be neglected because they reflect the fundamental ethical values of each country, such as reducing health inequities and development and support of ethical policies and regulations that avoid exploitation and promote justice and fairness of organ allocation decisions. On the organ recipient's side, ethical factors assume responsible adherence to medical recommendations, justifying the organ allocation decision.

(4) *Economic factors* include organized efforts to maximize the number of successful organ transplantations, as well as efforts to combat illegal organ trade. In some cases, economic factors are combined with ethical factors, e.g., in efforts to prevent access to organ transplantation from being determined by wealth, privilege, or political power.

(5) *Political factors* are visible in the transparency and formal public control of the process of creating policies and regulations for organ transplantation, as well as in openness and clarity of the process of recipient selection and organ allocation. The frequent use of simple additive scoring for decision making can be considered primarily a political factor because the ultimate simplicity is a prerequisite for including general public in the process of organ allocation decision-making.

(6) *Decision engineering aspects* investigate relationships between the existing organ transplantation priority scoring practice, and the state-of-the-art status in multiple criteria decision methods. According to [OPT24], the organ allocation decisions can be based on adding five or eight strictly medical scores. These scores are carefully selected and fully justified, but nevertheless it is obvious that such decision models do not benefit from the sophisticated methodology that decision engineering can currently offer.

The social, ethical, economic, and political factors include important components that should not be neglected because they naturally exist everywhere in real life. Therefore, we believe that organ allocation decisions should not be based only on medical priority factors. Generally, various social and ethical priority factors should also be included, making them explicitly visible and quantifiable. This approach

1 PRIORITY FACTORS FOR ORGAN TRANSPLANTATION
11 MEDICAL PRIORITY FACTORS
111 Medical urgency for organ transplantation
112 Medical prerequisites for organ transplantation
113 Expected recovery time after organ transplantation
114 Medical benefits of organ transplantation
12 SOCIAL AND ETHICAL PRIORITY FACTORS
121 Priority factors based on recipient's contribution to society
122 Priority factors that reflect expected social benefits for recipient
123 Priority factors that support ethical aspects of organ transplantation

Fig. 19.1 The main groups of organ transplantation priority factors

yields the initial hierarchical structure of organ transplantation priority factors shown in Fig. 19.1. Each of the seven basic factors (111, 112, 113, 114, 121, 122, 123) can be further decomposed until we reach the level of transplantation priority attributes. The presented initial structure in Fig. 19.1 is general, and it can be customized for each specific organ. In this chapter we will develop the priority attribute tree customized for the case of liver transplantation.

Organization of Organ Procurement, Distribution, and Allocation

Each country has policies and rules for organ procurement, distribution, and allocation. These policies have both similarities and differences. From a general perspective of decision-support systems, the organ transplantation prioritization problem can be defined for a database that contains records of N patients who wait for a specific organ. Each patient (potential recipient) is characterized by n priority attributes that include medical components (organ-dependent and organ-independent) and non-medical (social and ethical) components. Let k denote the number of attributes that characterize the donor organ and the donor/recipient compatibility. Therefore, there are $n - k$ attributes that characterize each patient independently of the unknown donor organ that can become available in the future (in Fig. 19.4, we have n = 35, k = 8, and 27 attributes characterize the patient and patient's environment).

We assume that each country has a specialized *organ procurement, distribution and allocation system* (OPDAS). OPDAS maintains the patient database with $n - k$ periodically updated attributes that characterize each patient who waits for a specific organ. For each organ, OPDAS should provide an official priority criterion function $p : \mathbb{R}^n \rightarrow [0, 100\%]$ which uses the complete list of priority attributes $a_1, \ldots, a_n$ for computing the *Transplantation Priority Score* (TPS), $P = p(a_1, \ldots, a_n)$, as a numeric indicator of patient transplantation priority. The TPS should reflect expected overall benefits of specific organ transplantation, and thus it must be a comprehensive indicator based on all relevant medical, social, ethical, and other necessary inputs. Then, TPS can be used with justification and confidence for offering available organs to patients according to decreasing TPS values. By definition, TPS is the degree of truth of statement "*the patient has the highest organ transplantation priority*."

When an organ becomes available in a specific geographic location, the donor organ characteristics are used to complete the priority attribute list of each patient.

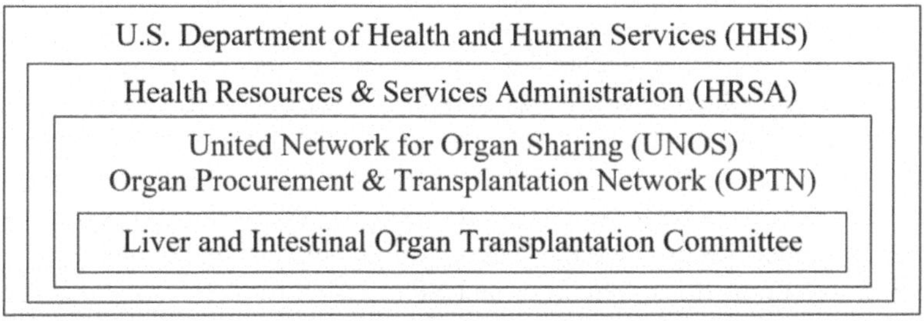

Fig. 19.2 The OPDAS structure for liver and intestinal organ allocation currently used by the U.S. federal government

The next step is the transplantation priority processing of *all* N patients, computing their priority degrees, and creating the ranking $P_{(1)} \geq P_{(2)} \geq \cdots \geq P_{(N)}$, $P_{(i)} \in [0, 100\%]$, $i = 1, \ldots, N$. . OPDAS can create this initial ranking as the first urgent step of prioritization. In the next step, the ranking can be used to determine which transplantation center, and which patient(s) will be selected for the transplantation procedure. That can include additional decision criteria.

The described general prioritization decision problem is different for each specific country and for each specific organ. The goal of this chapter is to propose a general set of priority attributes $a_1, \ldots, a_n$ and a general structure of the LSP priority criterion function p. The proposed general criterion should have a flexible logic structure, so that it can be customized for most countries and organs. Our goal is to show how to customize the general criterion in the case of computing the TPS of liver transplantation patients.

Organ procurement, distribution, and allocation activities are regularly controlled by governmental and specialized professional organizations [WIK24]. Each country has a specific approach to organ transplantation problems and specific quantitative criteria that are used for selecting the highest priority recipient for each specific organ [ABA21, WIK24a, SHI20, EUR20, KIA18, OPT24, OPT24a, KIM21].

In the case of liver transplantation, the current U.S. OPDAS structure is shown in Fig. 19.2. At the top level is the U.S. Department of Health and Human Services (HHS), which is a branch of the U.S. federal government. HHS has various agencies and one of them is HRSA (Health Resources & Services Administration). Next levels include the United Network of Organ Sharing (UNOS) [WIK24a], the Organ Procurement and Transplantation Network (OPTN), and the Liver and Intestinal Organ Transplantation Committee that develops methodology for computing the patient priority scores (TPS).

Organ donors can either be living or deceased. In many cases an organ becomes available for donation and transplantation after donor's death, and that is the most common form of liver transplantation. If that happens at an unpredictable time (e.g. after traffic accidents), then the allocation decision must be made rather quickly because each organ can be viable for transplantation only for a limited time.[3] In all cases, biological properties of available organ must be quickly investigated, and the organ must be transplanted as soon as possible. Thus, it is necessary to match the organ with the whole database of potential recipients and create the TPS ranking of patients for that specific organ. Since each organ creates a different ranking of patients, it is necessary to have a carefully prepared TPS criterion and a decision support system that can produce the ranking of patients fast and at any time.

The first step in the development of an LSP criterion for computing the TPS is to create the set of necessary and sufficient suitability attributes. Then, we need attribute criteria and a logic aggregation structure for computing the TPS. The final step is the verification of the created LSP criterion. This procedure is presented below.

Attribute Tree and Attribute Criteria for Liver Transplantation Priority Evaluation

All organ transplantations have common properties, and solutions in a specific organ area can be partially reused in other areas. Thus, we are primarily interested in the general structure of TPS criteria. Based on this goal, an attribute tree for prioritizing liver transplantation patients, developed using LSP.NT [SEA17], is proposed in Fig. 19.3. The corresponding list of 35 priority attributes (14 medical and 21 social and ethical) is shown in Fig. 19.4. A detailed description of all attribute criteria is shown in Table 19.1, and the LSP logic aggregation structure, based on UGCD.15 aggregators, is presented in Fig. 19.5.[4]

The selection of attributes shown in the attribute tree (Fig. 19.3), the parameters of attribute criteria shown in Table 19.1, and the weights and andness of aggregators (Fig. 19.5) reflect the collective expertise accumulated in a liver transplantation community. In this situation LSP criteria have only predictive character. After a liver has been transplanted into a specific patient, it is not possible to know what would happen if that organ were transplanted into somebody else. Therefore, there is no historical data that could be used for reliable training of the whole logic aggregation structure or the whole LSP criterion. Of course, GCD aggregators can be used as preferential neurons and individually trained as introduced in [DUJ91] and shown in

[3] Unsurprisingly, organ transplantations are the area of very intensive medical research characterized by fast changes and frequent improvements. Among the most important research areas are organ preservations techniques that reduce the risk of ischemic injury (injury caused by the absence of blood flow) and extend the organ viability time [NIH19]. Modern machine perfusion methods increased the liver viability up to 27 h. The organ preservation time depends on the organ; for kidney it is currently up to 36 h and for heart and lungs 4 to 6 h. These values can increase in the future.

[4] The LSP.NT notation of aggregation structure can be explained in the case of aggregator #112, which is a hard conjunction HC + that aggregates the donor-recipient compatibility #1121 (relative importance 60%), the organ delivery time #1122 (relative importance 20%), and the organ recipient overall health status and comorbidities #1123 (relative importance 20%). The conjunctive partial absorption root aggregator #1 aggregates as mandatory inputs medical priority factors #11, and as optional inputs social and ethical priority factors #12, using the proposed penalty of 20% and the reward of 15%.

1 TRANSPLANTATION PRIORITY SCORE (TPS)
- **11 MEDICAL PRIORITY FACTORS FOR LIVER TRANSPLANTATION**
  - 111 Medical urgency for organ transplantation
    - 1111 Disease severity
      - 11111 MELD score for liver disease severity evaluation
      - 11112 MELD exceptions: complications of liver disease
    - 1112 Waiting time for organ transplantation
    - 1113 Expected time of life without organ transplantation
  - 112 Medical prerequisites for organ transplantation
    - 1121 Donor-recipient compatibility
      - 11211 Donor-recipient blood compatibility
        - 112111 Donor-recipient ABO compatibility
        - 112112 Donor-recipient HLA compatibility
      - 11212 Donor-recipient age compatibility
      - 11213 Donor-recipient organ size compatibility
        - 112131 Donor-recipient height compatibility
        - 112132 Donor-recipient weight compatibility
    - 1122 Organ delivery time
    - 1123 Organ recipient overall health status and comorbidities
  - 113 Expected recovery time after organ transplantation
  - 114 Medical benefits of organ transplantation
    - 1141 Expected time of life after organ transplantation
    - 1142 Utility: collective benefits of allocation optimization
- **12 SOCIAL AND ETHICAL PRIORITY FACTORS**
  - 121 Priority factors based on recipient's contribution to society
    - 1211 Governmental role/responsibility
    - 1212 Important professional status or responsibility
  - 122 Priority factors that reflect expected social benefits for recipient
    - 1221 Priority factors based on recipient's family and social support
      - 12211 Number of family members of the organ recipient
      - 12212 Number of dependents of the organ recipient
      - 12213 Social support network of the organ recipient
      - 12214 Parenthood status of the organ recipient
    - 1222 Priority factors based on organ recipient residential environment
      - 12221 Living in non-retirement homes
      - 12222 Living in retirement and nursing homes
      - 12223 Degree of urbanity in the patient residential area
      - 12224 Density of population in the patient residential area
  - 123 Priority factors that support ethical aspects of organ transplantation
    - 1231 Priority factors based on organ allocation policies and regulations
      - 12311 Donor selection policy
      - 12312 Fairness of the organ allocation process
      - 12313 Transparency of the organ allocation process
      - 12314 Accountability of the organ allocation organizations and individuals
      - 12315 Data protection, privacy, and security
      - 12316 Public feedback during the development of organ allocation policies
    - 1232 Priority factors based on reducing health inequities
      - 12321 Protection of disadvantaged groups
      - 12322 Protecting patients with low economic status
      - 12323 Educational, cultural, and language priority factors
      - 12324 Promoting frameworks that foster a fair society
    - 1233 Recipient's responsible adherence to medical recommendations

Fig. 19.3 An attribute tree for the liver transplantation priority evaluation

1. MELD score for liver disease severity evaluation
2. MELD exceptions: complications of liver disease
3. Waiting time for organ transplantation
4. Expected time of life without organ transplantation
5. Donor-recipient ABO compatibility
6. Donor-recipient HLA compatibility
7. Donor-recipient age compatibility
8. Donor-recipient height compatibility
9. Donor-recipient weight compatibility
10. Organ delivery time
11. Organ recipient overall health status and comorbidities
12. Expected recovery time after organ transplantation
13. Expected time of life after organ transplantation
14. Utility: collective benefits of allocation optimization

15. Governmental role/responsibility
16. Important professional status or responsibility
17. Number of family members of the organ recipient
18. Number of dependents of the organ recipient
19. Social support network of the organ recipient
20. Parenthood status of the organ recipient
21. Living in non-retirement homes
22. Living in retirement and nursing homes
23. Degree of urbanity in the patient residential area
24. Density of population in the patient residential area
25. Donor selection policy
26. Fairness of the organ allocation process
27. Transparency of the organ allocation process
28. Accountability of the organ allocation organizations and individuals
29. Data protection, privacy, and security
30. Public feedback during the development of organ allocation policies
31. Protection of disadvantaged groups
32. Protecting patients with low economic status
33. Educational, cultural, and language priority factors
34. Promoting frameworks that foster a fair society
35. Recipient's responsible adherence to medical recommendations

Fig. 19.4 The list of attributes for liver transplantation priority evaluation

Chaps. 10 and 13. That is used wherever a reliable training set is available. Generally, in all nontrivial LSP criteria, the indispensable expert opinions and experiences penetrate in LSP analytic models in a parallel way, affecting all decisions, including the selection of attributes, the definition of attribute criteria, and the justifiable selection of numerous parameters of aggregation structures.

The basic idea of our TPS criterion is that the organ transplantation priority decisions should be based on both medical and non-medical (social and ethical) components. That should be done in an explicit, justifiable, and quantitative way. The LSP method is a convenient framework to satisfy these requirements. Of course, our proposed criterion is an initial step in that direction, and the suggested values of

Table 19.1 Attribute criteria for liver transplantation

11111		**MELD score for liver disease severity evaluation**
Value	%	In the case of liver transplantation, the disease severity for adult patients is primarily assessed using the Model for End-Stage Liver Disease (MELD 3.0 score). MELD score is used as a predictor of the three-month mortality risk for patients without access to liver transplant. For children below 12 years of age, the Pediatric End-Stage Liver Disease (PELD) score is used instead of the MELD score, but the range [6, 40] remains the same.
0 6 40	0 10 100	

11112		**MELD exceptions: complications of liver disease**
Value	%	MELD 3.0 is the primary indicator of liver transplantation urgency. MELD exceptions are the liver disease severity scores based on complications of liver disease. OTPN identifies 14 diagnoses that can be presented to the Review Board with a request to be used instead of the MELD score. E.g., the hepatocellular carcinoma (HCC) currently accounts for 20-25% of liver transplants. The liver disease severity scores used as MELD exceptions can be evaluated using the following rating scale: 1=lowest, 2=very low, 3=low, 4=mid-low, 5=medium, 6=mid-high, 7=high, 8=very high, 9=highest.
0 9	0 100	

1112		**Waiting time for organ transplantation**
Value	%	According to donors1.org the average wait time for liver transplant in the USA is 11 months. This criterion prioritizes patients who spent longer time waiting for transplant. [The wait time is measured in months]
1 12	20 100	

1113		**Expected time of life without organ transplantation**
Value	%	This criterion prioritizes patients with a short expected lifespan without liver transplantation. The expected time of life without liver transplantation is evaluated by a team of experts in the range from Tmin to Tmax. The evaluation is based on the following ratio: R = 100*Tmean/T [%] , where Tmean = (Tmin+Tmax)/2 denotes the mean expected remaining lifetime, and T denotes the current recipient age.
5 50	100 20	

112111		**Donor-recipient ABO compatibility**
Value	%	Blood type compatibility is a critical factor in liver transplantation. Rh compatibility is not a primary concern and blood types A, B, AB, and O are evaluated as follows: 1 = compatible donor->recipient blood types for adults: A->A, O->A B->B, O->B A->AB, B->AB, AB->AB, O->AB (AB = universal recipient) O->O (O = universal donor) 0 = incompatible blood types (can be used in special cases)
0 1	0 100	

(continued)

Table 19.1 (continued)

<table>
<tr><th colspan="2">112112</th><th>Donor-recipient HLA compatibility</th></tr>
<tr><td>Value</td><td>%</td><td rowspan="2">Human leukocyte antigen compatibility (HLA) plays a crucial role in some organ transplantation. It is used in simultaneous liver-kidney transplantation. In strictly liver transplantation it is not used.
Evaluation is based on the donor->recipient HLA Matching Score that is normalized to range [0,100]. Higher scores describe higher HLA compatibility. In some cases, a partial match can be sufficient for a successful transplantation.</td></tr>
<tr><td>0
100</td><td>0
100</td></tr>
<tr><th colspan="2">11212</th><th>Donor-recipient age compatibility</th></tr>
<tr><td>Value</td><td>%</td><td rowspan="2">This criterion gives moderate credit to donors and recipients that are of similar age. Evaluation is based on the following age compatibility (AC) indicator:
AC = 100 |AD - AR| / (AD + AR) [%]
Inputs: AD = age of donor ; AR = age of recipient .</td></tr>
<tr><td>0
50</td><td>100
50</td></tr>
<tr><th colspan="2">112131</th><th>Donor-recipient height compatibility</th></tr>
<tr><td>Value</td><td>%</td><td rowspan="2">Evaluation is based on the following height compatibility (HC) indicator:
HC = 100 |HD - HR| / (HD + HR) [%] .
Inputs: HD = height of donor ; HR = height of recipient.
Low values of HC denote good match.</td></tr>
<tr><td>0
40</td><td>100
0</td></tr>
<tr><th colspan="2">112132</th><th>Donor-recipient weight compatibility</th></tr>
<tr><td>Value</td><td>%</td><td rowspan="2">Evaluation is based on the following weight compatibility (WC) indicator:
WC = 100 |WD - WR| / (WD + WR) [%] .
Inputs: WD = weight of donor ; WR = weight of recipient.
Low values of WC denote good match.</td></tr>
<tr><td>0
30</td><td>100
0</td></tr>
<tr><th colspan="2">1122</th><th>Organ delivery time</th></tr>
<tr><td>Value</td><td>%</td><td rowspan="2">Geographic location of recipient affects the organ delivery time.
Evaluated as the following relative organ delivery time:
Trel = 100*Tdel/Tmax [%].
Here we use the following variables:
Tdel = organ delivery time (transport time from current organ location to recipient's surgery room).
Tmax = maximum organ viability time.
In the case of liver transplantation, this criterion can be based on Tmax=24 h (or 27 h), or defined as the cold ischemia time criterion, e.g. {(0,100), (12,85), (24,0)}</td></tr>
<tr><td>0
100</td><td>100
0</td></tr>
<tr><th colspan="2">1123</th><th>Organ recipient overall health status and comorbidities</th></tr>
<tr><td>Value</td><td>%</td><td rowspan="2">A patient's overall health status includes kidney functions, cardiovascular health, and comorbidities that affect the likelihood of successful surgery and post-transplant recovery. The overall health rating scale is 0=very poor, 1=below average, 2=average, 3=good, 4=excellent</td></tr>
<tr><td>0
4</td><td>0
100</td></tr>
</table>

(continued)

Table 19.1 (continued)

113		**Expected recovery time after organ transplantation**
Value	%	Patient ability to recover after the organ transplantation can be measured as the expected patient recovery time (time necessary to reach a steady health state). Evaluation can be based on the expert opinion provided for a specific organ and a specific patient, using a rating scale, as follows: 0=very long, 1=long, 2=medium, 3=short, 4=very short. This criterion can be precisiated using 5 components: (1) hospital stay after the transplant surgery, (2) first few weeks after the surgery (immunosuppressive medications with side effects), (3) first few months, (4) first year, and (5) long-term. In the case of liver transplants the criterion can be {(3 month, 100), (12 months, 50)}.
0 4	10 100	

1141		**Expected time of life after organ transplantation**
Value	%	The main medical benefit is the patient's time of life after an organ transplantation. The life expectancy after an organ transplantation can vary in a wide range and is influenced by many factors, including age, healthy lifestyle, management of comorbidities, adherence to medication regimens, and other factors. In the case of liver transplantation, UNOS data for USA show the one-year survival rate around 85% to 90%, and the five-year survival rate around 70% to 80%. The proposed rating scale reflects this situation. The time is measured in years.
0 1 5 10	0 40 75 100	

1142		**Utility: collective benefits of allocation optimization**
Value	%	This criterion balances the limited supply of organs with the long-term survival and quality of life of patients. The contribution to optimizing the allocation of available donor organs is evaluated using the following contribution rating scale: 0=no contribution, 1=low contribution, 2=average, 3=high, and 4=very high contribution
0 4	0 100	

1211		**Governmental role/responsibility**
Value	%	This criterion gives credit to patients of any age who have high responsibility and roles that are indispensable for functioning of society and directly or indirectly affect the lives of most people. Evaluated using 20 rank steps based on the scoring criterion SCORE = 100-5(RANK-1) as follows: 1 = The top rank (score = 100%) 2 = Next to the top rank (score = 95%) ... 20 = Lowest rank (score = 5%) 21 = Not in this group (score = 0)
1 21	100 0	

(continued)

Table 19.1 (continued)

1212		**Important professional status or responsibility**
Value	%	This criterion gives credit to patients who have significant professional status and/or important roles in various areas, such as industry, medicine, business, entertainment, research, sports, education, military, police, etc. Evaluated using 20 rank steps based on the scoring criterion SCORE = 100-5(RANK-1) as follows: 1 = The top rank (score = 100%) 2 = Next to the top rank (score = 95%) ... 20 = Lowest rank (score = 5%) 21 = Not in this group (score = 0)
1	100	
21	0	

12211		**Number of family members of the organ recipient**
Value	%	This criterion gives credit to organ transplantation patients who are members of large families. All family members benefit if one of them (the organ recipient) significantly improves her/his health status. The family is defined in a traditional way, or as any strongly connected social unit, primarily those that share residential units and maintain permanent contacts.
1	20	
2	50	
4	80	
6	100	

12212		**Number of dependents of the organ recipient**
Value	%	Adult patients with more dependents deserve more help because positive health effects caused by organ transplantation are shared by all dependents. Dependents can be children or adults. Pediatric patients are assumed to be counted as dependents and can be evaluated using the same criterion.
0	0	
1	50	
3	100	

12213		**Social support network of the organ recipient**
Value	%	This criterion evaluates the number of members of the organ recipient social support network who can regularly assist with the post-transplant recovery process. This includes adult family members, friends, and other people (or organized support groups) who can help with adherence to follow-up care and medications.
0	0	
10	100	

12214		**Parenthood status of the organ recipient**
Value	%	This criterion gives credit to organ recipients who are parents. Measured as the number of recipient's children. For pediatric patients, this criterion can be based on the number of siblings and the rating 0 => 60 2 => 100
0	40	
3	100	

12221		**Living in non-retirement homes**
Value	%	This criterion gives credit to patients who live in groups with other people. Evaluated using the following (or similar) group living options: Life in family, group homes/dorms, or homeless shelters with the total of n members: SCORE = min[100, 20(n-1)]
0	0	
100	100	

(continued)

Table 19.1 (continued)

12222		**Living in retirement and nursing homes**
Value	%	This criterion gives credit to patients who live in retirement homes and assisted living institutions. Evaluation is based on the following rating scale: 0 = excellent conditions 1 = very good conditions 2 = average conditions 3 = poor conditions 4 = very poor conditions
0 4	0 100	

12223		**Degree of urbanity in the patient residential area**
Value	%	This criterion gives credit to patients who live in areas with a low degree of urbanity (e.g., in underdeveloped areas), regardless of the density of population. Such areas are assumed to have limited access to medical services. Evaluated using the following rating scale for the degree of urbanity: 0 = very low (isolated rural areas) 1 = low 2 = medium 3 = high 4 = very high
0 4	100 0	

12224		**Density of population in the patient residential area**
Value	%	This criterion gives credit to patients who live in areas with low density of population, which are presumably less developed and offer less health protection opportunities. Evaluated using the following rating scale for the density of population: 0 = very low density of population 1 = low 2 = medium 3 = high 4 = very high
0 4	100 0	

12311		**Donor selection policy**
Value	%	This criterion gives credit to social policies that contribute to the availability of donor organs. Evaluation based on the following scale: 1 = "opt-out" organ donation system: all individuals are presumed to consent to be organ donors unless they explicitly opt out. 2 = "opt-in" organ donation system: individuals who voluntarily agree to donate their organs register their consent. 3 = insufficient legal and organizational support for recruiting organ donors.
1 3	100 0	

12312		**Fairness of the organ allocation process**
Value	%	Ensuring that the organ procurement organizations are not biased and do not discriminate against individuals or groups based on religion, race, gender, socioeconomic status, or ethnicity. Existing policies and practice are evaluated for fairness using the following fairness rating scale: 0 = unknown, 1 = low, 2 = medium, 3 = high, 4 = very high
0 4	0 100	

(continued)

Table 19.1 (continued)

12313		**Transparency of the organ allocation process**
Value	%	Providing clear and publicly available explanations of allocation rules, decisions, and practice. Transparency builds trust and allows patients and their families to understand and challenge outcomes. Transparency is evaluated using the following transparency rating scale: 0 = unknown, 1 = low, 2 = medium, 3 = high, 4 = very high.
0 4	0 100	

12314		**Accountability of the organ allocation organizations and individuals**
Value	%	Holding organ procurement and allocation organizations and leading individuals responsible for the development of organ allocation policies and the practice of organ procurement and allocation (serving the maximum number of patients in equitable and efficient way). Accountability is evaluated using the following accountability rating scale: 0 = unknown, 1 = low, 2 = medium, 3 = high, 4 = very high.
0 4	0 100	

12315		**Data protection, privacy, and security**
Value	%	This criterion evaluates organized activity for protecting the privacy of organ transplantation recipients: (1) Implementing strong data protection measures. (2) Obtaining patient's informed consent for the collection, sharing, and use of personal information. Security includes the following activities: (1) Preventing unauthorized access to transplantation database (2) Preventing manipulation or malicious use of sensitive data Privacy and security are jointly evaluated using the following rating scale: 0 = unknown, 1 = low, 2 = medium, 3 = high, 4 = very high.
0 4	0 100	

12316		**Public feedback during the development of organ allocation policies**
Value	%	In the process of development of the organ transplantation policy and prioritization criteria, there is a period reserved for collecting and considering public feedback (comments, suggestions, concerns) from professionals, patients, caregivers, and general public. 0 = not available 1 = available
0 1	0 100	

(continued)

Table 19.1 (continued)

12321		Protection of disadvantaged groups
Value	%	This criterion evaluates organized protection of ethnic/social/racial minority groups disadvantaged in relation to organ transplantation and health care (based on religion, place of residence, race, ethnicity, education, occupation, etc.). That can include selected people from homeless shelters, people with intellectual or development disabilities, substance use disorder, and sexual and gender minorities. Disadvantaged groups protection is evaluated using the following rating scale: 0 = no protection, 1 = low, 2 = medium, 3 = high, 4 = very high.
0 4	0 100	

12322		Protecting patients with low economic status
Value	%	This criterion evaluates protection of people living in poverty with poor access to health care. People in this group usually also experience the problem in necessary time and travel cost to go to places offering health protection. Evaluation is based on the following rating scale: 0 = no protection, 1 = low, 2 = medium, 3 = high, 4 = very high.
0 4	0 100	

12323		Educational, cultural, and language priority factors
Value	%	This criterion evaluates protection of people with insufficient education, low socioeconomic status, family problems (no parents, single parents, family conflicts), cultural and language isolation, and similar factors that reduce health care equity. Health risks for people in this group are evaluated using the following rating scale: 0 = no risk, 1 = low, 2 = medium, 3 = high, 4 = very high.
0 4	0 100	

12324		Promoting equity, inclusivity, and diversity
Value	%	Evaluation of existing policies and regulations that explicitly promote equity, inclusivity, and diversity for organ transplantation recipients. Equity, inclusivity, and diversity are evaluated using the following rating scale: 0 = unknown, 1 = low, 2 = medium, 3 = high, 4 = very high.
0 4	0 100	

1233		Recipient's responsible adherence to medical recommendations
Value	%	This criterion evaluates the expected patient's contribution to success of organ transplantation. Potential recipients of transplanted organs are expected to be responsible and cooperating in health protection: (1) Actively engaged in maintaining their health (2) Adhering to medical recommendations (e.g., no substance abuse, healthy lifestyle, sustenance, etc.) Evaluation is based on the following expected responsibility rating scale: 0 = very low, 1 = low, 2 = medium, 3 = high, 4 = very high
0 4	0 100	

(continued)

Table 19.1 (continued)

12321		**Protection of disadvantaged groups**
Value	%	This criterion evaluates organized protection of ethnic/social/racial minority groups disadvantaged in relation to organ transplantation and health care (based on religion, place of residence, race, ethnicity, education, occupation, etc.). That can include selected people from homeless shelters, people with intellectual or development disabilities, substance use disorder, and sexual and gender minorities. Disadvantaged groups protection is evaluated using the following rating scale: 0 = no protection, 1 = low, 2 = medium, 3 = high, 4 = very high.
0 4	0 100	
12322		**Protecting patients with low economic status**
Value	%	This criterion evaluates protection of people living in poverty with poor access to health care. People in this group usually also experience the problem in necessary time and travel cost to go to places offering health protection. Evaluation is based on the following rating scale: 0 = no protection, 1 = low, 2 = medium, 3 = high, 4 = very high.
0 4	0 100	
12323		**Educational, cultural, and language priority factors**
Value	%	This criterion evaluates protection of people with insufficient education, low socioeconomic status, family problems (no parents, single parents, family conflicts), cultural and language isolation, and similar factors that reduce health care equity. Health risks for people in this group are evaluated using the following rating scale: 0 = no risk, 1 = low, 2 = medium, 3 = high, 4 = very high.
0 4	0 100	
12324		**Promoting frameworks that foster a fair society**
Value	%	Evaluation of existing policies and regulations that explicitly promote systems that ensure equal access and fair treatment for all. Frameworks that foster a fair society are evaluated using the following rating scale: 0 = unknown, 1 = low, 2 = medium, 3 = high, 4 = very high.
0 4	0 100	
1233		**Recipient's responsible adherence to medical recommendations**
Value	%	This criterion evaluates the expected patient's contribution to success of organ transplantation. Potential recipients of transplanted organs are expected to be responsible and cooperating in health protection: (1) Actively engaged in maintaining their health (2) Adhering to medical recommendations (e.g., no substance abuse, healthy lifestyle, sustenance, etc.) The evaluation is based on the following expected responsibility rating scale: 0 = very low, 1 = low, 2 = medium, 3 = high, 4 = very high
0 4	0 100	

parameters (weights, andness, penalty and reward) are expected to be adjusted to satisfy specific needs of a given liver transplantation system of a specific country. The proposed parameters in Table 19.1 and Fig. 19.5 reflect a general domain expertise in this area: medical attributes are a superset of attributes that are currently used in various countries, and social and ethical components are those that are currently frequently encountered in the social media and literature. However, it is easy to understand that each social/political/religious environment can expand or reduce the proposed attributes, the attribute criteria, and the aggregation structure. In particular,

```
1 [CPA: P=20%, R=15%] TRANSPLANTATION PRIORITY SCORE (TPS)
  11 [MAN; HC] MEDICAL PRIORITY FACTORS FOR LIVER TRANSPLANTATION
    111 [30%; HC-] Medical urgency for organ transplantation
      1111 [75%; D] Disease severity
        11111 [50%] MELD score for liver disease severity evaluation
        11112 [50%] MELD exceptions: complications of liver disease
      1112 [15%] Waiting time for organ transplantation
      1113 [10%] Expected time of life without organ transplantation
    112 [30%; HC+] Medical prerequisites for organ transplantation
      1121 [60%; HC] Donor-recipient compatibility
        11211 [60%; HC] Donor-recipient blood compatibility
          112111 [70%] Donor-recipient ABO compatibility
          112112 [30%] Donor-recipient HLA compatibility
        11212 [15%] Donor-recipient age compatibility
        11213 [25%; SC+] Donor-recipient organ size compatibility
          112131 [40%] Donor-recipient height compatibility
          112132 [60%] Donor-recipient weight compatibility
      1122 [20%] Organ delivery time
      1123 [20%] Organ recipient overall health status and comorbidities
    113 [20%] Expected recovery time after organ transplantation
    114 [20%; CPA: P=60%, R=50%] Medical benefits of organ transplantation
      1141 [MAN] Expected time of life after organ transplantation
      1142 [OPT] Utility: collective benefits of allocation optimization
  12 [OPT; SC] SOCIAL AND ETHICAL PRIORITY FACTORS
    121 [20%; D] Priority factors based on recipient's contribution to society
      1211 [50%] Governmental role/responsibility
      1212 [50%] Important professional status or responsibility
    122 [40%; SC-] Priority factors that reflect expected social benefits for recipient
      1221 [70%; SC-] Priority factors based on recipient's family and social support
        12211 [30%] Number of family members of the organ recipient
        12212 [30%] Number of dependents of the organ recipient
        12213 [20%] Social support network of the organ recipient
        12214 [20%] Parenthood status of the organ recipient
      1222 [30%; SD] Priority factors based on organ recipient residential environment
        12221 [30%] Living in non-retirement homes
        12222 [20%] Living in retirement and nursing homes
        12223 [30%] Degree of urbanity in the patient residential area
        12224 [20%] Density of population in the patient residential area
    123 [40%; SC] Priority factors that support ethical aspects of organ transplantation
      1231 [40%; SC-] Priority factors based on organ allocation policies and regulations
        12311 [18%] Donor selection policy
        12312 [18%] Fairness of the organ allocation process
        12313 [18%] Transparency of the organ allocation process
        12314 [18%] Accountability of the organ allocation organizations and individuals
        12315 [18%] Data protection, privacy, and security
        12316 [10%] Public feedback during the development of organ allocation policies
      1232 [35%; A] Priority factors based on reducing health inequities
        12321 [30%] Protection of disadvantaged groups
        12322 [30%] Protecting patients with low economic status
        12323 [20%] Educational, cultural, and language priority factors
        12324 [20%] Promoting frameworks that foster a fair society
      1233 [25%] Recipient's responsible adherence to medical recommendations
```

Fig. 19.5 Logic aggregation structure of the TPS criterion function the social and ethical criteria are a sensitive area. In many countries, such criteria are not included in the organ transplantation priority decision making, and decision making is based (at least theoretically) only on medical priority factors that give equal opportunities to all patients regardless of their social status. To suggest an evolutionary change of the current practice, we propose that the satisfaction of medical criteria should be mandatory, and the satisfaction of non-medical criteria should be optional. In GL and the LSP method that can be achieved using the partial absorption function. These concepts are used in the LSP criterion presented in Figs. 19.3–19.5 and Table 19.1.

Attribute criteria listed in Fig. 19.4 are functions $g_i : \mathbb{R} \rightarrow [0, 1], i = 1, ..., n$. They use the attribute values $a_1, \ldots, a_n$ to create the attribute priority scores $x_i = g_i(a_i), i = 1, \ldots, n$. It is important to emphasize that all priority scores are degrees of truth: x_i is the degree of truth of statement "a_i *completely satisfies the requirements of the* i^{th} *attribute criterion.*" These functions are presented in Table 19.1 in vertex notation. For example, the criterion #1112 is the waiting time for organ transplantation. If $a_i \leq 1$ month then $x_i = 20\%$ and if $a_i \geq 12$ months then $x_i = 100\%$. Between these extreme values we use the linear interpolation $x_i[\%] = 20+80(a_i - 1)/11$; so, if $a_i = 6.5$ months, then the corresponding attribute priority factor is $x_i = 60\%$. The attribute priority scores are aggregated using the logic aggregation structure $L : [0, 1]^n \rightarrow [0, 1]$, shown in Fig. 19.5, to compute the TPS: $X = L(x_1, \ldots, x_n) = L(g_1(a_1), \ldots, g_n(a_n))$.

The aggregation structure shown in Fig. 19.5 aggregates the suitability of 35 input attributes listed in Fig. 19.4. Both the aggregation structure and the attribute criteria in Table 19.1 include a significant number of parameters. The selection of attributes and all LSP criterion parameters reflect the expert opinions accumulated in the liver transplantation community, combined with the expertise in GL and experience with LSP method and corresponding software tools. That combined expertise gives credibility to LSP logic structures and criteria. More precisely, the applications of GL and the LSP method assume the interaction of three participants:

- *Stakeholder*: an organization that is using the LSP method for decision making.
- *Evaluator*: a decision engineer familiar with GL, LSP, and LSP tools.
- *Domain expert*: the provider of expertise with the object of evaluation.

These three participants can be individuals, or organizations, or any combination of individuals and organizations. In all medical applications, domain experts are medical doctors.

LSP Criterion Tuning and Validation Process

Before using a proposed LSP criterion it is necessary to verify and adjust the criterion properties. The goal of validation is to confirm that the proposed criterion generates acceptable (explainable and justifiable) results. The first step is to define a test patient data and to compute the patient's priority score. This is presented in Fig. 19.6 where we investigate the medical part of the TPS criterion and its interaction with the social and ethical factors. A typical test patient is denoted P1, and Fig. 19.6 shows four variations of patient's attributes (P2, P3, P4, and P5). Test patients P2 and P4 are

Id	Attribute	P1	P2	P3	P4	P5
	Cost	1.0000	1.0000	1.0000	1.0000	1.0000
11111	MELD score for liver disease severity evaluation	35	35	35	35	35
11112	MELD exceptions: complications of liver disease	0	0	0	0	0
1112	Waiting time for organ transplantation	10	10	10	10	10
1113	Expected time of life without organ transplantation	10	10	10	10	10
112111	Donor-recipient ABO compatibility	1	1	1	1	1
112112	Donor-recipient HLA compatibility	80	80	80	80	80
11212	Donor-recipient age compatibility	10	10	10	10	10
112131	Donor-recipient height compatibility	5	5	5	5	5
112132	Donor-recipient weight compatibility	8	8	8	8	8
1122	Organ delivery time	10	10	10	10	50
1123	Organ recipient overall health status and comorbidities	3	3	3	3	3
113	Expected recovery time after organ transplantation	3	3	3	3	3
1141	Expected time of life after organ transplantation	8	8	8	8	8
1142	Utility: collective benefits of allocation optimization	3	3	3	3	3
12	SOCIAL AND ETHICAL PRIORITY FACTORS	70	0	*****	100	70

Fig. 19.6 A test patient P1 and four variations of his/her attributes

equal to P1 except that they have extreme contributions (0 and 100) from social and ethical factors. P3 denotes the same patient in the case of unknown contribution of social and ethical factors.[5] Finally, the test patient P5 is equal to the patient P1 in the case where the liver delivery time is increased from 10 to 50% of the organ viability time.

The computation of TPS for all variants of a typical test patient is shown in Fig. 19.7. Initially, P1 achieves $TPS = 80.28\%$. In the case of P2, without support of social and ethical factors, the TPS is reduced to 66.20%. In the case of P4, and the maximum support of social and ethical factors, the TPS value increases to 85.88%. Consequently, for a typical patient P1, the absolute impact range of social and ethical factors is $D_{soc} = 85.88 - 66.20 \cong 20\%$. A detailed presentation of the impact of social and ethical factors, in the full range of impact, is shown in Fig. 19.8. This impact range should be analyzed by the stakeholder organization, resulting in either acceptance or modification. In situations where this influence range must be modified, that can be easily achieved by changing/adjusting the penalty and reward parameters of the partial absorption aggregator 1 shown in Fig. 19.5.

The computation of TPS in the case of missing/unknown attribute values is shown for variant P3 in Fig. 19.7. The missing value is denoted with asterisks. If we don't know the value of social and ethical factors, then TPS should be affected only by medical priority factors. That is visible in Fig. 19.7 where $TPS = 82.67\%$ and that value is the aggregated priority score of the medical factors, without the impact of social and ethical components. Fig. 19.8

The investigation of nonlinear properties of the proposed criterion, for the selected typical patient, can be done using various forms of sensitivity analysis [DUJ18]. The

[5] LSP.NT supports criteria with unknown/missing attribute values [DUJ12a, SEA17, DUJ18]. In such cases the unknown attribute values are replaced by *concordance values* that are neutral in the sense that they neither increase nor decrease the PTS generated by other (known) attributes.

Id	Attribute	P1	P2	P3	P4	P5
1	ORGAN TRANSPLANTATION PRIORITY SCORE (TPS)	80.28	66.20	82.67	85.88	73.53
11	MEDICAL PRIORITY FACTORS FOR LIVER TRANSPLANTATION	82.67	82.67	82.67	82.67	74.34
111	Medical urgency for organ transplantation	86.98	86.98	86.98	86.98	86.98
112	Medical prerequisites for organ transplantation	83.53	83.53	83.53	83.53	62.59
114	Medical benefits of organ transplantation	81.28	81.28	81.28	81.28	81.28
1111	Disease severity	86.76	86.76	86.76	86.76	86.76
1121	Donor-recipient compatibility	86.63	86.63	86.63	86.63	86.63
11211	Donor-recipient blood compatibility	92.06	92.06	92.06	92.06	92.06
11213	Donor-recipient organ size compatibility	78.56	78.56	78.56	78.56	78.56
12	SOCIAL AND ETHICAL PRIORITY FACTORS	70.00	0.00	*****	100.00	70.00
1142	Utility: collective benefits of allocation optimization	75.00	75.00	75.00	75.00	75.00
1141	Expected time of life after organ transplantation	90.00	90.00	90.00	90.00	90.00
113	Expected recovery time after organ transplantation	77.50	77.50	77.50	77.50	77.50
1123	Organ recipient overall health status and comorbidities	75.00	75.00	75.00	75.00	75.00
1122	Organ delivery time	90.00	90.00	90.00	90.00	50.00
112132	Donor-recipient weight compatibility	73.33	73.33	73.33	73.33	73.33
112131	Donor-recipient height compatibility	87.50	87.50	87.50	87.50	87.50
11212	Donor-recipient age compatibility	83.33	83.33	83.33	83.33	83.33
112112	Donor-recipient HLA compatibility	80.00	80.00	80.00	80.00	80.00
112111	Donor-recipient ABO compatibility	100.00	100.00	100.00	100.00	100.00
1113	Expected time of life without organ transplantation	91.11	91.11	91.11	91.11	91.11
1112	Waiting time for organ transplantation	85.45	85.45	85.45	85.45	85.45
11112	MELD exceptions: complications of liver disease	0.00	0.00	0.00	0.00	0.00
11111	MELD score for liver disease severity evaluation	86.76	86.76	86.76	86.76	86.76

Fig. 19.7 Resulting TPS values for all variants of the test patient P1

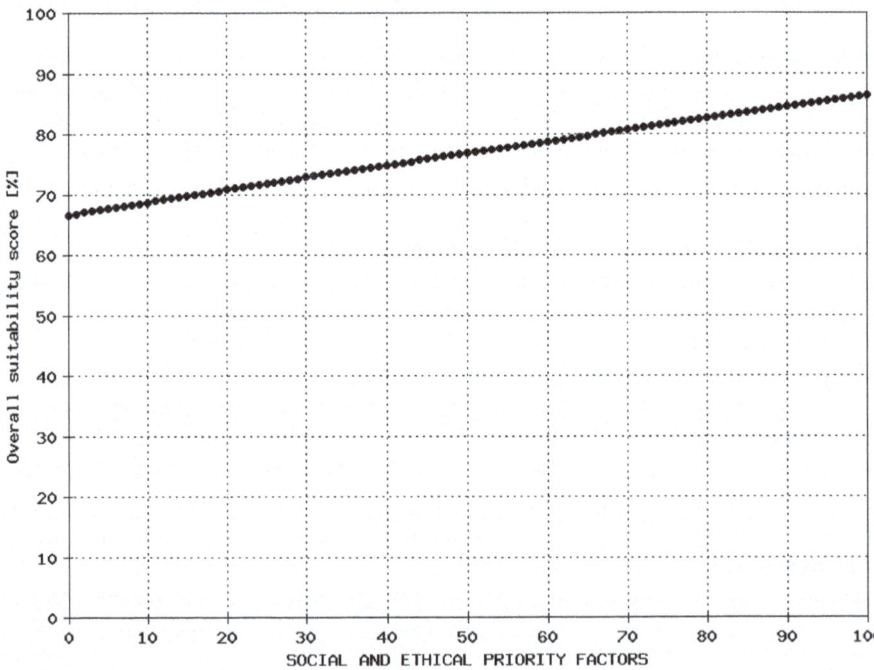

Fig. 19.8 TPS as a function of social and ethical priority factors

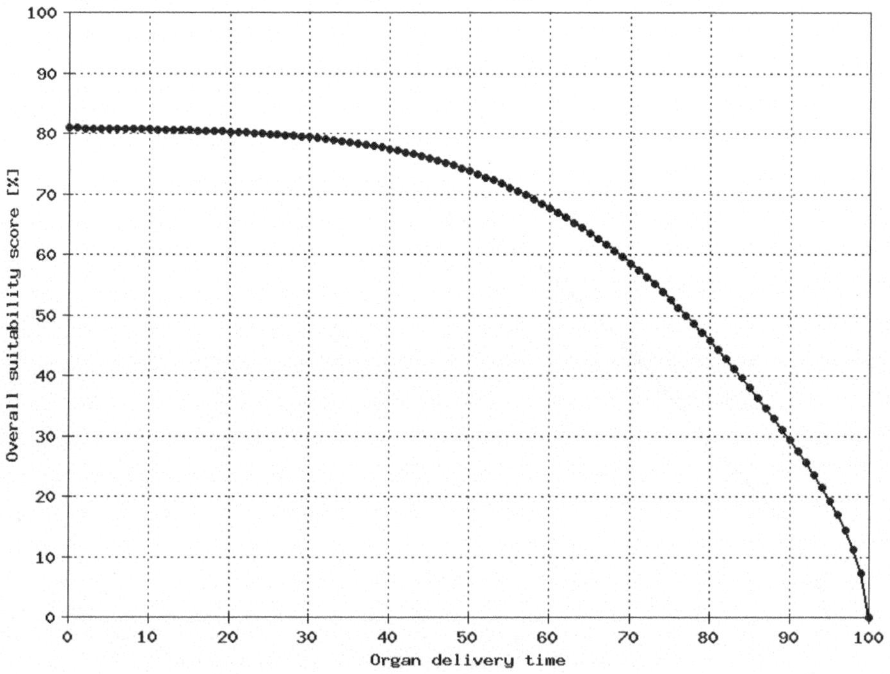

Fig. 19.9 TPS as a function of the organ delivery time

sensitivity analysis is used to investigate whether the variations of a selected priority attribute produce acceptable variations of the resulting TPS value. An example of sensitivity analysis for the organ delivery time is shown in Fig. 19.9. Unsurprisingly, it is very important to deliver liver (or any other organ) to the recipient as soon as possible. Therefore, the TPS must be a steep decreasing function of the organ delivery time, and patients who are close to the available organ should have priority compared to patients who are far and need a long delivery time. Our criterion provides such a function and, according to Figs. 19.7 and 19.9, an increase of delivery time from 10 to 50% reduces the TPS value from 80.28% to 73.53%. That decrease becomes much more significant for delivery times above 50%. The shape of the sensitivity function in Fig. 19.9 can be modified by adjusting the attribute criterion #1122 and its weight (20%) in the aggregation structure shown in Fig. 19.5.

The most important factor in the medical section of the TPS criterion is the MELD score [KIM21] which affects the medical urgency of transplantation three times more than the remaining attributes #1112 and #1113. Therefore, an increase of the MELD score should strongly positively affect the overall transplantation priority, as shown in the sensitivity curve in Fig. 19.10. If the MELD score approaches its maximum value 40, then the TPS does not approach 100% because other factors also affect and limit the TPS value. Unfortunately, some patients can die while waiting for a liver transplantation and the primary motivation for developing decision-support systems

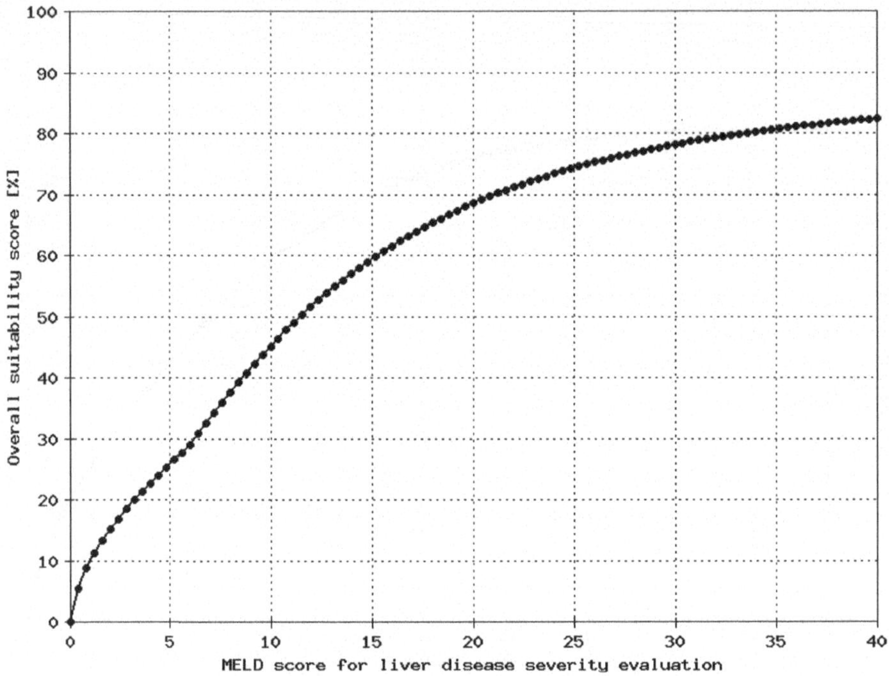

Fig. 19.10 TPS as a function of the MELD score

for prioritizing organ transplantation using the LSP method is to make decisions that can save more lives.

As a part of the LSP criterion tuning and validation process, the analysis of impact of priority attributes should be done for all attributes in the LSP criterion. This can be done for hypothetical test patients that have the same priority scores in all input attributes, typically 50% or 75%. The priority score of each input can vary from 0 to 100% and that causes variations of the overall TPS in the range $[P_{min}, P_{max}]$. The difference $D = P_{max} - P_{min}$ denotes the impact or importance of the analyzed priority attribute. The priority attributes can then be ranked according to decreasing impact, and this ranking can be used as one of components in the LSP criterion validation and explainability processes (see Chap. 18).

After the adjustment of selected priority attributes and tuning of the priority attribute criteria, aggregation functions and their parameters, the LSP criterion attains a validated final form and is ready for practical use and periodical updates. In this chapter, the proposed generic LSP criterion for organ transplantation priority evaluation is customized for liver transplantation. Organizations that develop prioritization criteria for various specific organs can consider modifying and expanding their criteria based on the proposed set of 35 transplantation priority attributes.

In the liver transplantation practice, some decision models are based on additive scoring, using less than ten carefully designed medical attributes. In addition, non-medical attributes are not formally included in the decision process. Considering the undeniable importance of some social and ethical factors, we believe that the most important such factors should be formally included in the organ allocation decision processes. The use of additive scoring is motivated primarily by the ultimate simplicity of that method, and the necessary political goal of making decision process easily understandable and controllable by the general public.

The spectrum of graded logic aggregators that are used in the LSP method can make the organ transplantation priority criteria significantly more credible than the simple additive scoring. The expressive power of LSP criteria is based on graded logic functions that can easily model graded conjunctive, disjunctive, symmetric and asymmetric logic relationships, differentiate mandatory, sufficient, and optional attributes, and precisely adjust their importance. The precision and justifiability of resulting TPS criteria can be maximized by both design and fine-tuning. Using more precise TPS criteria, the scarce organs can be allocated to the most deserving patients, maximizing the transplantation benefits for patients, organ transplantation community, and society.

References

ABA21 Abaalkhail, F.A., et al. 2021. Clinical practice guidelines for liver transplantation in Saudi Arabia. *Saudi Medical Journal*, 42 (9): 927–968.

ALL11 Allen, W.L., O.M. Amundsen, J. Dujmović, and K.D. Messer. 2011. Identifying and selecting strategic mitigation opportunities: criteria design and project evaluation using logic scoring of preference and optimization. *Journal of Conservation Planning*, 7: 61–68.

CHO12 Cholongitas E., and A.K. Burroughs. 2012. The evolution in the prioritization for liver transplantation. *Annals of Gastroenterology*, 25: 6–13.

COO90 Cook R.D., S. Staschak, and W.T. Green. 1990. Equitable allocation of livers for orthotopic transplantation: an application of the analytic hierarchy process. *European Journal of Operational Research* 48(1) 49–56, (1990).

DRA17 Dragićević, S., J. Dujmović, and R. Minardi. 2018. Modeling urban land-use suitability with soft computing: the GIS-LSP method. In *GeoComputational analysis of regional systems* 257–275, ed. Thill J.-C., S. Dragićević. Springer.

DRA21 Drackett, L., and S. Dragićević. 2021. Suitability analysis of acoustic refugia for endangered killer whales (*Orcinus orca*) using the GIS-based logic scoring of preference method. *Environmental Management*, 68: 262–278. https://doi.org/10.1007/s00267-021-01481-y.

DRA23 Drackett, L., S. Dragićević, A.K. Solomon. 2023. Integrating multiple perspectives in marine spatial planning using the GIS-based Logic Scoring of Preference method. *Ocean & Coastal Management*. 232 (1): 106423.

DUJ03 Dujmović, J. 2003. Optimizing computer system configurations. *Journal of Automatic Control*, 13 (2): 22–33.

DUJ06a Dujmović, J., and H. Nagashima. 2006. LSP method and its use for evaluation of Java IDE's. *International Journal of Approximate Reasoning*, 41 (1): 3–22.

DUJ08a Dujmović, J., J.W. Ralph, and L.J. Dorfman. 2008. Evaluation of disease severity and patient disability using the LSP method. in *Proceedings of the 12th Information*

Processing and Management of Uncertainty International Conference (IPMU 2008) 1398–1405. ed. L. Magdalena, M. Ojeda-Aciego, J.L. Verdegay Torremolinos (Malaga).

DUJ13 Dujmović, J., G. De Tré, N. Singh, D. Tomasevich, and R. Yokoohji. 2013. Soft computing models in online real estate. In *Advance trends in soft computing, WCSC 2013. studies in fuzziness and soft computing* vol 312, 77–91. ed. M. Jamshidi, V. Kreinovich, and J. Kacprzyk. Springer.

DUJ21a Dujmović, J. and D. Tomasevich. 2021. COVID-19 vaccination priority evaluation. In *Explainable AI and other applications of fuzzy techniques (Proceedings of the North American Fuzzy Information Processing Society, NAFIPS 2021)*. ed. Julia Rayz, Victor Raskin, Scott Dick, and Vladik Kreinovich . Springer 2021.

DUJ21f Dujmović, J., and W. Allen, III. 2021. Soft computing logic decision-making in strategic conservation planning for water quality protection. *Ecological Informatics*, 61: 101167.

DUJ22a Dujmović, J., and W.L. Allen. 2022. Explainable decision-making for water quality protection. *Information*, 13 (12): 551. https://doi.org/10.3390/info13120551.

DUJ25 Dujmović, J., and N. Mehta, LSP models for organ transplantation priority evaluation. In *Proceedings of the 2024 International Conference on Computational Science and Computational Intelligence*. Springer (in print).

DUJ73c Dujmović, J. 1973. Mixed averaging by levels (MAL)—A system and computer evaluation method, (in Serbo-Croatian). In *Proceedings of the Informatica Conference*, paper d28, Bled, Yugoslavia.

DUJ80 Dujmović, J. 1980. Computer selection and criteria for computer performance evaluation. *International Journal of Computer and Information Sciences*, 9 (6): 435–458.

EUR20 European Parliament. 2020. *Organ donation and transplantation. Fact, figures, and European Union action.* https://www.europarl.europa.eu/RegData/etudes/BRIE/2020/649363/EPRS_BRI(2020)649363_EN.pdf. Retrieved Feb 02, 2024.

HUN14 Hunink, M.G.M., and M.C. Weinstein, et al. 2014. *Decision-making in health and medicine: integrating evidence and values*, 2nd ed. Cambridge University Press.

KIA18 Kiani, M., M. Abbasi, M. Ahmadi, and B. Salehi. 2018. Organ transplantation in iran; current state and challenges with a view on ethical consideration. *Journal of Clinical Medicine*, 7: 45. https://doi.org/10.3390/jcm7030045.

KIM21 Kim, W.R., et al. 2021. MELD 3.0: the model for end-stage liver disease updated for the modern era. *Gastroenterology*, 161 (5): 1887–1895.

KLI95 Klir, G.J., and B. Yuan. 1995. *Fuzzy sets and fuzzy logic*. Prentice-Hall.

MES17 Messer, K.D. and W.L. Allen, III. 2017. *The science of strategic conservation: Protecting more with less*. Cambridge University Press, 2017.

NIH19 National Institutes of Health, *Scientists triple storage time of human donor livers*. News Releases, September 9, 2019. https://www.nih.gov/news-events/news-releases/scientists-triple-storage-time-human-donor-livers. Last reviewed on March 8, 2024.

OPT24 OPTN. 2024. *OPTN policies effective as of January 10*. https://optn.transplant.hrsa.gov/media/eavh5bf3/optn_policies.pdf. Retrieved February 26, 2024.

OPT24a OPTN. 2024. *Guidance to liver transplant programs and the national liver review board for: adult MELD exception review*. https://optn.transplant.hrsa.gov/media/2847/liver_guidance_adult_meld_201706.pdf. Retrieved February 15, 2024.

PIT10 Pittas, M.A. 2010. *Location suitability maps based on the LSP method*. San Francisco State University, Department of Computer Science, Report SFSU-CS-CE-10.08.

REB14 Rebolledo, B. 2014. *Spatial vulnerability to nitrate contamination from agricultural origin: development and application of parametric model IVNA-LSP (in Spanish)* (Doctoral thesis). University of Zaragoza, Instituto Universitario de Investigación Mixto CIRCE, http://zaguan.unizar.es.

REB16 Rebolledo, B., A. Gil, X. Flotats, and J.A. Sanchez. 2016. Assessment of groundwater vulnerability to nitrates from agricultural sources using a GIS-compatible logic multicriteria model. *Journal of Environmental Management*, 171: 70–80.

RIC21 Richards, E.G., R.M. Farrell, S. Ricci, U. Perni, C. Quintini, A. Tzakis, and T. Falcone. 2021. Uterus transplantation: state of the art in 2021. *Journal of Assisted Reproduction and Genetics*, 38: 2251–2259.

SAA80 Saaty, T.L. 1980. *The analytic hierarchy process*. McGraw-Hill.

SCH86 Schwartz. S., and T. Griffin. 1986. *Medical thinking: the psychology of medical judgement and decision making*. Springer.

SEA17 SEAS Co. 2017. *LSP.NT—LSP method for evaluation over the Internet*. LSP.NT User Manual. SEAS.

SHE21a Shen S., S. Dragićević, and J. Dujmović. 2021. GIS-based logic scoring of preference method for urban densification suitability analysis. *Computers, Environment and Urban Systems*, 89: 1016542021.

SHE21b Shen S., S. Dragićević, and J. Dujmović, GIS.LSP: a new soft computing method and tool for geospatial suitability analysis. *Transactions in GIS*, 25 (3): 1228–1248.

SHI20 Shi, B.Y., Z.J. Liu, and T. Yu. 2020. Development of the organ donation and transplantation system in China. *Chinese Medical Journal (Engl)*, 133(7): 760–765 (2020).

SOX13 Sox, H.C., M.C. Higgins, and D.K. Owens. 2013. *Medical decision making*. Wiley.

STR89 Streiner, D.L., and G.R. Norman. 1989. *Health measurement scales: a practical guide to their development and use*. Oxford University Press.

SU87 Su, S.Y.W., J. Dujmović, D.S. Batory, S.B. Navathe, and R. Elnicki. 1987. A cost-benefit decision model: analysis, comparison, and selection of data management systems. *ACM Transactions on Database Systems*, 12 (3): 472–520.

TAK22 Takherkhani, N., M.M. Sepehri, R. Khasha, and S. Shafaghi. 2022. Ranking patients on the kidney transplant waiting list based on fuzzy inference system. *BMC Nephrology*, 23: 31.

WIK24 Wikipedia. *Organ transplantation*. https://en.wikipedia.org/wiki/Organ_transplantation. Retrieved February 02, 2024.

WIK24a Wikipedia. *United network for organ sharing*. https://en.wikipedia.org/wiki/United_Network_for_Organ_Sharing. Retrieved February 02, 2024.

YAG94 Yager, R.R., and D.P. Filev. 1994. *Essentials of fuzzy modeling and control*. Wiley.

Chapter 20
The Road Ahead

We can only see a short distance ahead, but we can see plenty there that needs to be done.
—Alan Turing [TUR50].

At each point of development, for each scientific area, there is a road ahead. For Graded Logic, the road ahead can go in various directions, based on possible cooperation with various bordering areas. In this book we have shown that the Graded Logic is inseparably connected to decision making. On the other hand, decision making is a part of various disciplines where GL can (and should) be used. Such disciplines create bordering areas of GL, and the road ahead can go in the direction of any of bordering areas, or in all of them.

The Bordering Areas of Graded Logic

Human logical reasoning and decision making is a scientific area that is studied in psychology, economics, philosophy/logic, mathematics, computer science, data science, business administration, management, medicine, geography, urban studies, and other disciplines. Every discipline has natural interactions with its bordering areas. Graded Logic can be interpreted as an applicable decision-support methodology that occupies a space between several scientific areas shown in Fig. 20.1. At each border between GL and another area, GL can be used to affect and expand existing methods of the primary area, and the primary area can affect the further development of GL. Therefore, such interactions are desirable and mutually beneficial.

Graded Logic relies on scientific methods that are similar to physics: observations, formulation of hypotheses, experimental verification, and the development of mathematical models. The physical world is "hard" in the sense that fundamental properties and behaviors of matter and energy are stable, precisely measurable, and modelable. GL operates in a world that is "soft" in the sense that the properties of natural human logical reasoning are less stable, less precisely measurable, and not easily modelable. In both cases, mathematics is a tool for describing observed phenomena. In the case of GL, mathematics is also a natural environment. However, GL has interests and goals that are different from traditional goals of logic as a branch of philosophy and a

J. Dujmović, *Graded Logic*, Intelligent Systems Reference Library 273,
https://doi.org/10.1007/978-3-031-88558-7_20

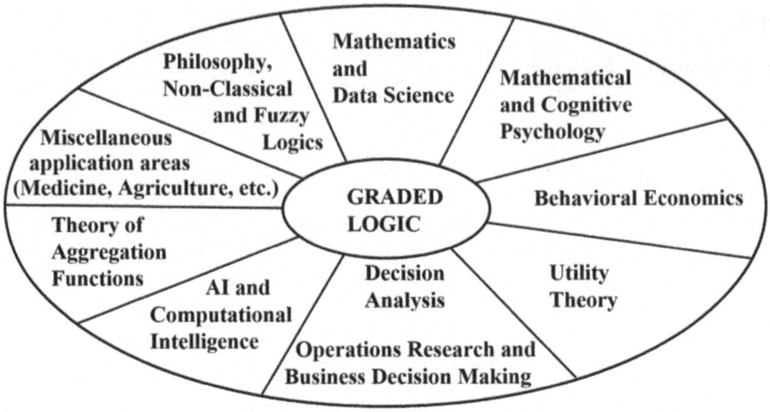

Fig. 20.1 Bordering scientific areas that interact with Graded Logic

branch of mathematics. First, the main interest of GL is the commonsense reasoning. Second, GL is currently restricted to propositional logic because the most frequent applications are at the propositional level, and that level is visible in human decision making and communication using natural languages. Third, the main goal of GL is to create decision-making applications that are consistent with the commonsense logical reasoning, but don't have inherent human limitations.

In mathematics and philosophy, logic is usually defined as the science that investigates principles of correct inference, or, more precisely, as the science of the formal principles of reasoning. To study the formal principles of reasoning, both mathematicians and philosophers create formalized axiomatic deductive theories that are usually characterized by the absence of humans and the absence of applications. That is normal because axiomatic theories are self-contained, and the presence of specific human thinkers is not needed.

On the other hand, reasoning and inferring are (primarily human) cognitive processes. So, we can use a human-centric approach that assumes the explicit presence of a specific human (a stakeholder/decision-maker) and observe/analyze/model the stakeholder's logical reasoning. Obviously, such a reasoning has the status of *commonsense reasoning*, i.e., the reasoning that is natural and based on everyday knowledge that humans find intuitive and straightforward. Fundamental properties that are provably present in natural human logical reasoning can be specified as a set of GL postulates (shown in Chap. 3) that are necessary and sufficient to build mathematical models, specify useful conjectures, and make successful applications. Obviously, the role of GL postulates is the same as the role of axioms in formal logic. Since the GL postulates are derived from properties that are provably present in natural human reasoning, all mathematical infrastructure built using these postulates is also consistent with natural human reasoning. Graded logic relies on that approach and currently supports a graded propositional calculus. Therefore, it is possible to have desirable interactions between mathematical and human-centric approaches where both approaches can benefit: the mathematical and philosophical approach can move towards the realities of human mental activities, and the human-centric approach can move towards the theoretical purity of the mathematical and philosophical approach. The road ahead is expected to go in that direction.

The mathematical theory of aggregation functions [FOD94, CAL02, BEL07, GRA09, BEL16] is a discipline closest to Graded Logic. All aggregation functions (including means) can always be interpreted as logic functions, but so far this interpretation has not been used in the mathematical developments of the theory of aggregators. On the other hand, there is plenty of space for mathematically expanding and improving the functionality of graded propositional calculus presented in [DUJ18] and in this book. So, in the future theoretical work, the cooperation between the Graded Logic and the theory of aggregation functions is expected to produce useful theoretical results and successful applications.

In its basic role of the mathematical infrastructure for decision methods, GL can be used in a wide spectrum of applications [DUJ18]. These include applications in AI systems, strategic conservation in ecology, suitability maps in geography and agriculture, selecting strategies for protection of endangered species, decisions related to urban development, development of personalized recommender systems, applications in computer science, medicine, and many others.

Decision making in AI is an important area necessary for autonomous performing complex tasks. It is based on methods developed in computational intelligence, fuzzy systems, and soft computing [KAC15, RAY15]. In such applications machines are expected to behave similarly as humans and consequently GL models of commonsense reasoning can be useful for supporting such systems. Of course, there is the eternal question "how does the human mind work," and there are countless attempts to answer that question. There are attempts at the micro level (neurons in human brain), at the level of neural networks, natural language models, attempts in the framework of medicine, philosophy, and other disciplines, most notably in current AI research. Based on our efforts to model natural commonsense reasoning, we cannot say that we have no interest in this eternal question. However, it is easy to see that there are many levels of complexity in problems that are analyzed and solved by the human mind. We believe that human commonsense decision making is either the simplest, or among the simplest such problems. Fortunately, the intuitive solutions of such problems are both observable and explainable at the "macro level," i.e., using a black-box approach, where it is possible to reliably identify the inputs and the output of the corresponding mental process.

The simplicity and observability of the commonsense decision making resulted in the quantitative model presented in Fig. 6.1. Graded Logic is the indispensable component of that model, and the mental model can be easily expanded in the LSP decision method and corresponding software support that can be used to solve a variety of delicate decision problems, such as the patient priority for organ transplantation, presented in Chap. 19. Of course, simple problems have the highest likelihood of getting effective solutions. However, regardless of the simplicity of the commonsense decision making, there are many opportunities for improvement of existing models and experimental verification of proposed methodology. That work, inside the global framework of AI, is the primary road ahead for Graded Logic refinement and improvement.

One of the most visible AI problems is the explainability of AI-assisted decisions about individuals [ICO22, LES24]. To extract explanations from the "black box" AI systems we need sophisticated techniques. On the other hand, the main advantages of GL and the LSP method are their inherent interpretability and explainability. For example, if we use the preferential neuron training process for a GCD aggregator, we provide a training set that contains inputs and desired outputs (e.g., see Figs. 13.11–13.13). The resulting aggregator is a function which in addition to the analytic form automatically explains the basic logic properties of the GCD aggregator:

- Is the aggregator conjunctive or disjunctive?
- Is the aggregator hard or soft?
- What is the necessary andness/orness of the GCD aggregator?
- What is the degree of importance of each input?
- What is the precision of the GCD aggregator?

Similarly, the result of training partial absorption neurons (Figs. 16.4 and 16.5) can be used to find the resulting penalty and reward indicators.

Any GL propositional calculus formula can be trained sequentially, aggregator by aggregator. In that process, all input and output values have clearly interpretable semantic identity. These properties directly contribute to the explainability of GL/LSP models and their results. In GL/LSP models, we can simultaneously generate the results, explanations, and the degrees of confidence in the obtained results. As opposed to that, the "black box" AI models create results, but to extract explanations from such models we need separate methods. So, the use of automatic explainability of GL/LSP models in AI-assisted decision-making is an attractive direction for the future work. The development of social and conversational humanoid robots [FUR25] assumes that robots behave, reason, and communicate like humans. If the robot is used to help in explainable decision making, the decision models based of GL and LSP are suitable for generating results close to commonsense human reasoning. In such applications, the robot can provide not only the results of evaluation and comparison decisions, but also the explanations that verbally justify the obtained results.

Applications of AI decision systems are possible in many areas, including autonomous vehicles, humanoid robotics, recommender systems, and healthcare. To exemplify opportunities for future GL/LSP applications, let us briefly investigate the area of medical decision making.

Decision-making in health and medicine is an area that has significant potential for the applications of Graded Logic. Most authors of medical decision-making books explicitly claim that decision making is the key activity in the practice of health care [SCH86, GRO99, SCH08]. Indeed, decision-making in health and medicine is an area that deserves high attention, and there is a rich literature devoted to medical decision-making. In most cases the primary users of decision methods and tools are physicians and decisions are related to selecting diagnostic and therapeutic options

[HUN14]. More generally, medical decision making includes three major participants: patients, physicians, and health care policymakers. Patients regularly have the role of stakeholder, and the first step in most decision problems is to understand and specify the goals and objectives of each specific patient. Since medical care is directly related to health care policy, medical decision problems are frequently related also to patient's family, community, and society [SCH08].

Medical decisions naturally occur in an environment of uncertainty, risk, imprecise probability of outcomes, and limited confidence. So, medical literature primarily exposes probabilistic aspects of medical decision-making [HUN14, GRO99]. Areas of medicine and public health policy where GL and the LSP method can contribute improvements include the following:

- Disease severity evaluation
- Patient disability evaluation
- Risky therapy decision making
- Organ transplantation priority evaluation
- Vaccination priority evaluation
- Evaluation and monitoring of the effects of a therapy
- Selection of medical equipment and software tools
- Optimum location of hospitals, pharmacies, labs, and medical services.

Let us briefly discuss each of these areas. Disease severity evaluation is a typical evaluation problem. This problem is significant because a quantitative indicator of disease severity is critical for deciding about the most appropriate therapy, and as the communication means between physicians (when a physician refers a patient to another physician). This is the reason why in many cases disease severity evaluation is based on rating scales [HER06, DYC93, DYC05] and on variety of scoring techniques [MER06, GRA06a, GRA06b]. In many cases it is possible to seriously criticize simple additive scoring used for disease severity evaluation and improve evaluation results using the LSP method [DUJ08a].

Patient disability evaluation can significantly contribute to personalized medicine. Indeed, there are disabilities that are insignificant and negligible for a specific patient, but crucial for another patient (e.g. a minor injury to a left-hand finger can be insignificant for a math professor but disabling for a violin player). Reducing patient disability is the primary goal of all therapies, and in many cases that goal is only partially achieved. In many cases an overall disability degree can be computed using questionnaires posted on the Internet. Then both patients and physicians can monitor the status of disability as a function of time, assess the effects of a therapy, and make various other decisions. Such a system for monitoring the development of peripheral neuropathy can be found in [DUJ18]. There are many specific diseases that can benefit from LSP models used for disability evaluation, and that is one of the areas for future research and development.

Risky therapy decision-making is an optimization problem. Some therapies are both beneficial and harmful. Many medications come to patients with a standard claim that "your doctor prescribed the therapy believing that benefits are greater than negative effects and risks." For example, if a disease is increasingly developing, then it is desirable to start therapy as soon as possible. On the other hand, to reduce various negative effects of the therapy, it is desirable to postpone the therapy as late as possible. Obviously, between these extremes there is an optimum point that maximizes the patient's quality of life. Such problems exist for many diseases and many therapies and deserve serious attention. A sample solution of such a problem can be found in [DUJ18].

One of the permanently present medical decision problems is the patient prioritizing problem that occurs whenever there are limited resources and many patients waiting for them. In such cases it is necessary to develop a justifiable criterion for computing the priority of each patient and then allocate resources according to decreasing priority degrees of patients (starting with the highest priority patient). Two typical such problems are the organ transplantation priority evaluation (Chap. 19) and vaccination priority evaluation [DUJ21a]. Of course, there are many similar problems of allocation of scarce resources, and the allocation of resources according to justifiable priority criterion yields the optimum use of resources and maximizes positive medical effects. Generally, the criterion cannot be justifiable if it is incomplete. Thus, priority criteria should include both medical and social (and/or ethical) priority factors, as suggested in the liver transplantation prioritizing problem discussed in Chap. 19.

Evaluation and monitoring of the effects of physical (or another) therapy is necessary in almost all therapies. Physical (or other) therapies can prove their success by quantifying the attained decrease of patient disability. That is possible only if there is a quantitative disability criterion function that is updated during the therapy and used for explainable measurement of the degree of success. In the case of peripheral neuropathy, the use of such a system is described in [DUJ18]. Similar systems can be developed for many other therapies.

Selection of medical equipment and software tools is a problem like any other equipment or software selection problem. The LSP method is routinely used for evaluation, comparison and selection of complex equipment in electrical, computer, and software engineering [DUJ06a, DUJ06c, DUJ80, SU87]. The method determines the overall value of equipment/software as a logically justified combination of the overall suitability of equipment/software and its total cost. Evaluation, comparison, and selection of expensive medical equipment can be based on the same LSP methodology.

Optimum location of hospitals, pharmacies, labs, and medical services is the same problem as the optimum location of public objects (stores, theatres, sports stadiums, airports, etc.). Each point in a selected urban area has a degree of suitability as a location of a new medical institution. The LSP method provides location suitability maps, necessary for optimum decisions [DUJ10a, DUJ10b, DUJ18, SHE21a, SHE21b].

Experiments with human subjects are the prerequisites for the development of all human-centric logical reasoning models. Organization, realization, and interpretation of such experiments should be done in cooperation with psychologists and that is another area of expected mutual benefits for both GL and psychology. Experimental psychology and cognitive psychology [MYE06, KAN09, SOL98, STE06] study selected aspects of human perception, judgement, and reasoning. However, so far there has been no research in the aera of human natural logic, and modeling based on experiments. So, that is an area that can contribute to logic models that we currently use. In particular, problems related to aggregated commonsense percepts of overall importance [DUJ12] and their decomposition in relative and andness/orness-based logic importance deserve experimental research and interest of psychologists. The use of hyperconjunction and hyperdisjunction was experimentally studied in the context of fuzzy logic [THO79, ZYS79, ZIM79, ZIM80, ZIM87, ZIM96, KOV92, DUJ22] but not in psychology, and that opportunity is still available.

Human commonsense decision making is based on adjustable, continuum-valued simultaneity, substitutability, and importance. Therefore, logic functions must be andness-directed in the whole range from drastic conjunction to drastic disjunction. Decision models used in the utility theory [FIS64, FIS70, KEE76, KEE92, FIG05a, HEN15], economics, traditional operations research, and decision analysis [MIL70, MOR77, BUN84, OLS96, BEL02, ISH13] are mostly additive and less frequently multiplicative or based on t-norms, but in all cases they have fixed andness. Therefore, such models are not fully consistent with human reasoning and can be improved by using andness-directed GL models. So, business decision making is another area of natural expanding the applicability of GL and the LSP method.

In this book, the study of GL is focused on two topics: (1) understanding the anatomy of natural intuitive decision making and the need for fully continuum-valued logic, and (2) using the current knowledge of natural human logic to propose how we ought to organize decision-making processes. Our interest is focused on professional/industrial decision problems where the complexity of decisions is beyond the limits of commonsense reasoning. To solve such problems, we need a complete propositional logic armamentarium, necessary and sufficient for precise quantifying of all logic operations, as well as providing explainability, and confidence. Our approach to GL is based on a sequence of five activities, typical for science and engineering:

(1) Observe phenomena of natural evaluation and decision-making.
(2) Develop experiments with human subjects.
(3) Measure responses from human subjects.
(4) Develop mathematical models of measured phenomena.
(5) Apply mathematical models in decision engineering.

Then, the same sequence repeats in cycles, in attempts to increase the precision, confidence, explainability and applicability. Over the last fifty years of GL's development and use, several cycles have occurred through this classical sequence, and this book aims to help initiate a new one.

Takeaway

The last words in this book are devoted to a brief takeaway, specifying the key facts, points, and ideas to be remembered. The main idea is that Graded Logic is the logic of human commonsense logical reasoning. Logical reasoning and decision making are human mental activities. They don't exist in a vacuum. They are inseparably related to a specific stakeholder/decision-maker who is human, either an individual or an organization.

Natural human commonsense logic is observable, measurable, modelable, and explainable. Observations and experiments convincingly prove that human decision-making uses graded propositional logic that is based on graded percepts of *truth*, *importance*, and *logic operations* of simultaneity and substitutability, which are also graded. All graded percepts have adjustable intensity in a continuous range from a minimum value to a maximum value. In decision making we are primarily interested in the percepts of suitability, importance, and satisfaction of requirements. Any percept of partial satisfaction can be interpreted as the degree of truth of the statement that claims the full (maximum) satisfaction. That makes the evaluation-based decision making an application based on graded propositional calculus, which is the main result of Graded Logic. We must be able to evaluate a single object, and the evaluation, comparison, and selection in the case of m objects can be reduced to m evaluations of a single object.

In mathematics and philosophy logic can be built as a formalized axiomatic deductive system. Such systems are self-contained and generally not related to humans, and human decision making. In addition, formal logic theories are not related to specific decision-making processes or other applications. Thus, there are two characteristic approaches to the development of logic systems:

- Human-centric approach, based on experimentally verified postulates, derived from observable natural human logical reasoning.
- Formal axiomatic approach, based on axioms and deductive methods for development of a mathematical theory.

These two approaches are different because they have different goals. The goal of human-centric approach is to model actual commonsense human natural logical reasoning, evaluation, and decision making. These results are then applied for building analytic logic models that are necessary for the development and use of decision-support systems, and other applications. The goal of an axiomatic approach is to develop valid mathematical theories. If these two approaches coexist as separate entities, there are possibilities for mutually beneficial cross-fertilization. However, it is advisable to avoid situations where one approach claims the achievement of goals of the other approach.

This book promotes a human-centric approach to propositional calculus. The result is Graded Logic that offers a propositional calculus with graded percepts that include the following necessary fundamental properties, provably present in human commonsense logical reasoning:

- Graded truth (continuum-valued truth of statements).
- Semantic identity of each degree of truth.
- Graded importance and noncommutativity (importance weighting of statements).
- Graded simultaneity/substitutability (adjustable intensity of graded conjunction and graded disjunction).
- Andness-directedness (using andness as explicitly visible and easily adjustable graded parameter).

- The continuum of the simultaneity/substitutability logic operations in the full range from the drastic conjunction to the drastic disjunction.
- Complementary unification of simultaneity and substitutability in a single logic operation, Graded Conjunction/Disjunction (GCD).
- Selectability of idempotence (supported or not supported) in basic logic operations.
- Selectability of annihilators (supported or not supported) in models of idempotent graded logic operations (hard and soft graded conjunction and graded disjunction).
- Duality of simultaneity and substitutability for all possible values of andness.

These properties are indispensable if logic is expected to describe human commonsense reasoning and be used in decision-making applications. This functionality is necessary for building professional decision methods, such as the Logic Scoring of Preference (LSP).

Professional evaluation-based decision methods cannot have credibility if they are not compatible with human commonsense logical reasoning and related decision making. Compatibility with natural human decision-making is the prerequisite for the explainability of decision method results and for achieving the degree of confidence necessary to implement proposed decisions.

Logic systems that lack support for continuum-valued truth, importance weighting of statements, idempotence and annihilator selectability, and andness-directedness—enabling precisely controlled continuous transitions across the full range from drastic conjunction to drastic disjunction—fail to provide sufficient functionality for effective use in human-centric logic, decision-making, and AI applications.

Starting from Graded Logic and the LSP method, many paths extend in different directions. The author hopes that the reader will choose one of these paths and take the next steps forward. Good luck!

References

BEL02 Belton, V., T.J. Stewart, *Multiple criteria decision analysis: an integrated approach.* Kluwer Academic Publishers.

BEL07 Beliakov, G., A. Pradera, and T. Calvo. 2007. *Aggregation functions: a guide for practitioners.* Springer.

BEL16 Beliakov, G., H. Bustince Sola, and T. Calvo Sanchez. 2016. A practical guide to averaging functions. Studies in Fuzziness and Soft Computing 329, Springer 2016.

BUN84 Bunn, D.W. 1984. *Applied decision analysis.* McGraw-Hill Book Company.

CAL02 Calvo, T., A. Kolesarova, M., Komornikova, and R. Mesiar. 2002. Aggregation operators: properties, classes and construction methods. In *Aggregation operators. New trends and applications*, 3–104. ed. T. Calvo, G. Mayor, and R. Mesiar. Heildeberg: Physica-Verlag.

DUJ06a Dujmović, J., and H. Nagashima. 2006. LSP method and its use for evaluation of Java IDE's. *International Journal of Approximate Reasoning*, 41 (1): 3–22.

DUJ06c Dujmović, J., and H. Bai. 2006. Evaluation and comparison of search engines using the LSP method. In *ComSIS* vol. 3, No. 2, 31–56.

DUJ08a Dujmović, J., J.W. Ralph, and L.J. Dorfman. 2008. Evaluation of disease severity and patient disability using the LSP method. In *Proceedings of the 12th Information*

Processing and Management of Uncertainty International Conference (IPMU 2008) 1398–1405. ed. L. Magdalena, M. Ojeda-Aciego, J.L. Verdegay. Torremolinos (Malaga).

DUJ10a Dujmović, J., G. De Tré , and N. Van de Weghe. 2010. LSP suitability maps. Published online June 5, 2009, ISSN 1433-7479, by Springer. Soft Computing, ISSN 1432-7643 Vol 14, No 5 421–434. Springer.

DUJ10b Dujmović, J., and D. Scheer. Logic aggregation of suitability maps. In *Proceedings of the 2010 IEEE World Congress on Computational Intelligence*, 2222–2229. Barcelona, Spain, July 18–23, 2010. ISBN 978-1-4244-6920-8.

DUJ12 Dujmović, J.2012. Andness and orness as a mean of overall importance. In *Proceedings of the IEEE World Congress on Computational Intelligence, June 10–15, 2012 -Brisbane, Australia* 83–88.

DUJ18 Dujmović, J. 2018. *Soft computing evaluation logic*. Wiley.

DUJ21a Dujmović, J., and D. Tomasevich. 2021. COVID-19 vaccination priority evaluation. In *Explainable AI and Other Applications of Fuzzy Techniques. (Proceedings of the North American Fuzzy Information Processing Society, NAFIPS 2021)*. ed. Julia Rayz, Victor Raskin, Scott Dick, and Vladik Kreinovich. Springer.

DUJ22 Dujmović, J., and D. Tomasevich. 2022. Experimental analysis and modeling of human conjunctive logic aggregation. In *2022 IEEE International Conference on Fuzzy Systems (FUZZ-IEEE)*, Padua, Italy 1–8. https://doi.org/10.1109/FUZZ-IEEE55066.2022.988 2665.

DUJ80 Dujmović, J. 1980. Computer selection and criteria for computer performance evaluation. *International Journal of Computer and Information Sciences*, 9 (6): 435–458.

DYC05 Dyck, P.J., R.A.C. Hughes, and P.C. O'Brien. 2005. Quantitating overall neuropathic symptoms, impairments, and outcomes. In *Peripheral neuropathy*. 4th ed. 1031–52. ed. P.J. Dyck, P.K. Thomas. Philadelphia: Elsevier.

DYC93 Dyck, P.J. 1993. Quantitating severity of neuropathy. In *Peripheral neuropathy*, 3rd ed. 686–697. ed. Dyck P.J., P.K. Thomas J.W. Griffin. and J.F. Poduslo. Saunders Philadelphia.

FIG05a Figueira, J., S. Greco, and M. Ehrgott (eds.) *Multiple criteria decision analysis*, Springer.

FIS64 Fishburn, P.C. 1964. *Decision and value theory*. Wiley.

FIS70 Fishburn, P.C. 1970. *Utility theory for decision-making*. Wiley.

FOD94 Fodor, J., and M. Roubens. 1994. *Fuzzy preference modelling and multicriteria decision support*. Dordrecht, The Netherlands: Kluwer Academic Publishers.

FUR25 The Furhat Robotics. 2025. *The Furhat Robot*. https://www.furhatrobotics.com/furhat-robot.

GRA06a Graham, R.C., and R.A.C. Hughes. 2006. A modified peripheral neuropathy scale: the overall neuropathy limitations scale. *Journal of Neurology, Neurosurgery Psychiatry*, 77: 973–976.

GRA06b Graham, R.C., and R.A.C. Hughes. 2006. Clinimetric properties of a walking scale in peripheral neuropathy. *Journal of Neurology, Neurosurgery Psychiatry*, 77: 977–9.

GRA09 Grabisch, M., J-L. Marichal, R. Mesiar, and E. Pap. 2009. *Aggregation functions*. Cambridge University Press, Cambridge.

GRO99 Gross, R. 1999. *Making medical decisions: an approach to clinical decision making for practicing physicians*. American College of Physicians, Philadelphia.

HEN15 Hensher, D.A., J.M. Rose, and W.H. Greene. 2015. *Applied choice analysis*. Cambridge University Press.

HER06 Herndon, R.M. 2006. *Handbook of neurologic rating scales*. Second edition. Demos Medical Publishing, 2006.

HUN14 Hunink, M.G.M., and M.C. Weinstein, et al. 2014. *Decision-making in health and medicine: integrating evidence and values*, 2nd ed. Cambridge University Press.

ICO22 ICO22Information Commissioner's Office and The Alan Turing Institute. *Explaining decisions made with AI*. https://ico.org.uk/media/for-organisations/guide-to-data-pro tection/key-dp-themes/explaining-decisions-made-with-artificial-intelligence-1-0.pdf. Accessed October 17, 2022.

ISH13 Ishizaka, A., and P. Nemery. 2013. *Multi-criteria decision analysis*. Wiley.
KAC15 Kacprzyk, J., and W. Pedrycz (eds). 2015. *Springer handbook of computational intelligence*. Springer.
KAN09 Kantowitz, B.H., and H.L. Roediger III. 2009. *Experimental psychology*. Wadsworth, Cengage Learning.
KEE76 Keeney, R.L., and H. Raiffa. 1976. *Decisions with multiple objectives/preferences and value tradeoffs*. Wiley.
KEE92 Keeney R.L. 1992. *Value-focused thinking*. Harvard University Press.
KOV92 Kovalerchuk, B., and V. Taliansky. 1992. Comparison of empirical and computed values of fuzzy conjunction. *Fuzzy Sets and Systems*, 46: 49–53 (North-Holland).
LES24 Leslie, D., C. Rincón, M. Briggs, A. Perini, S. Jayadeva, A. Borda, S.J. Bennett, C. Burr, M. Aitken, S. Mahomed, J. Wong, M. Waller, and C. Fischer. 2024. *AI explainability in practice*. The Alan Turing Institute.
MER06 Merkies, I.S.J., and P.I.M. Schmitz. 2006. Getting closer to patients: the INCAT Overall Disability Sum Score relates better to patients' own clinical judgement in immune-mediated polyneuropathies. *Journal of Neurology, Neurosurgery Psychiatry*, 77: 970–972.
MIL70 Miller, J.R., III. 1970. *Professional decision-making*. Praeger, New York.
MOR77 Morris, W.T. 1977. *Decision analysis*. Grid Inc.
MYE06 Myers, A., and C. Hansen. 2006. *Experimental psychology*. Thomson Wadsworth.
OLS96 Olson, D.L. 1996. *Decision aids for selection problems*. Springer.
RAY15 Ray, K.S. 2015. *Soft computing and its applications*. Apple Academic Press.
SCH08 Schwartz, A., and G. Bergus. 2008. *Medical Decision-making. A Physician's Guide*. Cambridge University Press.
SCH86 Schwartz. S., and T. Griffin. 1986. *Medical thinking: the psychology of medical judgement and decision Making*. Springer.
SHE21a Shen S., S. Dragićević, and J. Dujmović. 2021. GIS-based logic scoring of preference method for urban densification suitability analysis. *Computers, Environment and Urban Systems*, 89: 1016542021.
SHE21b Shen S., S. Dragićević, and J. Dujmović. 2021. GIS.LSP: a new soft computing method and tool for geospatial suitability analysis. *Transactions in GIS*, 25 (3): 1228–1248.
SOL98 Solso, R.L., H.H. Johnson, and M.K. Beal. 1998. *Experimental psychology*. Longman.
STE06 Sternberg, R.J. 2006. *Cognitive psychology*. Thomson Wadsworth.
SU87 Su, S.Y.W., J. Dujmović, D.S. Batory, S.B. Navathe, and R. Elnicki,. 1987. A cost-benefit decision model: analysis, comparison, and selection of data management systems. *ACM Transactions on Database Systems*, 12 (3): 472–520.
THO79 Thole, U., H.-J. Zimmermann, and P. Zysno. 1979. On the suitability of minimum and product operators for the intersection of fuzzy sets. *Fuzzy Sets and Systems* 2: 167–180.
TUR50 Turing, A.M. 1950. Computing machinery and intelligence. *Mind*, 49: 433–460.
ZIM79 Zimmermann, H.-J., and P. Zysno. 1979. Decisions and evaluations by hierarchical aggregation of information. *Fuzzy Sets and Systems*, 10: 243–266.
ZIM80 Zimmermann, H.-J., and P. Zysno. 1980. Latent connectives in human decision-making. *Fuzzy Sets and Systems*, 4: 37–51.
ZIM87 Zimmermann, H-J. 1987. *Fuzzy sets, decision-making, and expert systems*. Kluwer Academic Publishers.
ZIM96 Zimmermann, H-J. 1996. *Fuzzy set theory and its applications*. New York: Springer.
ZYS79 Zysno, P. 1979. One class of operators for the aggregation of fuzzy sets. In *EURO III Congress*, Amsterdam.

Index

J. Dujmović, *Graded Logic*, Intelligent Systems Reference Library 273,
https://doi.org/10.1007/978-3-031-88558-7

H

I

J

T

U

V

W

The manufacturer's authorised representative in the EU is Springer Nature Customer Service Centre GmbH, Europaplatz 3, 69115 Heidelberg, Germany. If you have any concerns regarding our products, please contact ProductSafety@springernature.com

Printed and bound by CPI Group (UK) Ltd, Croydon, CR0 4YY
30/09/2026
02225588-0012